Human Nature

IN GEOGRAPHY

Human Nature IN GEOGRAPHY

FOURTEEN PAPERS, 1925–1965

John Kirtland Wright

HARVARD UNIVERSITY PRESS
CAMBRIDGE, MASSACHUSETTS

1966

Distributed in Great Britain by Oxford University Press, London
Library of Congress Catalog Card Number 66–10128
Printed in the United States of America

To My Children

Preface

IN 1953 my friend Dr. J. Russell Whitaker, professor of geography at George Peabody College for Teachers, Nashville, Tennessee, honored me by suggesting that I get out a book of this sort. Such a "little book in one's own library," he wrote, "is worth a great number of articles scattered through magazines . . . In the language of my boyhood, I 'double-dare' you." Now, more than a decade later, I have taken his dare.

Of the fourteen papers herein, four (Chapters 8, 10, 13, 14) are making their debuts in print, one (Chapter 1) was first published when I was thirty-four, and one (Chapter 12) in 1962 when I had just turned seventy; four (Chapters 2, 3, 4, 5) appeared when I was in my fifties and four (Chapters 6, 7, 9, 11) when I was in my sixties (the median year was 1956 and median age 65, facts that the reader may interpret as he wishes). The papers are arranged chronologically according to when they were first written or delivered as addresses. The title of the third paper suggested that of the fourth, and the latter, in turn, suggested that of the volume as a whole. "Human nature in geography" refers to the impact upon geographical awareness (perception, cognition, knowledge, belief, study) of human emotions, motives, and behavior, and all of the papers have to do with this rather large subject. Furthermore, except for the third and fourth, they are all mainly concerned with past times and hence with the subject conventionally known as the "history of geography."

In the title and throughout this book the word "geography" should be understood as referring to geographical awareness rather than to the objective realities upon which such awareness is focused. These objective realities consist in the self-evident circumstance that the features of this earth and its inhabitants are not homogeneous and uniformly distributed in space, but differ from place to place. This circumstance might be called either "the spatial diversity of the earth and its inhabitants," or "terrestrial diversity in space," or, more concisely, "*geodiversity*," and, accordingly, geography might be defined as "awareness concerning geodiversity"—which, in fact, is what we often have in mind when we say or write the word "geography." (Obviously, "awareness concerning geodiversity" would hardly do for a definition of geography as an academic discipline today, since academic geography involves considerably more than such awareness, alone. I refrain, however, from attempting to define academic geography.)

The term *geodiversity* and the concept that it designates call for special comment here for the benefit of those interested in *geosemantics* (the semantics of geography). Geodiversity is more or less identical with Richard Hartshorne's well-known "areal differentiation of the earth's surface" in *The Nature of Geography* (p. 413 [237]; see Chapter 10, n. 29). It would seem desirable, however, to distinguish between diversity and differentiation. Diversity is the product of the diversifying actions of God, Nature, and Man. Differentiation results, rather, from human contemplation and study of the "given" facts of diversity, as when statesmen draw boundary lines on maps, surveyors mark them out on the ground, and geographers divide up the earth's surface into regions of divers kinds. (Differentiation, of course, may lead to diversity, as in the case of boundary lines.) There are three kinds of diversity: "topical," temporal, and spatial; and spatial diversity is an attribute not only of areas but of points, lines, and volumes. The opposites of spatial and topical diversity are spatial and topical homogeneity (or uniformity); the opposite of temporal diversity is monotony (or uniformity). "Geodiversity" is a coined term defined arbitrarily as designating the *spatial* diversity of the earth, "the earth" to be understood here as this terrestrial spheroid of ours with its enveloping atmosphere and all that is contained therein and thereon. (The beauty of coined terms is that we may define them as we please, whereas established words and phrases when used to designate special concepts often carry overtones of implication that we do not wish them to have and that may confuse the reader.)

The relation between geography, as I have defined it (strictly for present purposes only), and human nature is comparable with that between Maine and America. One might study what is American as distinguished from what is distinctively Mainelike about Maine. Analogously, these papers deal with what is, or has been, "humanly natural" as distinguished from what is or has been distinctively "geographical" about geography, although in a larger sense geography is no more separable from human nature than Maine is from America. Indeed, the fibers that bind this book together in a rather loose bundle are primarily those of interest in what is *not* geographical about geography and in what is geographical about nongeography.

Although not so interpreted here, the phrase "human nature in geography" might with perfect propriety be used with reference to manifestations of human nature in geodiversity. This could be called "the geography of human nature" and would be analogous to "the geography of rivers" or "of animals." Cultivated under such labels as "human geography," "cultural geography," "political geography," and the like, the study of the geography (that is, the *geodiversity*) of human nature lies outside the scope of this book.

The two different meanings of "human nature in geography" arise, of course, from the ambivalence of the word "geography." Like "history," this noun is employed interchangeably both for a species of human awareness and for the actualities, real or imagined, with which such awareness has to do, the actualities that Hartshorne called "areal differentiation" and for which I have coined the term "geodiversity." This ambivalence has given rise to a great deal of obscurity, misunderstanding, and argument. In ordinary speech we do not call ornithology "birds" nor birds "ornithology," but we *do* use the same word ("geography") for geograph*ic* actuality that we do for geograph*ical* awareness (discourse, knowledge, belief) pertaining to such actuality.

To guard, insofar as possible, against misunderstanding, I try not to refer to geographic actuality (geodiversity) in these pages as "geography" except where the meaning is unmistakable, as, for example, when we say, "Let us study the geography of Maine" (what we mean here, in my terminology, is, "Let us study the geodiversity of Maine"). Furthermore, an arbitrary and, so far as I know, unprecedented distinction is drawn between the adjectival forms "geograph*ical*" and "geograph*ic*," the former being used solely with reference to geograph*ical* awareness and the latter solely with reference to geodiversity, or geograph*ic* actuality (thus, "Maine's geograph*ic* features," "geograph*ic* distribution," "geograph*ic* location," but "geograph*ical* knowledge," "geograph*ical* society").

There is more unity and coherence in this volume than might meet the eye upon a first glance over the table of contents. The latter suggests the miscellany of a rummage sale, and, indeed, some of the papers are not unlike old clothes dug out of a trunk and patched up a bit. Actually, as Dr. Whitaker explains in the Foreword and as will be further explained in the Introduction, they are all offshoots of a long-sustained and gradually developing interest in the history of geography. But, like most things, this book is incomplete. Not only does every paper fall far short of "covering" the topic with which it nominally deals, but, in preparing the volume as a unit, I was constantly tempted by alluring byways, highways, and superhighways that further explorations might follow. Where some of these could lead is hinted at in the Epilogue.

The previously published papers have been revised to eliminate repetition, to bring nonbibliographical information from the footnotes into the text, and to shift the bibliographical notes to the end of the volume. Otherwise, except for a few rather inconsistent attempts to bring things up to date, the papers have been reprinted essentially as first published.

In these papers comments in the text and references (other than cross references) in the notes added subsequently are, in many instances, enclosed in brackets and indicated thus: (1965). Some of the references given in the notes to the 1926 version of the first paper have been omitted in cases where the works to which they refer are referred to elsewhere in the notes to this volume, and cross-references are given to pages where ideas sketched in that paper are more fully developed.

I am enormously indebted to Dr. Whitaker both for the boost he gave my spirits in 1953 and for writing the Foreword. Others from whose suggestions and criticism I have benefited are Professor Clarence J. Glacken, University of California; Mr. Martyn J. Bowden, Clark University; Mr. R. A. Skelton, British Museum; my brother-in-law, the Rev. Dr. A. C. McGiffert, Jr.; Mr. J. D. Elder, the editor; and my wife, who has helped me in countless ways, including the preparation of the index. Several of the papers were typed by Mrs. Doris G. Michael. Living, as I do, in a tiny village in northern New Hampshire, it would have been impossible for me to prepare this book for publication had it not been for privileges and courtesies extended by the director and staff of the Baker Library of Dartmouth College. I am also most appreciative of help cheerfully given on many occasions by Miss Nordis Felland and Mrs. Mollie Friedman of the American Geographical Society's staff.

I acknowledge with gratitude the following permissions: by the American Geographical Society, the Association of American Geographers, and the publishers of *Isis*, *Science*, and *The American Quarterly*, to include in this volume revisions of papers first published by them, as specifically indicated in the table of contents; and by Doubleday and Company and Harcourt, Brace and Company to quote on pages xviii and 216 from poems published by them, as indicated specifically in the notes.

It would be ungracious not to acknowledge my especial debt to the two institutions mainly responsible for this book: Harvard University, which not only first aroused my interest in human nature in geography but also, through its Press, is publishing these papers, and the American Geographical Society, where I actually learned most of what I know about the subject.

John K. Wright

Lyme, N. H.
September 17, 1965

Contents

ILLUSTRATIONS

Foreword

HUMAN NATURE IN GEOGRAPHY is a selection from the professional papers of a lifelong student of the history of geography. Concerned with widely varying aspects of the field, these papers are threaded together by the life and thought of one who has for more than fifty years studied and observed the development of geographic thought, and whose constant concern has been to contribute to that development. They are part of the harvest of these years, and reflect the insights gained in work as student, productive scholar, librarian, editor, and director of an outstanding geographical society. The volume deals with geography "in the sense of all that has been written and depicted and conceived on the subject," and with human nature, not as expressed in the face of the earth, but rather as it has affected the facts, ideas, truths, and myths brought together by geographically minded men.

The fourth paper identifies and illustrates the role in geographical scholarship—and in scholarship in general—of the human qualities of scholars. The effect in geography of such personal traits as originality, open-mindedness, accuracy, and intellectual curiosity appears in numerous places in this volume. Here, too, we see geography as it reflects the capacities and motives of many noteworthy geographers, such as Jedidiah Morse, William Morris Davis, Isaiah Bowman, and Ellen Churchill Semple. And from the opening paragraphs of the introductory essay to the end of the volume, the reader is aware that he is in the presence of a gentle scholar through whose spirit runs a puckish vein, an active worker in the field who shares with his readers his mature reflections on the role of human nature in the history of geography.

Although written over a number of years, these essays make up a symphony of thought, with various themes coming in and dropping out, so that the reader comes to identify them, to relish them, and to anticipate their reappearance. Included are such themes as the constant struggle to supplant error and myth with thoroughly substantiated facts, the problem of map making and map interpretation, explorers and their discoveries, measurement and the development of mathematical geography, creative imagination and its place in the discipline, and the relation of geography to other fields of thought and creativity, including poetry, religion, and mathematics. Indeed, we have here a personally conducted tour of the cathedral of learning that geographers have built, by one who has had a significant part in its construction. This recital of accomplishments and of failures may well arouse in fellow geographers both a sense of achievement and some concern lest the mistakes and biases of the past are present today in new guises. And the Introduction shows John K. Wright as ever the historian of geographical thought, of his own geographical thinking as well as that of others.

In the Preface, the author credits me with having stirred him to assemble this collection of papers. I am glad to share in the credit for the appearance of the volume, but I am painfully aware that I may seem, to some readers, to be casting myself in the role of the fly in La Fontaine's fable, "The Coach and the Fly." The fly, you may remember, buzzed around the team of six horses as they pulled a heavily loaded wagon to the top of the hill, whereupon he advised them to rest while he claimed credit for the achievement. As Philip Wayne translates La Fontaine's verses for us,

> After much strain the party tops the ridge.
> "Now we can take a breather," says our midge,
> "The job I had, to bring you up the slope!"[1]

[1] From "The Coach and the Fly" from the book *A Hundred Fables from La Fontaine* by Philip Wayne. Copyright © by Philip Wayne. Reprinted by permission of Doubleday & Company, Inc.

Disclaiming all credit except that due me for buzzing and stinging until the author settled to his work of selecting essays for this volume, I am pleased to emulate the fly by calling attention both to the deed and to the doer while he rests a bit before the next long pull.

For in a work like this the text and the writer are inseparable. As the famous eighteenth-century French geographer and naturalist, Comte de Buffon, put the matter for all later students of writing, "Le style est l'homme même." Clearly we are in the presence of scholarship that is profound, elaborate, and meticulous. Moreover, the play of mind of a historically inclined person is, surprisingly, combined with that of a mathematically minded one. Dr. Wright's interest in quantities ranges from the influence of mathematical concepts on geography-as-actuality (as in the locating of boundary lines) to the manner in which geographical writers have used mathematical data both in medieval and in more recent times. One finishes the essay on "Measuring and Counting in Early American Geography" with an uneasy feeling that the manipulation of numbers shown there as having polished over and concealed crudities in the basic data a century and a half ago may also be doing much the same today.

Throughout the volume, we recognize the work of a free-ranging mind. Could this volume have emerged from a strictly academic environment, where there is a serious dedication to the field of geography as an autonomous discipline? Fortunately, the American Geographical Society and, in recent years, the home study in Lyme, New Hampshire, fostered in Dr. Wright the scholarly daring and detachment that breathe through these pages. Serious reflection on these essays is guaranteed to restore a measure of mental balance to youth too soberly absorbed in the current brand of "new" geography, and to console and cheer the aging practitioner who has been, perforce, concerned with geography in its broadest sense.

We see here, then, a geographical scholar at serious work and, at the same time, at play, his findings shot through with the joyful antics of an adventurous, imaginative mind. We find ourselves a little envious of the man who can take his professional cares so lightly, who can do his work and still have time, energy, and inclination for some of the embroidery of life.

In truth I cannot always tell where the serious aspects of these essays end and the playful antics begin, where Dr. Wright is really serious and where he is just spoofing his colleagues. How concerned should we get, he may be asking, with elaborate computations based on data that are themselves only approximate? How far can symbols take us when the words they represent are few and clear, whereas the symbols used are easily confused and quickly forgotten? Carried along as we are from the description of a nonexistent country, in the Introduction, to the mapping of an unmappable concept, in the twelfth paper, we can expect almost anything from this man. Dr. Wright's plea for the free play of the imagination finds full expression in this selection of his essays.

John K. Wright provides weighty evidence in this volume that "the history of geographical knowledge and belief should be viewed as part of the larger history of civilization," and that cross-fertilization of disciplines is of critical importance to the development and uses of geographical scholarship. This book abundantly supports his thesis, moreover, that it is possible and desirable "broadly to conceive and broadly to treat of geography and its variegated history."

J. Russell Whitaker

Nashville, Tennessee
January 1964

Human Nature IN GEOGRAPHY

'Tis evident, that all the sciences have a relation, greater or less, to human nature; and that however wide any of them may seem to run from it, they still return back by one passage or another.

David Hume, *Treatise of Human Nature*, Introduction

In a word, Geography will prove an useful amusement to every curious and inquisitive mind.

A. F. Büsching, *A New System of Geography* (London, 1762)

Introduction

After making some futile attempts to arrange and explain the contents of this book according to systematic principles, I finally decided that it would be simpler and better to put the papers in chronologic order and tell in the Introduction how they came to be written. Naturally, this has involved autobiographic reminiscence, for which my apologies are offered, since scholarly mores call for the appearance, at least, of impersonality.

This book is a mixture of history and geography, two subjects in which I developed an interest about the time of the Spanish-American War. That conflict occurred soon after I had learned to read and had grown old enough to notice events of the kind. The war enlarged the historical and geographical outlook of the American nation—and also mine. But, like my father and brother and many other persons when they were children, I focused my interests upon an imaginary country, called "Cravay." I mapped Cravay and nearly put my eyes out. There were several successive small-scale maps of the whole country; there were also maps of parts of it on medium and large scales, of which the large-scale series resemble the topographic sheets of the United States Geological Survey and may be fitted together so as to present outlines conforming, as they should, with those shown in lesser detail on the maps of smaller scale. About 1903 I wrote a history of Cravay with the aid of friends whom I had roped in. It began in medieval times—1862 to be exact—with a period of geographical discovery and exploration which lasted for about a decade and was punctuated and followed by many wars, including one of liberation from the Mother Country. My brother, Austin, who was eight years older, encouraged me in this and helped illustrate the history. By being mysterious about his submarine boat and his island, both of which he refused to let me visit, Austin may have been partly responsible for the inception of Cravay. I think, however, that it was due principally to spontaneous combustion, though the flames, once ignited, were fanned by map study and history study in school. Professor William Morris Davis[1] added realistic fuel to the fires and did much to bring them down to earth from outer space (Cravay was on a planet of Sirius). At the age of 14 and 15, fifteen months in Europe, mostly in Greece, completed the transfer by making this terrestrial habitation seem more fascinating than anything that the imagination could conjure up among the stars; and thus Cravay passed into oblivion not long before Austin's island was beginning to take shape as *Islandia*.[2]

During my college years (1909–1912) the geographical fire flared higher than the historical. In the spring and summer of 1911 I served as rodman on a surveying expedition along the Atlantic coast between the Florida keys and the New Hampshire beaches, and many vacations then and later were spent with friends exploring and mapping the northern half (in Maine) of what was later named the Mahoosuc Range and was at that time a miniature *terra incognita* (see Chapter 5). Since no courses in the geography of man were given at Harvard in those days, I got my geography through geology, physiography, meteorology, and climatology, and nearly became a geomorphologist. Man, however, interested me more than did nature-minus-man, and the historical fire, though only smoldering, had not died out. So, when it became necessary to make a decision in the summer of 1912, it was to "go into history." This was radical, for the only college course in history that I had taken was one on Napoleon (you could do things like that at Harvard when the freedom of the will instituted by President Eliot still prevailed). "Going into history" meant graduate study and teaching, and finally, after interruptions due to the First World War, a doctorate in European history in 1922. But from the beginning in 1912 my resolve was to make my history geographical. I considered two alternatives: (1) study of history-as-actuality in a geographical manner (that is, "geographical history") à la Buckle or Turner or Ratzel or Semple, or (2) study of the history of geographical knowledge and belief as part of the larger history of civilization. The former had a strong lure, especially after I had read Miss Semple's *Influences of Geographic Environment* in the summer of 1912 (see Chapter 12) and had audited Professor Turner's course on the American frontier. But, for a variety of reasons (the year in Europe at an impressionable age being one), I chose European rather than American history and was thus precluded from becoming a Turnerian. Also, the philosophy of history that animated those who taught me at Harvard was (or I felt it to be) one of disapproval of the espousal of any particular philosophy of history. Hence, with the connivance of my professors I adopted the second alternative; the history of geography then became my main scholarly interest in 1912–1913 and has remained so ever since.

I took no courses in the history of geography because there were none to take. Indeed, the Harvard historians, other than Channing, would seem to have known little about the subject. But my teachers looked with indulgence on my writing term papers on ancient and medieval geographers and exploration, and they gave me the rare (perhaps unique) privilege of offering the history of geography as my special field for the doctor's degree.

Since my professors probably knew little about the history of geography, I did not receive much guidance from them with regard to that subject. An interest in geographical explorations and the nature of geographical ideas prevalent during the Middle Ages was aroused by Professor H. L. Gray's course on the Renaissance (embracing also much that was medieval) which I took in 1913–1914, and in 1919 Professor Charles H. Haskins persuaded me to write my doctoral thesis on the geographical science and lore of Western Europe in the twelfth and early thirteenth centuries. This I did under his supervision, and Gray and Haskins introduced me to many books relating to medieval civilization, which, in turn, made it comparatively simple to track down materials of geographical interest. Of the larger subject of the history of geography as a whole I was self-taught, except that Professor Channing helped me become acquainted with maps and books bearing on the discovery of America. In the course of this self-education I regret to say that

I overlooked von Humboldt's immortal *Kosmos.*[3] I did, however, "discover" Peschel's and Günther's histories of geography,[4] and for a long while they were my bibles, Peschel for its breadth of scope and easily comprehensible German and Günther for its bibliographic references. Peschel stressed the role of mathematics and statistics in the evolution of geography, and his book aroused a lasting interest, kept alive and augmented in later years by the necessity of dabbling in cartography while I was a member of the staff of the American Geographical Society and by finding in O. M. Miller of that staff an ingenious mathematically minded friend. In this present book the papers on the "Heights of Mountains" (1958) and on "Measuring and Counting in Early-American Geography" (1964) more or less in their entirety, as well as parts of the ones on "Map Makers are Human" (1942) and on "Miss Semple's 'Influences of Geographic Environment' " (1961), bear witness to this interest.

After a year in Europe (1919-1920) traveling, working on the thesis, and attending lectures in Paris by the geographers Gallois, Demangeon, De Martonne, and Brunhes—which gave me some feeling for the nature of French geography in those days—I became Librarian of the American Geographical Society of New York, an immediate result of Professor Haskins's association at the Paris Peace Conference with Dr. Isaiah Bowman, the Director of the Society. Thus, having intended for eight years to teach history, I threw in my lot with the geographers and since October 1, 1920, have been professionally associated with them. Yet, in "going into" geography, I resolved to keep my geography as historical as possible, much as I had done the reverse in 1912. This was easy. The position at the Society gave ample opportunity for indulgence in study of the history of geography and often made such indulgence a pleasant duty as well, for though I was titularly a librarian, my work was for the most part that of an editor and writer, and I was often called upon, and could often decide for myself, to perform tasks of a historical nature. Thus, by the mid-twenties I had come to feel that my ideas concerning the history of geography were ripe enough to warrant presentation before professional societies, and I proceeded so to present them in papers read during the Christmas holidays of 1925 before the History of Science Society and the Association of American Geographers.[5] One of these, "A

Plea for the History of Geography" (1925), is included in this volume as a result of a plea made to me by my friend, Martyn J. Bowden, early in 1964, and it represents a link between the thoughts expressed as follows in the opening paragraph of the published version of my doctoral thesis and thoughts further developed in many places later on in this book:

> When viewed historically, geographical concepts are seen to have come from an immense variety of sources. They have sprung partly from activities that cause men to travel over the surface of the earth: war, commerce, pilgrimage, diplomacy, pleasure. They have also sprung from the accumulated learning and lore of preceding ages and to no small extent from unfettered flights of the imagination. The history of geography, therefore, leads its students into many fields, affording them a key by means of which they may gain a sounder understanding of the extensive ranges of human activity and of the evolution of important phases of intellectual life.[6]

It is one of the minor laws of human nature that in educational and research institutions the less time one has to study and think owing to administrative responsibilities, the greater the quantity of public pontification one is tempted and obliged to engage in. This is particularly true of university presidents. It was also true of me, though to a far lesser degree, while I was serving as Director of the American Geographical Society (1938–1949), and it accounts for three of the four papers dating from that period—"Where History and Geography Meet" (1940), "Human Nature in Science" (1943), and "Terrae Incognitae" (1946)—all of them addresses before learned societies. The fourth, "Map Makers are Human" (1942), though not an address, is akin to the other three in spirit and subject and, like them, was called forth by the special circumstances of my position.

"Where History and Geography Meet," as its subtitle ("Recent American Studies in the History of Exploration") specifies, actually deals with only a very small portion of the gigantic meeting place implied. It advocates a "geographical approach" to the history of geographical discovery, suggests how such an approach differs from the more frequent nongeographical approaches, and intimates what it might entail. Since the paper was prepared for a Pan-American group, "recent American studies" were cited to illustrate these general ideas, all of which were adaptations of ideas expounded in 1925.

"Map Makers are Human," "Human Nature in Science," and "Terrae Incognitae" are loosely allied. During the Second World War, both before and after Pearl Harbor, the American Geographical Society did much work for various agencies of the United States Government. We were asked, among other things, to compile from highly generalized maps even more highly generalized ones. This struck me as like making impressionistic copies of impressionistic landscape paintings, or writing generalized histories from generalized secondary sources. Skepticism regarding the latter procedure that Professor Haskins had engendered in me thirty years earlier, together with a certain familiarity with the ways in which maps were compiled, prompted the composition of "Map Makers are Human" (1942), in illustration of the propositions that since "maps are drawn by men and not turned out automatically by machines" they may reflect not only human abilities but human disabilities, and that, in facing the problems of amplifying insufficient and simplifying superabundant evidence, the cartographer is confronted with difficulties not unlike those that every historian faces. When, two years later, it became my duty to address a group of geologists and geographers on the assigned theme of "the indispensability of science for the future of civilization," all I had to do (in "Human Nature in Science") was to turn the tables and speak, rather, on the indispensability of civilization for the future of science and use, in illustration, ideas of the same sort as those expressed in Chapter 3, "Map Makers are Human."[7]

When the War was over and one could relax despite the Bomb, I addressed the Association of American Geographers on "Terrae Incognitae: the Place of the Imagination in Geography" (1946). This was Cravay once more, except that it urged geographers to look for their "Cravays" in the unknownness that still lurks within the lands and seas about them. Perhaps it was also motivated in part by the ghost of the rebelliousness that in 1912 had made me turn from the rocky trails of geomorphology to the more human pathways of history. "Terrae Incognitae" begins with the proposition that the words "*terra incognita*" have an appeal to the imagination. Then, after considering the nature of various kinds of *terrae incognitae*, both literal and figurative, the paper discusses some of the functions of and needs for the exercise of imaginative faculties in the pursuit of geography. The thought is developed that geographical writing and teaching could be made more interesting, inspiring, and generally effective were there at least a few scholars who could and would treat geography as one of the humanities rather than exclusively as a natural or social or socionatural science. As a means to this end the cultivation was suggested of a field for which I ventured (with some hesitation) to coin the term *geosophy*, a field that I defined as "the study of geographical knowledge from any or all points of view" as distinguished from *geography*, or the study of the realities with which geographical knowledge has to do.

After ceasing to be Director in 1949, I continued at the American Geographical Society as a Research Associate for seven years. During this period and since my retirement (1956) my concern has been mainly with the history of geography in this country. Hence, the papers dating from after 1949 are all products of a transatlantic crossing of the focus of my interests, a change that may have been initiated in the early thirties by the editing of Dr. Paullin's *Atlas of the Historical Geography of the United States*[8] and a symposium on New England,[9] and was certainly confirmed by my first work at the Society as Research Associate. The latter was the congenial task, made possible by the late Archer M. Huntington, Esq., of writing a history of the Society.[10] Institutional histories are proverbially among the dullest of books, especially to readers not already familiar with the institutions to which they relate. But, other things being equal, their dullness is in direct mathematical ratio to the relative amount of detailed information that they purvey concerning the internal affairs of the institutions treated, as compared with what they have to say about broader matters. In writing the Society's history I tried to keep this mathematical ratio in mind and to avoid the ultimate in dullness by giving approximately equal attention to the Society as a society and to the growth of American geography since 1850 as illustrated in the Society's interests, contacts, activities, and publications. Thus I was led to make many detours off the main road of the institutional record and in the process to encounter much of interest, some of which I did and some of which I did not report upon in the book.[11] In the present volume the papers on "The Open Polar Sea" (1953) and on "Daniel Coit Gilman" (1961) are immediate by-products of studies pursued in connection with the history, and my personal experience in carrying on these studies gave me an illustration of (and perhaps the inspiration for) one of the two leading ideas suggested in "Medievalism and Watersheds" (1960).

In 1949 Dr. Stefansson called my attention to the seemingly strange belief, in which Kane and other American Arctic explorers of the mid-nineteenth century had shared, that a great ice-free sea surrounds the North Pole. Several pages are devoted to this in the history of the American Geographical Society, and I later made a more detailed examination of the origins, rise, and fall of this long-lived and highly influential error. Although geographical errors of divers kinds, both in method and in fact, are touched upon in nearly every paper in this volume, "The Open Polar Sea" is the only one for which the nature and history of such error provides the central theme.

While investigating the Society's institutional history I learned that, as a young man, Daniel Coit Gilman (later the first President of the Johns Hopkins University) served for a few months as General Secretary of the Society and also that he was a geographer. This helps explain the paper devoted to him as a geographer and historian and also, incidentally, as one who, through his creative leadership in the establishment of postgraduate education in American universities, exerted a powerful indirect influence, still felt today, upon the study and teaching of geography.

Four of the seven papers that remain to be mentioned are revisions of talks originally given to divers groups and hence are adaptations to the interests of those groups of special interests of my own. Two were first prepared for, and were read and discussed at, sessions of the Columbia University Seminar in American Civilization. This consists of professors (not all from Columbia), museum curators, librarians, and others who meet regularly to consider various aspects of the civilization of the United States. "What's 'American' About American Geography" (1956) was presented at a time when the Seminar was devoting itself to the elusive question of whether or not different manifestations of American civilization are distinctively or characteristically "American," a question bristling with substantive and semantic problems upon which my paper touches, none too heavily, I hope. Having thus been impelled to consider the "Americanness" of American geography, I turned at a later time to the opposite, its "non-Americanness," on which I wrote "On Medievalism and Watersheds in American Geography" (1960), the latter for the Seminar at a time when it was examining the question of "continuity versus change in American history." In a sense, the two Seminar papers were made to order, as was also "Miss Semple's Influences . . ." (1962), the latter for a special session on Miss Semple arranged by the program committee for the annual meeting of the Association of American Geographers at East Lansing in 1961. One of the talks, however, was on a topic wholly of my own choosing. "The Heights of Mountains" (1958), originally read before the New York Chapter of the History of Science Society, is the resultant of four different motivations: an athletic interest in climbing mountains, an aesthetic interest in their grandeur, a statistical interest in their comparative heights, and a historical interest in the manner in which knowledge of those heights has been acquired.

To complete the roster, mention must be made of "From 'Kubla Khan' to Florida" (1956), "Measuring and Counting in Early American Geography" (1964), and "Early American Geopiety" (1964). Although none of these was prepared as a talk, part of the second was so used. All three relate to periods in the development of American geography that preceded the founding of the American Geographical Society in 1851 and were written with a view to possible use in a book on the history of Early American geography —a book still little more than a dream. "Kubla Khan" is concerned with the realm where poetry and geography meet, "Measuring and Counting" with that of geography and mathematics, and "Geopiety" with the meeting ground of geography and religion.

The appearance of a northern landscape in the early springtime is often determined by three factors: the major, enduring contours of the terrain; snowbanks flecked across open fields; and young growths just springing to life. The dominant aspect of Early American Geography was similarly determined by enduring characteristics that it shared with all geographical knowledge in the Western cultural tradition, by characteristics that were waning and disappearing like the snowbanks, and by characteristics which, like the young growths, were first appearing and destined for more luxuriant blossoming and increase in later times. Traditional, authoritarian, and relics of the past, the "snowbanks" in Early American geography in some ways resembled West European geography during the Middle Ages, of which they were in part a direct heritage, as suggested in the tenth paper. Hence they may be classed as "medieval." In contrast, the "young growths" were modern, in that they more nearly resembled and foreshadowed the geography of today. Like the tenth, the last paper ("Early American Geopiety") dwells upon lingering snows, whereas the next to the last one ("Measuring and Counting") is concerned rather with harbingers of spring and of a summer that promises to be hot and thundery.

The seemingly preposterous diversity of the times and matters considered in this book exemplify what might be called "foolrushery"—what fools do when they rush in where angels fear to tread. My belief in this has been growing on me. Were I a sensible angel I should have been much too afraid of the classicists and medievalists, the Americanists and statisticians, the logicians and theologians, the literary critics and psychologists, the geologists and methodologists, to write three-quarters of this volume. I believe in foolrushery especially on the part of scholarly persons over sixty-five years of age, because it promotes the cause of interdisciplinary cross-fertilization. Were the latter to perish from the earth, science and scholarship would shrivel up and die of sterility, just as our gardens would if all the bees were exterminated by pesticides. Interdisciplinary cross-fertilization is as indispensable to the balance of scholarship as the bees are to the balance of nature. But, however that may be, this book, I hope, exemplifies the chief thought expressed in "Terrae Incognitae"—that it is possible and perhaps desirable broadly to conceive and broadly to treat of geography and its variegated history.

CHAPTER 1

A Plea for the History of Geography

THE HISTORY of geography as a whole and in its wider bearings has been neglected in America, or at least it has not received the attention which an enthusiast may, perhaps, be permitted to regard as its due. Though they refer to very different matters, the terms *history of geography* and *historical geography* are often loosely employed even by geographers and historians themselves. Lest there be any confusion, at the outset a working definition may be given of each. *Historical geography* is the study of geographic facts as they have existed in historical times—boundary changes, former distributions of population, the development of trade routes, and so on. The *history of geography*, on the other hand, is the history of geographical ideas.

HISTORICAL SKETCH OF STUDIES OF THE HISTORY OF GEOGRAPHY

Investigations into the history of the geographical ideas of earlier ages were made even in classical times. An academic controversy was waged over the reliability of geographical data in Homer's *Odyssey*.[1] Strabo, who believed the *Odyssey* to be authentic and reliable, in a long and controversial passage leveled criticism against Eratosthenes for holding that Homer should be read as a poet and not as a scientific authority.[2] All through the Middle Ages students were interested in what classical writers had written on geographical subjects, thus, in a certain sense, in the history of ancient geography. Ancient geography, however, was of little moment to medieval scholars from the historical point of view. They studied it, rather, for the light that it might shed on the geography of their own time. The revival of the Greek and Roman classics in the Renaissance brought with it enthusiasm not only for classical history, literature, and archaeology, but also for classical geography and geographers.[3] But this enthusiasm was more for the historical geography of antiquity than for the history of geography in antiquity. Ptolemy's treatise with the atlases based upon it and Strabo's great work were translated into Latin and published early in the fourteenth century.[4] The humanists, however, looked to Strabo and Ptolemy for what they actually told of classical lands and countries, not for what they revealed of stages in the growth of geographical knowledge. As had been the case with Pliny, Aristotle, and others in the Middle Ages, Ptolemy's *Geography* until the latter part of the sixteenth century was used as a source for geographical descriptions and as a basis for maps of the contemporary world. Thus the impress of Ptolemy's overestimate of certain longitudes persisted until the close of the seventeenth century.[5] Nevertheless, it was partly to the classical revival of the Renaissance that we may trace the origins of modern research in the history of geography. The interest in historical geography that was then kindled has lasted ever since and has led many students to delve into the geographical works of earlier times; and from the investigations thus inspired many of the data have been assembled upon which studies in the history of geography have been and will be based.

Researches in the history of geography during the Age of Discovery were also stimulated by the explorations themselves. The navigators of the time believed that the writings of earlier geographers contained information of practical value. Columbus was a serious student of the scientific opinions of Aristotle, Seneca, and Ptolemy, and of the travels of Marco Polo. The acceptance of ancient theories of the distribution of land and water over the earth's surface persisted until as late as the middle of the eighteenth century. If Columbus' glory rests upon the discovery of continents not known to exist before his time, that of Captain Cook rests upon the sweeping away of an imaginary continent that for centuries had filled most of the Southern Hemisphere.[6]

Then again, the romantic interest of the great discoveries, no less than their political and commercial aspect, early led to the compilation of collections of voyages.[7] Among these those of Ramusio, De Bry, and the volumes of the English Hakluyt and Purchas are the best known. The work of compiling, editing, and commenting on the narrations of voyages has been continued ever since, and the results are now being embodied in such monumental series as the publications of the Hakluyt and Linschoten Societies, or the *Recueil de voyages et de mémoires* of the Société de Géographie, or the *Recueil de voyages et de documents pour servir à l'histoire de la géographie*, or the *Library of the Palestine Pilgrims Text Society;* and in America by the publications of the Champlain Society and Thwaites's *Jesuit Relations.*

Before the nineteenth century examination into the history of geography was devoted almost exclusively to the regional phases of the subject—to voyages and explorations—and indeed at all times this has been by far the most intensively cultivated portion of the field. The development of modern scientific geography in Europe, however, has been accompanied by a growth of interest on the part of a relatively few students in the evolution of geographical theories and methods, in the history, that is, of mathematical and physical geography, cartography, and of bio- and anthropogeography.[8]

[While the foregoing is probably more or less well founded, I have learned since 1925 that there was considerably more interest in certain phases of the history of geography—notably those relating to cartography, mountain altitudes, theories concerning the effects of the Deluge, and so forth—during the Renaissance and on down into the eighteenth century than I had suspected. (1965)]

The nineteenth and twentieth centuries have also seen the production of a limited number of synthetic works by European scholars relating to the development of geographical knowledge and belief as a whole and over long periods of time. Among these Oscar Peschel's *Geschichte der Erdkunde* (1877)[9] should be given the first place as the only really adequate and satisfactory work in existence in which the attempt is made to trace the entire development of geography in the Western World. Thoroughly documented, written in a pleasing style, giving due weight to the scientific as well as to the exploratory phases of the subject, this volume would seem to be a model of what a study in the history of science should be; but unfortunately it is now old and seems to be little known outside of

Germany. Günther's compact *Geschichte der Erdkunde* (1904)[10] is more in the nature of a reference work, encyclopedic in its style and full of bibliographical notes. Vivien de St. Martin, in his otherwise admirable *Histoire de la Géographie*,[11] which appeared at about the same time as the work of Peschel, tended to neglect the scientific side of the subject.[12] [In English there is still no major work of scholarship dealing comprehensively with the whole history of geography, although a number of lesser books provide competent introductions to the field.[13] (1965)]

Thus we see that in European scholarship, particularly on the continent, historical geography and the history of geography have long held honored places. One of the main reasons for this is the fact that geography has been associated in continental schools and universities with history and the humanities rather than with the physical and natural sciences. Indeed, Professor Emmanuel de Martonne of the Sorbonne, now one of the leading representatives of the modern French geographical school, would seem to regard the recent development of geography in France as something of an emancipation from the dominance of history.[14] In America, on the other hand, scientific geography has been evolved mainly out of geology and instruction in geography in the universities is usually given in close connection with geology departments. Until the past few years our geographers have devoted less attention than foreign students to the human and historical phases of their subject and almost none to its history. Indeed, most American geographers have been so intent upon building up the newer aspects of their specialty and upon struggling for its recognition as a topic worthy of a dignified and independent place in university curricula, that they have had little opportunity to interest themselves in its past development. Yet it is pleasing to note that a course in the history of geography is offered by Professor J. Paul Goode at the University of Chicago. So far as I know, this is the only college course in the field now given in the United States.

[Since the above was written there has no doubt been a considerable increase in the amount of attention given to the history of geography, both in the teaching of geography in graduate schools and in the production by geographers of publications on the subject, but as compared with the gains in other branches of geographical study the gains have been trifling. In the *Handbook-Directory* of the Association of American Geographers for 1961 only 14 members of that association are listed as having specified the history of geography, of geographical thought, or of geographical exploration among their special fields. During the last two or three years, however, there would seem to have been a marked freshening of interest in the subject, if we may judge from the number of papers devoted to it read at meetings of the Association. (1965)]

The beginnings of American history are but a chapter in the history of geographical exploration, and consequently our historians have interested themselves not only in the history of the exploration of the Western Hemisphere but also—though to a lesser degree—in those earlier phases of the history of geography which bear directly or indirectly on the discovery of America. We need but mention the names of Winsor, Fiske, Thacher, Harrisse, and Vignaud (the last two American by birth but French by residence) among the older generation of students whose interest led them to investigate the relations of ancient and medieval ideas regarding the size and shape of the earth and distribution of land and water to the more immediate problems which faced Columbus and other navigators. An American, the late W. H. Tillinghast, had to his credit an unusually profound study of ancient geographical thought.[15] Moreover, the interest not only of genuine historians but of a formidable array of hack writers and cranks has at all times been concerned with Columbus and with the more nebulous questions of pre-Columbian voyages and cultural connections across the Atlantic and Pacific. The decade of the 1890's, including as it did the quadricentennial year, was notable for the vast volume then poured out of Columbian literature both good and bad. Probably one of the most important recent American contributions to Columbian studies is a volume on the geographical conceptions and aims of Columbus by a young Californian scholar, George E. Nunn.[16] A valuable service to the history of geography has been performed by the Hispanic Society of America in publishing many early maps and by Professor E. L. Stevenson in stressing the importance of maps as historical sources. California and Texas historians are doing very important work on the early Spanish discoveries. The publication of the *Jesuit Relations* by Thwaites has opened the door to the history of explorations in the interior of North America. The *Relations* were used, for instance, as the primary source for an investigation of the geographical knowledge acquired by the Jesuits, recently published by Nellis M. Crouse as a Cornell doctoral dissertation.[17] But on the whole, though much important special work has been done, almost no American historian has devoted himself to the history of geography as it bears upon other parts of the world than America or to its broader evolution. [This was something of an overstatement when first written in 1925 and would certainly call for qualification as of 1965.]

THE SCOPE OF THE HISTORY OF GEOGRAPHY

The foregoing sketch of what has actually been accomplished in the history of geography must serve as the background for a statement of the possible scope of future studies in this field.

Brunhes asserts that the history of geography is nothing more than a part of the history of the sciences.[18] This is certainly true of the history of *scientific* geography. Yet scientific geography is merely one of a large group of consciously geographical activities and interests, the history of which is an important element in intellectual and social history. I should include within the history of geography not only the history of scientific geography, but the history of these other geographical interests and activities as well. That many of the latter may be approached historically from wholly different points of view does not render them any less fitting subjects for examination from the point of view of geography. Geography in its essence is a sphere of ideas relating to the regional groupings of phenomena on the earth's surface. The fact that it overlaps and borrows from other spheres of ideas bothers us no longer; we are fortunately outgrowing the tendency to mark off sharp and exclusive boundaries between the different domains of scholarship. [The undiscriminating all-inclusive "we" in the preceding sentence reflects the optimism of youth. (1965)]

Human beings possess in varying degrees a geographical sense. The habit of thinking in geographical terms is widespread among many peoples.[19] The Eskimos, some of the Bedouins, some of the Polynesians, have truly marvelous topographical powers and are able to draw outline maps of very complicated tracts of country.[20] There are a great many civilized individuals without notable geographical training and without any reason for professional interest in the subject who nevertheless find an atlas among the most absorbing of books[21] and enjoy works of travel quite as much for the geographical meat which they contain as for their exotic or adventurous flavoring. The geographical sense is an intellectual response to the environmental *milieu*. It leads to the acquisition of geographical ideas and to their expression in a multitude of forms. If the history of historiography is defined as the history of man's consciously expressed intellectual interpretations of events in their chronological order, the history of geography might be defined as the history of man's consciously expressed intellectual interpretations of his terrestrial environment; or, more simply, as the history of geographical ideas as they find spoken, graphic, or written expression. Nor need all these ideas be accurate, systematic, logical, or reasoned; for is not the history of error, folly, and emotion often as enlightening as the history of wisdom?

If we thus define the history of geography, what may we include within its limits?

We may include the consideration of ideas that for want of a better term we shall call "scientific." This does not mean that these ideas are necessarily scientific in the modern, which some would consider the absolute, sense of the word, that is, that they are necessarily based on accurate observation, logically developed and critically controlled. I use the term, rather, in a relative sense to mean ideas that are systematically worked out in conformity with the best intellectual standards of their age and that find expression in maps or formal scientific treatises. In addition to these, we may also include within the study of the history of geography the consideration of ideas that we shall call "nonscientific," not implying thereby that they are necessarily "unscientific" or erroneous, but, rather, that they are not expressed in scientific form.

THE ACQUISITION OF GEOGRAPHICAL IDEAS

Before we turn to the character of geographical ideas, both scientific and nonscientific, a few words must be said about the manner in which they are acquired at first hand; for the history of thought can hardly be understood without some knowledge of its origins and stimuli.

Geographical ideas are obtained at first hand from what is usually called exploration when conducted in little-known countries and field work when conducted in the better-known parts of the world

—whether this exploration or field work be strictly geographical, or geodetic, topographical, geological, ethnographical, biological, or otherwise.

Only within the last two or three centuries have explorations and field work been carried out exclusively for scientific purposes. One of the first scientific expeditions of the modern type, accompanied by specialists such as a botanist, a zoologist, a surveyor, an artist, was that of the Dane, Carsten Niebuhr, to the Yemen in 1761-1763. "Peter Forskall, a Swede by birth and a pupil of the great Linnaeus, was a physician with special knowledge of botany; Christian Charles Cramer, a surgeon and zoologist; Frederick Christian von Haven, a philologist and Oriental scholar; George William Baurenfeind, an artist; and lastly, Carsten Niebuhr, lieutenant of engineers, a mathematician and practical surveyor . . . Two of the party died in Yemen, one (and the Swedish servant) at sea on the voyage to India, and one on arrival there: none by violence, but all by the poison of the Yemen air. Niebuhr alone brought his report and the incomplete notes of his comrades to Denmark again."[22] Vitus Bering's two voyages (1725–1730, 1733–1742)[23] were also in the nature of scientific explorations, as were the famous French expeditions for the measurement of arcs of meridian in Peru and Ecuador and in Lapland (1736-1743).[24]

Nearly all earlier explorations and most of the more recent ones have been undertaken primarily for commercial or political reasons, though incidentally geographical knowledge has been acquired through them. Adequately interpreted, the history of exploration should be more than a dry catalogue of dates and names and routes, or a romantic but unsubstantial chronicle of adventures. It should involve some examination of the complex factors that lie back of exploration in any given age or region, and, in turn, it should throw some light on the effects of the expansion of regional knowledge upon economic, political, social, spiritual, and intellectual conditions. The progress of exploration is meaningless unless viewed against a wider historical setting.

Nonscientific geographical ideas are derived at first hand through travel, whether it be for commercial, military, political, or administrative ends or whether it fall into the category of recreational travel. The latter may range from "tripping" and "joy riding" at the lower end of the scale up through various gradations of tourist travel, yachting, student wandering, and so on, to those more serious levels, such as exploratory mountaineering, which merge into genuine field work and scientific exploration. Travel in all its forms is an enormously important social activity. Its history and philosophy have been all too little studied and interpreted.

[As of 1965 this last assertion would be highly misleading with respect to the history of travel[25] and more or less so with respect to its "philosophy,"[26] about which considerable interest is being developed on the part of American and British geographers of the younger generations and presumably by those of other countries as well.]

THE CHARACTER AND EXPRESSION OF GEOGRAPHICAL IDEAS

Scientific Geographical Ideas. Now let us consider briefly something of the character and expression of geographical ideas themselves, and first of those that we call scientific.

It should be observed that scientific geographical ideas are not necessarily expressed exclusively through publications devoted in name to geography. One of the most interesting trends in modern scholarship is the ever-increasing manifestation of the geographical spirit in all the natural and social sciences,[27] a development that no serious student of recent intellectual progress may well overlook. Nor in the study of earlier periods is it possible to avoid taking into consideration the geographical ideas revealed in nongeographical writings.

Among the various media for the specifically geographical expression of geographical ideas, the map since very early days has been the most graphic. The history of cartography is an integral part of the history of geography. But the history of cartography should always be more than an antiquarian study. A comparative examination of maps of different ages gives a remarkably clear view of varying intellectual qualities and technical abilities. A medieval map often reveals an atmosphere of credulity and respect for authority, but withal a love of beauty in form and detail. What better way is there of grasping some of the essential differences between the spirit of medieval and of modern science than to set side by side maps of the two ages? The evolution of maps is intimately associated with the development of astronomy, geodetics, and trigonometry; with the growth of spherical geometry in the projections adopted; and with the evolution of the technical arts in the draftsmanship, engraving, and printing. The manner in which the subject matter on maps is selected and represented sheds light on the critical acumen of the map makers.[28] The origin of most large-scale topographical maps of modern Europe is to be sought in military necessities and it is significant that nearly all the great topographical surveys have been made by war departments, except in the United States and Canada, where economic considerations have been foremost.[29] The history of topographical maps should always be linked with the history of military or administrative economic policy.

The different characters that geographical ideas have assumed in different times and countries have been conditioned partly by environmental and partly by human factors.[30]

The impress of geographic conditions themselves upon the nature of geographical thought is an alluring subject which has not been much investigated. For instance, in parts of Greece, Dalmatia, and Asia Minor, there are extensive limestone regions in which the process of solution has produced systems of subterranean caverns and watercourses. This not only gave rise to the old story that the River Alpheus in the Peloponnesus passes under the Ionian Sea to spring forth again in the fountain of Arethusa in Syracuse, but was the basis of a hydrographic theory that prevailed through the Middle Ages—the theory that the entire globe is seamed with watercourses which derive their ultimate source of supply either from vast interior reservoirs or from the sea.[31]

Another example of the same thing is to be found in the amazingly clear understanding of the processes of weathering and erosion held by certain Arabic and Persian writers of the Middle Ages, obviously facilitated by the aridity and bareness of the soil in the East, which permitted better observations of these processes than were possible in contemporary Europe.[32] Similarly, in our own West—where the physiographic nature of a region is frequently apparent even from the train window—aridity and lack of vegetation cover has contributed largely to the preeminence of American geographers in the study of land forms, or geomorphology.[33]

As a final example we may refer to the often obvious but often less apparent relation of geographic facts to the progress of exploration in the Age of Discovery. Nunn believes that Columbus deliberately chose a southern route sailing westward and a northern route sailing homeward in order to take advantage respectively of the northeasterly trade winds and the prevailing westerlies.[34]

More potent, however, than the direct effects of environmental conditions on the evolution of geographical ideas has been the influence of the general level of culture and intellectual life. The character of geographical writing in different countries and at different ages illustrates in a striking manner the outstanding qualities of contemporary or national thought. The scientific geography of the Greeks of the Hellenistic period is primarily speculative, theoretical, experimental; the geography of the Romans was essentially descriptive and practical. Early medieval geography was dominated by theology and based largely on scriptural exegesis.[35] The geographical works of the Renaissance reveal the newly revived interest in classical antiquity and something of the humanist's pagan love of the world about him. Work of the modern French school of human geographers reflects the severely critical quality of the French genius, its logical procedure of thought, its clarity and perfection of expression. The German anthropogeographers have been far more prone to build theoretical structures and systems. In Great Britain geographical interests (with notable exceptions) have tended to center on explorations and on economic geography. [As to "what's 'American' about American geography," see Chapter 8. (1965)]

Nonscientific Geographical Ideas. Geographical ideas of varying power and purity are expressed nonscientifically (and, alas, often unscientifically) in books and magazines of travel. The total number of these constitutes by no means an inconsiderable proportion of the total annual output of printed matter, other than newspapers and advertising, in all civilized states. Figures given in the *Publishers' Weekly* for January 31, 1925, show that 9012 books were produced in the United States in 1924. Of these, 1226 were fiction. Of the remaining 7786, 445 (or 4.9 percent of the grand total) are classified as geographical. They include, presumably, a small number of scientific treatises, a somewhat greater number of textbooks, and a preponderance of popular works of travel and description. The British statistics give an idea of the ratio of popular works of description and travel to works classified strictly as "geography": in 1924, 574 of the former as against only 97 of the latter.

Nor is the popular book of travel anything new: we need but think of Megasthenes' *Indica* or Giraldus Cambrensis' travels in Ireland and Wales, or Sir John Mandeville's delightful combination of fact and fancy. And one aspect of this almost universal interest in travel is the travelogue type of lecture so popular in the United States. It serves in this country much the same purpose served throughout parts of Europe by the local geographical society. Many of these are more like social clubs than scientific institutions in the strict sense of the word; their primary purpose is for the hearing of talks on travel.

There are many men of letters in whom the geographical sense is highly developed. In the works of Ruskin there are some extraordinarily interesting analyses of the influence of geographic features—particularly mountains—on artistic expression.[36] The evolution of the love of nature, the development of the appreciation of landscape—particularly of wild landscape and mountains[37]—and the outpouring of the emotions aroused by the contemplation of nature in poems[38] or prose may be regarded as an expression of geographical ideas. Some novelists have had an even clearer vision for the facts of geography that are of most significance to the average man than do professional writers on geographical subjects. One well-known bibliography by a geographer contains a list of geographical novels,[39] perhaps quite as valuable and even more reliable in their way than historical novels. The quest for local color in literature is also a quest for geographical expression. So is landscape painting, and it is perhaps not too farfetched to say that even program music may be of geographical interest provided one knows the title of the piece; Smetana's *Moldau* brings to mind the sweep of a great river; Sibelius' *Finlandia* evokes a melancholy vision of boreal moors. Ordinarily the approach to the study of the various ways in which the geographical sense has expressed itself is made—if made at all—through the study of the history of the *media* of expression—the history of science, literature, art, painting, or music. But will not the historian of civilization or the student of intellectual history find it possible and worth while also to approach these forms of expression from the geographical point of view?

Alexander von Humboldt in the second volume of *Kosmos* deals with the history of the response of the mind to the contemplation of the physical universe. He devotes stimulating chapters to the analysis of "incitements to the contemplation of nature" (that is, to the history of poetic descriptions of nature in early and modern times, and to the history of landscape painting). These are followed by a broad and sweeping review of the growth of human knowledge of the earth and the heavens.

[Upon rereading the foregoing remarks after nearly forty years, it occurs to me that geographers and other geographically minded persons might welcome having some of the following: (*a*) anthologies (or at least bibliographies) of selected geographically descriptive or responsive passages from poems, novels, short stories, books of travel, and so forth; (*b*) collections, bound or boxed in folders, of selected facsimiles in color of landscape paintings, past and present, illustrating scenes of geographical interest; (*c*) guidebooks to museums (or lists of exhibits therein) directing attention specifically to items, such as landscape paintings or maps, of geographical interest (what would I have given to have had such a guidebook when I was last in Rome!); (*d*) lists of musical compositions (including phonograph records) claimed by their composers or judged by competently musical geographers to convey a sense of the "personalities" of particular countries or of the *genius loci* of smaller areas. All this is not utterly fantastic. It would merely take some collaboration between geographically minded artists and artistically minded geographers, with the support of daring publishers. (1965)]

Some will think that this leads too far astray from the history of scientific geography, which, after all, is and should remain the central core of the history of geography. And, indeed, the writer is fully conscious that he has wandered widely and seemed to arrogate to the history of geography what would certainly appear to be a miscellaneous mass of dissociated ideas. But this has been done deliberately. The purpose has been to give an extensive view of possibilities in this field. The miscellaneous material over which we have roamed at least is united by a geographical thread. We are coming to recognize the importance of the "history of science and civilization,"[40] but, taken as a whole, this is immense and uncoördinated. It cannot well be studied or taught unless some unifying threads are selected and consistently followed. Whichever threads one may choose—and different scholars naturally will choose different threads—they should in any case be connected with widely varied portions of the whole. The geographical threads fulfill this requirement.

CHAPTER 2

Where History and Geography Meet

RECENT AMERICAN STUDIES IN THE HISTORY OF EXPLORATION

NATURE OF THE HISTORY OF GEOGRAPHICAL DISCOVERY

THE HISTORY of geographical discovery forms an important part of the zone where history and geography meet, and this paper will deal with recent studies in that field. I shall attempt the rather audacious task of presenting a sort of bird's-eye view of work that has been done of late, primarily by North Americans concerning the history of the exploration both of this continent and of other parts of the world. I shall also suggest certain areas in this field that might well be cultivated more intensively.

The history of geographical discovery is a fabric into which many colorful threads are interwoven: threads of adventure and hardship, of military conquest and political intrigue, of religious devotion, of commercial enterprise, of theoretical speculation, and of hard scientific thought.

Studies in this subject have been carried on in the United States to some extent by professional geographers, but in greater number by historians and other students of the civilizations of past times, and by popular writers. The geographers have devoted themselves largely to the developments of the last two centuries; the non-geographers to these as well as those of earlier periods. Geographers are mainly concerned with the world of today. Their interest in the growth of geographical knowledge tends to be restricted to the later phases of the record that are connected with the geographical thought of the present. On the other hand, the historians —and with them the orientalists, classicists, medievalists, students of comparative literature—tend to look upon geographical discoveries primarily as enterprises characteristic of the times in which they were made and to consider them in relation to other aspects of the life of those times. Some of the effects of these divergent points of view will be shown a little later.

Historians have devoted much labor to the preparation of the materials for synthetic and interpretative studies of the exploration of the Americas.

Controversial matters, as one would expect, have called forth a tremendous amount of scholarly activity. Some of this has been lavished upon attempts to solve problems of minor intrinsic historical and geographical importance, such as claims to priority in the discovery of particular regions, or the identification of the first landfall of Columbus, or the location of the exact spot where Sir Francis Drake harbored on the California coast. One hesitates, perhaps, to include in a similar category the research that has been devoted to pre-Columbian voyages in the Atlantic, for there is a romantic charm in these speculations that is its own justification. In a survey of studies of the Vinland voyages, W. S. Merrill points out that more than 400 writers have dealt with that subject alone. The literature on the career of Columbus would form a large library in itself and controversial questions connected with the nationality, education, supposed early voyages, and geographical ideas of Columbus are so thorny that a recent commentator has remarked that a definitive biography of the navigator would call for the efforts of a superscholar versed in many fields of learning other than history.

The skeleton that supports the whole structure of the history of exploration is the bare record of routes traversed in the opening to knowledge of previously unknown or little known regions. The original narratives and other documents relating to the outstanding explorations in the Americas were published in modern editions and translations before the First World War, but since then a great deal has been done to render available some of the less well-known materials. A large amount of space has been devoted in the *Revistas* and *Bolentines* of the historical and geographical societies of Hispanic America to the publication of such documents. In the United States and Canada societies comparable to the famous Hakluyt Society of England have been organized to promote the editing of the sources for explorations and discoveries in particular parts of North America—the Quivira Society, the Champlain Society, the Radisson Society. Professors, graduate students, and independent scholars have ransacked libraries and archives of the United States, Hispanic America, and Europe for rare printed and manuscript accounts of travels and have issued them in book form or as contributions to the periodicals of historical societies. [In 1960 a group of historians in the United States founded a Society for the History of (geographical) Discoveries (1965).]

Much, however, must be added to the skeleton of routes traversed if the record is to acquire vitality and meaning. North American writers of late have contributed a number of popular and semi-popular biographies of explorers as well as such histories of particular phases of the history of exploration as Leonard Outhwaite's *Unrolling the Map*, N. M. Crouse's *Search for the Northwest Passage*, or Felix Riesenberg's *Cape Horn*.[1] In these the bare bones become animated, clothed with the "human interest" in which the annals of geographical discovery are extraordinarily rich. Much is made of the experiences, the sufferings, and the adventures that the explorers have undergone, of the impress upon their minds and spirits of strange surroundings, of their friendships and enmities, of their feeling for beauty in wild nature. Yet, appealing though all this may be to the general reader—and to the scholar as well— the lure of the wild has been only a minor cause of geographical discovery and human interest, only one of many important elements in its history. The historian recognizes that the progress of exploration has been both the outgrowth and the cause of larger historical movements—colonial and commercial rivalries, politics and wars, missionary endeavors, the search for gold and slaves. Nearly every textbook on American history contains chapters on the early explorations along the shores and in the interior of the American Continent and on their far-reaching consequences. Every university student of European history is taught something of the social, political, and economic origins in and effects upon Europe of the explorations of the Great Age of Discovery. And when our historians deal with the history of exploration in their original investigations they have emphasized such relations.

This brings me to the main point that I wish to make—the somewhat paradoxical observation that geography has been more or less neglected in the majority of studies of the history of geographical discovery because the approach has been made more often from the point of view of human interest, or of political or economic or literary history, than from that of geography as such.

Therefore, with the indulgence of historians, I should like to indicate what a geographer might conceive some of the distinctively geographical values in the history of exploration to be. I am firmly convinced that the systematic study of these values is not only of absorbing interest in itself but absolutely essential to an adequate understanding of the enormously important part that exploration has played in human progress.

A GEOGRAPHICAL APPROACH TO THE HISTORY OF DISCOVERY

Briefly, in a geographical approach to the history of discovery one might seek to interpret, first, the influence of earlier geographical knowledge and belief upon the course of exploration; second, the actual relations between the course of exploration and the nature of the regions explored; and, third, the contributions made by exploration to subsequent geographical knowledge.

Influence of Earlier Geographical Knowledge. Explorers have seldom gone forth merely to probe about for whatever they may happen to discover. They have gone in quest of definite objectives believed to exist on the basis of such information as could be gathered from the geographical lore of their own and earlier times. When one explorer has failed in the quest, others have taken it up until the objectives have been attained or found to be illusory. Thus explorations have tended to run in cycles, within each of which there has been some degree of historical continuity and geographical unity. The great Asiatic journeys of Marco Polo and his contemporaries, the Portuguese voyages along the African coasts, the searches for the Northwest and Northeast Passages and for the Poles, have formed major cycles in which the geographical objectives were ultimately attained. The Spanish explorations in quest of the mythical land of Quivira in what is now the Southwest of the United States or of El Dorado in the remote interior of South America, the search for the Great Southern Continent that figures so enormously on maps of the sixteenth and seventeenth centuries, have formed similar cycles in which the geographical objectives proved to be will-o'-the-wisps. Throughout the history of geography erroneous notions have exerted a powerful fascination over men's minds and mistaken concepts have been hardly less influential than those finally found to be correct. Sometimes such errors have lured men on to disaster, as in the case of Colonel P. H. Fawcett, the English engineer who set out in 1925 to search for a mysterious lost city in the wilderness of Matto Grosso, a quest from which he never returned. Often, however, they have led to discoveries that might otherwise have been long postponed. The discovery of America itself was largely due to devoted belief in false geographical theories. In a suggestive paper, John Leighly has recently classified the types of error in geography and discussed the role of the simpler kinds of error concerning the distribution of lands and seas and habitable areas of the earth—errors that have been corrected as exploration has advanced. With "the flood of errors in the physical interpretation of terrestrial phenomena [which] form part of the general flux of scientific thought"[2] he does not attempt to deal.

Not only the routes of explorers but the character of their observations must be studied in the light of their preconceived ideas, both true and false. Columbus was constantly seeing things that to him betokened the nearness of Marco Polo's Cathay and the Grand Khan's realms; he played with the idea suggested by the geographical speculations of the Church Fathers, that the Orinoco might have its source in the Terrestrial Paradise. One of the oldest and most persistent of false geographical theories is derived from the Greeks—the doctrine that the polar regions are uninhabitable. Stefansson has shown how this theory colored the thought of certain nineteenth-century explorers who described the Arctic lands and seas as lifeless wastes, despite evidence to the contrary that they might have seen had their eyes been open to it. Similarly, things that would appear obvious in the light of later knowledge and experience have sometimes been completely overlooked by explorers. This may account for the fact pointed out by A. B. Hulbert that such sound and accurate observers as Lewis and Clark failed to perceive the economic opportunity latent in our Northwest—in the wealth of its soils, forests, and minerals.

Relations to the Regions Explored. This brings me to the second point, the actual relations between the course of exploration and the nature of the regions explored. It is not considered altogether good form among North American geographers today to define geography as the study of the influences of the geographic environment upon men or even as primarily the study of man's adjustments and adaptations to the environment. But however this may be, exploration is certainly one form of human enterprise in which such influences and adaptations have been of the utmost importance, and the history of exploration means little unless they are taken into account.

Guided by such information, true or false, as he may be able to gather, the explorer plans his methods of travel and sets his course, but only the successful aviator-explorer of today—a Wilkins or an Ellsworth—can follow the course exactly as set. In the case of the earthbound traveler, the nature of the environment into which he penetrates determines whether the methods are suitable and the approximate route can be maintained. Perhaps he will be turned back or deflected by obstacles: deserts, mountains, swamps, hostile natives; perhaps like Orellana, or La Salle, or Mungo Park, or Lewis and Clark, or Stanley, he will be swept forward by great rivers; perhaps, like Cabeza de Vaca in Texas or Doughty in Arabia, he may be forced to wander with nomad tribes; and even if he can keep to the general direction of his route as planned, its detailed trace will be controlled by the circumstances of the country, and he will be obliged to adapt his ways of living to these circumstances. Many explorers, unable or unwilling to make the necessary adaptations, have met misfortune. Some of the early conquistadores set out into the tropical wilderness of South America overwhelmingly burdened. For his expedition of 1541–1542 in quest of the Land of Cinnamon in the Oriente of Ecuador, Gonzalo Pizarro assembled 4000 Indians, 220 Spaniards, some 200 horses, more than 2000 hogs, and almost as many dogs. Yet in spite of, or perhaps because of, this burden Pizarro had to turn back and it was in a craft built from forest timbers with their own hands and by living off the country that Orellana and his party made the descent of the Napa and the Amazon. Polar explorers have also come to grief through rigid adherence to standards of living wholly unsuited to the conditions, and, as Stefansson has shown, through failure to adopt the expedient of living as the Eskimos do.

One approach to the interpretation of the relation between exploration and the nature of the regions explored is through geographical field work, a technique seldom employed by our historians. There are, however, notable exceptions. S. E. Morison has followed the routes of Columbus in yachts, seeking to undergo some of the experiences of the navigator and to supplement and correct by actual observations conclusions drawn from the examination of maps and documents. In tracing the routes of Father Kino and other explorers in Mexico and the southwestern United States, H. E. Bolton has had a similar purpose in view, and the same objective has prompted J. A. Robertson's field studies of Ponce de León's routes in Florida.

Subsequent Influence of Exploration. The record of exploring expeditions is only half of the history of geographical discovery. The other half is the record of their subsequent influence. As I have already remarked, the broader historical consequences of the enlargement of geographical horizons—the effects on commerce, land settlement, international relations, and so forth—have been frequently discussed in historical works. Here we are concerned, rather, with the immediate processes that have led up to these broader results—specifically, with the ways in which the information and misinformation acquired during explorations have been used in descriptive works and maps, have shaped geographical theories and the teaching of geography, and through these channels have been transmitted into general literature.

This subject has often been touched upon. In their biographies of explorers and their introductions and notes to the narratives of journeys our historians usually cite maps and other works in which the observations made during explorations have been incorporated. Subsequent influences have also been at least recognized in the histories of particular cycles of exploration, such as G. B. Manhart's *Search for the Northwest Passage in the Time of Queen Elizabeth* or Jeannette Mirsky's *To the North!* A few investigations have been made specifically of the contributions to geographical knowledge of certain large phases of exploratory activity, as in N. M. Crouse's thesis on the Jesuits in the interior of North America or F. W. Howay's paper on the early voyages to the Northwest coast. In their general reviews of the growth of scientific geography in the United States, W. M. Davis in 1924 and C. C. Colby in 1936 pointed to some of the connections between explorations and field surveys in this country and the development of North American geographical thought, particularly in its physiographic aspects. The influences of exploration have also been traced from a very different angle by students of the history of literature—J. L. Lowes, G. B. Parks, Geoffroy Atkinson, and others—who have sought to discover the origins in the records of travel and exploration of the geographical lore found in Chaucer, the Elizabethan dramatists, the French philosophers, nineteenth century poets and novelists, and the like.

RECONSTRUCTION OF PAST GEOGRAPHIC ACTUALITY

Another somewhat distinctive type of research should also be mentioned—the occasional attempts that have been made to reconstruct the geographic actuality of a particular region as it was in the past. A German scholar, Georg Friederici, has done this to some extent in a comprehensive three-volume work on the character of the discovery and conquest of America. A British geographer, E. W. Gilbert, has described the western part of the United States as portrayed in the narratives and reports of exploring expeditions of the period 1800 to 1850. To my knowledge, however, only one American geographer, R. H. Brown, has undertaken anything comparable, and his book does not deal with a previously unexplored or little-known region. In the light of all the data that he could gather from documents of the time, supplemented by his own personal knowledge of the region, Brown has painted a geographical picture of the landscapes, population distribution, and economic life of the Atlantic seaboard of the United States between 1780 and 1810. To accomplish their purpose Friederici, Gilbert, and Brown have examined not only the original records of explorers and travelers but also contemporary maps and secondary descriptive works derived from these as well as from lost records. In this way, incidentally to the main purpose, their studies shed light on the channels through which the results of geographical exploration have influenced subsequent geographical knowledge.

Although I have cited a number of examples of studies that seek to trace such influences, one cannot but feel that the work done so far is piecemeal. The general books are far from complete; the more detailed studies of particular explorations seldom follow the record through to the end; the work of the literary historians is fragmentary as regards exploration; and little in the way of systematic research has been devoted to the contributions of scientific explorations to scientific geography. Practically nothing, for example, is known of the explorations from which was derived the information appearing on the great general maps of Hispanic America prepared from time to time by the cartographic establishments of the Spanish and Portuguese Crowns during the period of discovery and colonization. And indeed the measure of scientific success attained by such a well-known expedition as that of Lewis and Clark has never been studied [this was remedied in 1954[3]].

In sum, it would appear that much virgin soil is still open for systematic cultivation in that alluring part of the broad domain where history and geography meet—the history of geographical discovery.[4]

CHAPTER 3

Map Makers are Human

COMMENTS ON THE SUBJECTIVE IN MAPS

CERTAIN implications of an obvious fact will be discussed in this paper—the fact that maps are drawn by men and not turned out automatically by machines, and consequently are influenced by human shortcomings. Although this fact itself is self-evident, some of its implications are often overlooked. The trim, precise, and clean-cut appearance that a well-drawn map presents lends it an air of scientific authenticity that may or may not be deserved. A map may be like a person who talks clearly and convincingly on a subject of which his knowledge is imperfect. We tend to assume too readily that the depiction of the arrangement of things on the earth's surface on a map is equivalent to a photograph—which, of course, is by no means the case. The object before the camera draws its own image through the operation of optical and chemical processes. The image on a map is drawn by human hands, controlled by operations in a human mind.

Every map is thus a reflection partly of objective realities and partly of subjective elements, a circumstance that has been dealt with implicitly in all works on the art of cartography, but seldom considered explicitly as a subject in itself.[1] No map, however, can be wholly objective. Even a photograph taken vertically downward (not a map but akin to one if the terrain imaged is flat) is subjective, in the sense that the photographer's choice determined the tract of country shown in it and also the time of day when the film was exposed and thus the aspect of the shadows appearing in the picture. Likewise, no map is wholly "nonobjective," as some forms of painting and sculpture are said to be. Even a map of an imaginary country is objective, in the sense that the mountains, roads, towns, and so on that it pictures were suggested by corresponding objective things in the real world.

The maps produced by governmental surveys or made in the field by explorers are more or less directly copied from nature. As in the case of the memoirs and letters written by those who have participated in historic events, their quality is influenced by the experience and powers of observation of their makers. A topographic map drawn by a man familiar with geology and physiography is likely to be far more expressive of relief and drainage than one drawn by an untrained observer, just as the memoirs of an experienced statesman are likely to present a more truthful record of an international conference than those of some inexperienced journalist. Many maps, however, are not drawn from nature but are compiled from such documentary sources as other maps, surveyors' notes and sketches, photographs, travelers' reports, statistics, and the like. As these sources are themselves man-made, the subjective elements they contain are carried over into the maps based on them. In the following paragraphs maps will be considered in the light of the effects on their sources and compilation of certain mental and moral qualities—scientific integrity, judgment, consistency, progressiveness, and their opposites—on the part of their makers.

MENTAL AND MORAL QUALITIES AFFECTING MAPS

SCIENTIFIC INTEGRITY

Fundamental among these qualities is scientific integrity: devotion to the truth and a will to record it as accurately as possible. The strength of this devotion varies with the individual. Not all cartographers are above attempting to make their maps seem more accurate than they actually are by drawing rivers, coasts, form lines, and so on with an intricacy of detail derived largely from the imagination. This may be done to cover up the use of inadequate source materials or, what is worse, to mask carelessness in the use of adequate sources. Indifference to the truth may also show itself in failure to counteract, where it would be feasible and desirable to do so, the exaggerated impression of accuracy often due to the clean-cut appearance of a map. Admittedly this is not always easy of accomplishment. A map is not like a printed text, in which statements can be qualified with fine shades of meaning. One cannot, on the face of a map, cite the evidence used and discuss its validity. A town or a mountain must be shown in one place, even though three sources of apparently equal validity may locate it at three different places. Nevertheless, there are certain ways in which the map maker can, within limits, modify the definiteness of his commitment. If he is not sure of the courses of rivers or if contours are approximate form lines only and are not based on actual field surveys, he can at least show them by broken lines. He can introduce question marks or, as on marine charts, such challenging letters as P.D. (position doubtful) or E.D. (existence doubtful) alongside islands and shoals. Another device is to include in the margins of maps small diagrams of the regions showing the character of the surveys and other sources on which the maps are based —"relative-reliability" diagrams—such as appear on the sheets of the American Geographical Society's "Map of Hispanic America, 1:1,000,000," and have now been accepted as standard for the "International Map of the World, 1:1,000,000." This accomplishes to some extent what the careful historian or economist does with qualifying phrases in his text and appended critiques of his sources. [Although this device is not often used at present, it may someday become standard practice in cartographical scholarship where maps, especially covering considerable tracts of territory, are based on different source materials of varying validity.]

Beware of maps prepared to substantiate a pet theory! There is a well-known type of reasoning that begins with a theory, gathers statistics and other data that seem to support it, makes a map on the basis of the statistics, and finally "proves" the theory by reference to the map. The dishonesty in such a procedure may be unconscious, but there is a large use of maps in propaganda with a view to conscious and deliberate deception in the service of special interests. The relative areas of different regions as disclosed on maps in railroad timetables are usually deliberately distorted so as to show particular railroad systems to best advantage. More subtle and dangerous is the type of deception found on maps designed for propaganda purposes—maps on which facts are played up or played down, omitted or invented, for nationalistic ends. Hans Speier has recently dealt with this subject in a paper entitled "Magic Geography":

> Today, maps are distributed on posters and slides, in books as propaganda atlases, on post cards, in magazines, newspapers and leaflets, in moving pictures and on postage stamps.
>
> Maps are not confined to the representation of a given state of affairs. They can be drawn to symbolize changes, or as blueprints of the future. They may make certain traits and properties of the world they depict more intelligible—or may distort or deny them. Instead of unknown relationships of facts they may reveal policies or illustrate doctrines. They may give information, but they may also plead. Maps can be symbols of conquest or tokens of revenge, instruments for airing grievances or expressions of pride.

Speier analyzes the ingenious manipulations of design and color to be found on some of the "geopolitical" maps issued in the interests of totalitarian propaganda. For example:

> On a map to illustrate the repatriation of Germans who had been living in Latvia, the German minority about to return to the fatherland is represented by a row of thirteen identical symbols, each standing for five thousand men. The symbols extend over the whole area of Latvia where it is widest, from Libau in the west to the eastern border. The size of the symbol is so chosen that the country seems to be populated by Germans, whereas, in point of fact, the German minority amounts to 3.7 percent of the total Latvian population.[2]

Speier exposes the more blatant forms of propagandistic maps intended to influence the masses. Nationalistic bias may also reveal itself in detailed maps published in serious books and learned periodicals. Superficially these may seem to have been prepared according to sound scientific principles. They may accurately record statistics, but with a symbolism devised to overemphasize the distribution of one people or language or religion or set of institutions. Assume, for example, that the frontier province of Pomeria, which formerly belonged to Sudia, was annexed by Nordia in the last war and that a recent census has shown that half of the population of Pomeria are Nordians and half Sudians. The Nordians are concentrated in the towns, the Sudians form the bulk of the rural population. On a detailed map of Pomeria in the *Bulletin of the Sudian Geographical Society* the areas where less than 10 percent of the total population is Sudian are left white and those where more than 90 percent is Sudian are shown in dark red, with intermediate gradations of lighter red to bring out intermediate percentages. Clearly most of the map would be red, giving an impression of a preponderant Sudian population. In 1925, in his great work on cartography, Max Eckert[3] discussed this particular type of deception, which he asserted had "recently been propagated on certain non-German maps," notably "maps showing the languages and the plebiscitary vote in Upper Silesia, which were drawn neither by German hands nor in the German scientific spirit."

JUDGMENT

Fully as important as scientific integrity in the making of maps—indeed, largely a function of scientific integrity—is judgment. This embraces critical acumen in the selection of source materials, discrimination in the use of techniques, taste in choice and arrangement of colors, symbols, lettering, and so forth, and, throughout, a feeling for consistency.

As maps are only rarely accompanied by relative-reliability diagrams, critiques of their sources, and explanations of the graphic techniques by which information is disclosed on them, it is not always possible to test the quality of the judgment that has gone into their construction. The essential accuracy of certain types of map, however, can be taken on faith. One hardly needs to question the basic reliability of the charts and topographic sheets issued by governmental institutions such as the United States Coast and Geodetic Survey[4] and the United States Geological Survey, whose high standards are well known. On the other hand, it is not safe to take on faith the reliability of the average reference map, atlas map, or statistical map. A general reference map can usually be spot-checked by comparison with good maps of the same region on larger scales, and a statistical map spot-checked against the sources if these can be identified and found. Whether or not such tests need to be carried out depends, of course, on the uses to which a map is to be put. A map exact enough for measuring the air-line distance from Boston to New York may be quite unsuitable for measuring the total length of the Maine coast line. In general, unless one has solid ground for confidence in the integrity and judgment of the makers, detailed information derived from maps should not be incorporated in other maps or used as a basis for conclusions and decisions of importance. The fallacy of overrefined inferences from maps should be carefully avoided. One can with good reason accept the general over-all picture that a map presents, but this does not mean that the map in its every particular presents the gospel truth.

Simplification and Amplification.[5] Although a map's reliability is no higher than that of its sources, it may be considerably lower unless good judgment is exercised in its compilation. In this process the "raw" information provided by the sources is transformed by the cartographer. Two operations may be carried out. If the "raw" information is too intricate or abundant to be fully reproduced to the scale of the map as it stands, it may be simplified and generalized. If it is too scanty, it may be amplified and elaborated. Similar operations are performed in all fields of investigation. For example, where the records yield scanty details, as for ancient times, the historian seeks to fill in the gaps on the basis of inference and conjecture. Where the information is superabundant, as for modern times, he selects. In both cases he does not copy the sources but gives a partly subjective interpretation. Frequently he must both amplify and simplify in the same study. So, too, one part of a map may be the result of amplification of the sources, another part the result of simplification. [It is my impression that the rather obvious concept of the historian's amplification of the evidence was first taught me by Professor C. H. Haskins, about 1913. On this subject R. G. Collingwood spoke as follows in a lecture in 1935: "I described constructive history as interpolating between the statements borrowed from our authorities, other statements implied by them. Thus our authorities tell us that on one day Caesar was in Rome and on a later day in Gaul; they tell us nothing about his journey from one place to the other, but we interpolate . . . with a perfectly good conscience . . . this activity which, bridging the gaps between what our authorities tell us, gives the historical narrative or description its continuity."[6] (1965)]

For most general reference maps the sources are simplified and generalized to some extent by the removal of minor irregularities in coast lines, railroads, slopes, and so on, as well as by the omission of many features—towns, hills, mountains, lakes, and the like. If the process of simplification and generalization is not carried out consistently, the results may be misleading. Sometimes one category of information is simplified to excess in relation to another. For instance, the omission of stream lines where form lines or contours are reproduced in detail may give an erroneous impression of dry watercourses. This is an inconsistency found even on standard atlas maps. There is also inconsistency where specific information is more radically simplified on one part of a map than on another. Such inconsistency is sometimes necessary on even the best of general reference maps to avoid overcrowding, for example, where, in the more densely settled regions, towns much larger than those shown in the sparsely settled regions are omitted because there is no room for them. In such cases the map maker must use his discrimination in deciding what to omit, and his map, although it may gain in legibility and beauty, will lose in reliability, at least with regard to the specific distributions that are inconsistently shown. On its "Map of the Americas, 1:5,000,000," the American Geographical Society attempted to reduce such inconsistency by showing all towns of more than a certain population selected on the basis of what it was possible to show on a map of that scale in the most congested areas and then adding elsewhere such places of smaller population as are important, notably road and rail centers and ports.

Amplification of the information provided by the source materials is the addition of details regarding the exact location of which the sources fail to provide precise evidence. Amplification is, of course, wholly unjustifiable where there is no sound evidence of the existence of the details added, as where an unsurveyed river is shown with sinuosities in the sincere belief that they are more "natural" than a straighter course would be. If, however, reliable observers report having crossed the river at several points, the careful cartographer need not omit it from his map because he has little to guide him in drawing its exact course. It is usually preferable to plot the river by conjecture, in the light of whatever evidence there may be. On maps designed for certain specific uses,

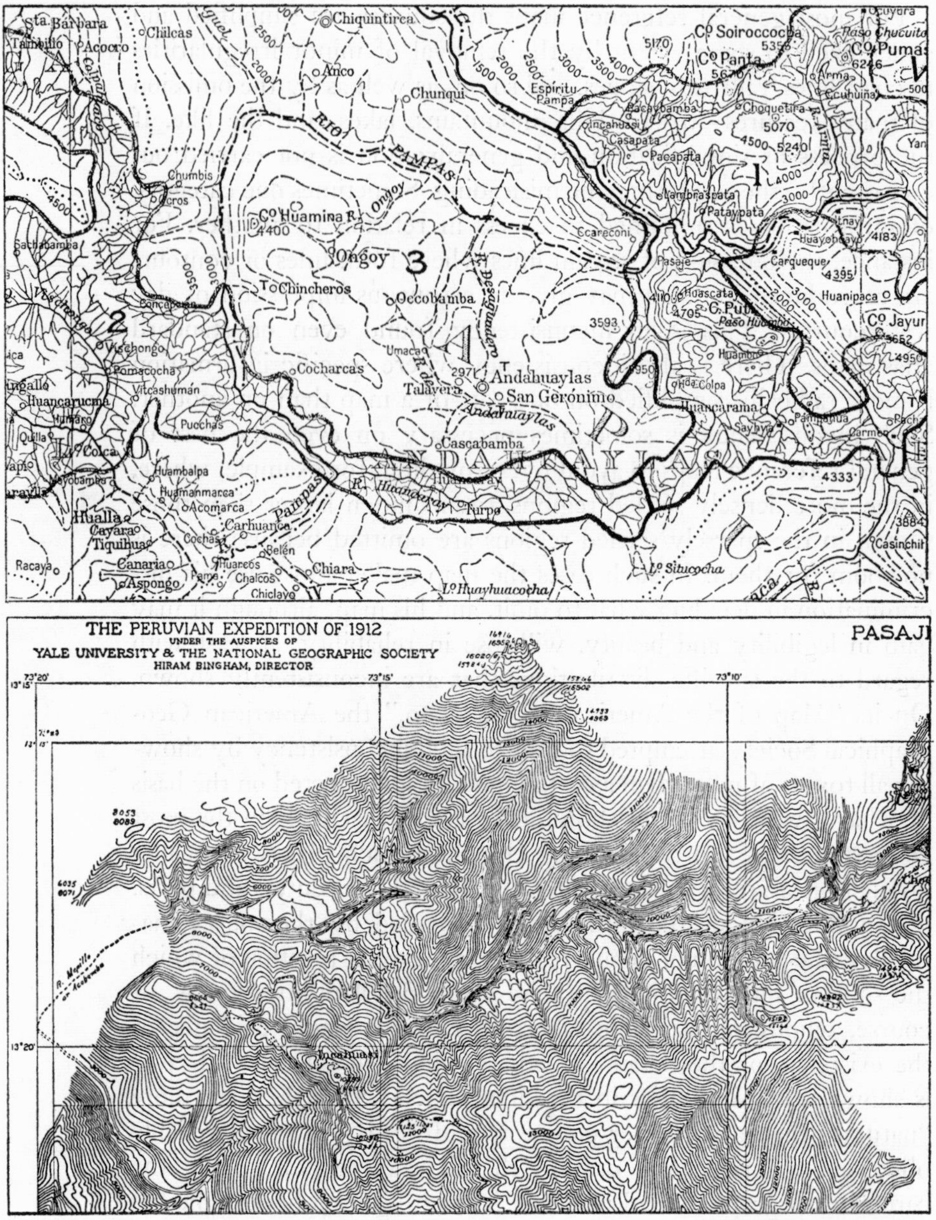

Fig. 1. These maps illustrate the processes of simplification and amplification of cartographic source materials (see text). The upper map, which shows a part of the Andean region of southern Peru, is reproduced from the Lima sheet (South D-18) of the American Geographical Society's

such as air-navigation charts, too much of the conjectural may be seriously confusing; but, in general, amplification of the sources, when done critically, is misleading only when the users of maps draw overprecise inferences from the features that have been amplified, and such danger is obviated if these features are represented by broken lines (Fig. 1).

Quantitative information shown on maps is peculiarly affected by the operations of amplification and simplification. By quantitative information I mean here information regarding the distribution of quantities of varying intensity. Contour lines, of course, provide such information. Statistical maps, or maps the purpose of which is to provide quantitative data, are essential tools in geography, climatology, oceanography, demography, and other branches of the natural and social sciences. They are also frequently consulted by the general public in atlases, magazines, and popular books. They are so widely used that it might be well if their comparatively high degree of subjectivity were more generally recognized than it is.

Certain terms must first be defined.[7] One way of setting forth quantitative data on a map is simply to copy from the sources numbers giving heights, depths, populations, and so forth. As this gives no visual impression of relative magnitudes, the more usual method is to show quantities by means of symbols. Three kinds of symbol may be used: point symbols, isarithms, or spatial symbols. A point symbol is a dot, a disk, a graphic representation of a cube or sphere,

---

Map of Hispanic America on the Scale of 1:1,000,000. The lower map is from one of the unpublished field sheets of the topographic survey made by the Peruvian Expedition of 1912 (scale 1:63,360; contour interval, 200 feet) used in the compilation of the area marked "1" on the upper map; this particular section lies to the northwest of the figure "1." Area 2 on the upper map was compiled from various large-scale topographic surveys made for the study of railway routes, including a survey on the scale of 1:48,000 carried out by the International Railway Commission in 1893. Area 3 was filled in from various compiled maps on which no contours appear.

As suggested by comparison with the lower map, the data appearing in the original sources for Areas 1 and 2 have been *simplified:* drainage patterns appropriate to the scale of the Millionth Map were selected; only the contours prescribed by the scheme of the Millionth Map are shown, the other contours being eliminated.

For Area 3 the details from the cartographic sources were adjusted to the adjoining topographic surveys covering Areas 1 and 2, which also furnished guidance for the pattern of the form lines shown by broken lines. The relief as represented thus reflects *amplification* of the sources.

or other geometric figure of conventional form that represents a specified quantity; an isarithm is a line representing the locus of a particular value of a quantity; and a spatial symbol is a color, shading, ruling, geometric pattern, or the like that represents either a single specified quantity (choropleth[8]) or a range of quantities between two specified limits.

Now it is obvious that the quantities the symbols represent—the "mapped quantities," that is—seldom correspond precisely to the quantities as given in the sources. As the latter guide or control the map maker in selecting the appropriate symbols and in placing them on his map, they may be called "control quantities." The mapped quantities are almost invariably simplifications or generalizations of the control quantities. For example, a point symbol may show that a city has a population between 500,000 and 600,000, whereas the population according to the census—a control quantity—may be 578,341. From isarithms one might determine that a given point was between 20 and 40 feet above sea level, whereas a bench mark at that point would show its altitude to be 32.5 feet.

A distinction must also be made between "locational quantities" and "spatial quantities." A locational control quantity is one that indicates the intensity of some condition at a "control point," such as the altitude of the point or the temperature or barometric pressure recorded there. Locational control quantities are usually determined by instruments and may be shown by point symbols of graded sizes, though more frequently isarithms (for example, contour lines, isotherms, or isobars) are used. A spatial control quantity is one that applies at least nominally to the whole of a given space, the "control space" (usually a political division)—for example, the population of a county. Spatial control quantities are usually determined by census or other enumerations rather than by instruments. Point symbols are properly employed (as on "dot maps") for showing total quantities; spatial symbols are more appropriate for showing spatial ratios. A spatial ratio may be the ratio of a certain quantity in a space either to the area of the space (for example, density of population) or to some other quantity in the same space (for example, income per capita, birth rate per 1000 persons).

In dealing with matters of this sort confusion may be avoided by calling any given part of the earth's surface a "space" rather than an "area" and using the latter term with reference to size only. In this way one avoids having to refer to the "area of an area" (see below, p. 210).

Spatial symbols may be bounded either by the limits of control spaces or by isarithms. Symbols of the former type, which represent quantities as actually determined for subdivisions of the region mapped, have been called "choropleths"; those of the latter might be called "chorisarithms" (for example, hypsometric tints, shadings bounded by isotherms). [In this paper as originally published in 1944, "isopleth," rather than "isarithm," was used as the generic term covering "iso-" lines of all kinds when used for plotting geographic distributions. Nowadays, at least in this country, "isopleth" would seem to be used more especially with reference to lines indicating the distribution of spatial ratios. I am following Professor Robinson here in favoring "isarithm" for the more inclusive concept.[9] Such terminology, however, is still fluid.]

In deciding what quantities his symbols are to designate, the cartographer must first study carefully the general character of the distribution he intends to map. Unless he does this, he may find, for example, that he has assigned too few people to each "dot" and consequently has to crowd so many dots into the more densely settled regions as to produce a solid black mass. Thus his map is inconsistent, since it fails to give any picture at all of the way in which the people are distributed within the black mass. Similarly, if he makes each dot represent too many people, he will meet with difficulties in placing the dots in the sparsely settled regions, where one dot may have to represent a population scattered over a very large tract. Much the same problem is encountered when spatial symbols are used. Many phenomena have a tendency toward extreme concentration in relatively small spaces; population is a notable example. If the class intervals of the densities of population represented by the spatial symbols are uniform (for example, 0–10, 10–20, 20–30, . . . to a square mile), there may be too few classes to show adequately the differences in density in the regions of sparse distribution and more classes than can be legibly differentiated for those of greater concentration. Hence it is customary in mapping distributions of this type to make the class intervals narrow at the lower end of the density scale and increase the width as the upper end is approached (Fig. 2, F, I).

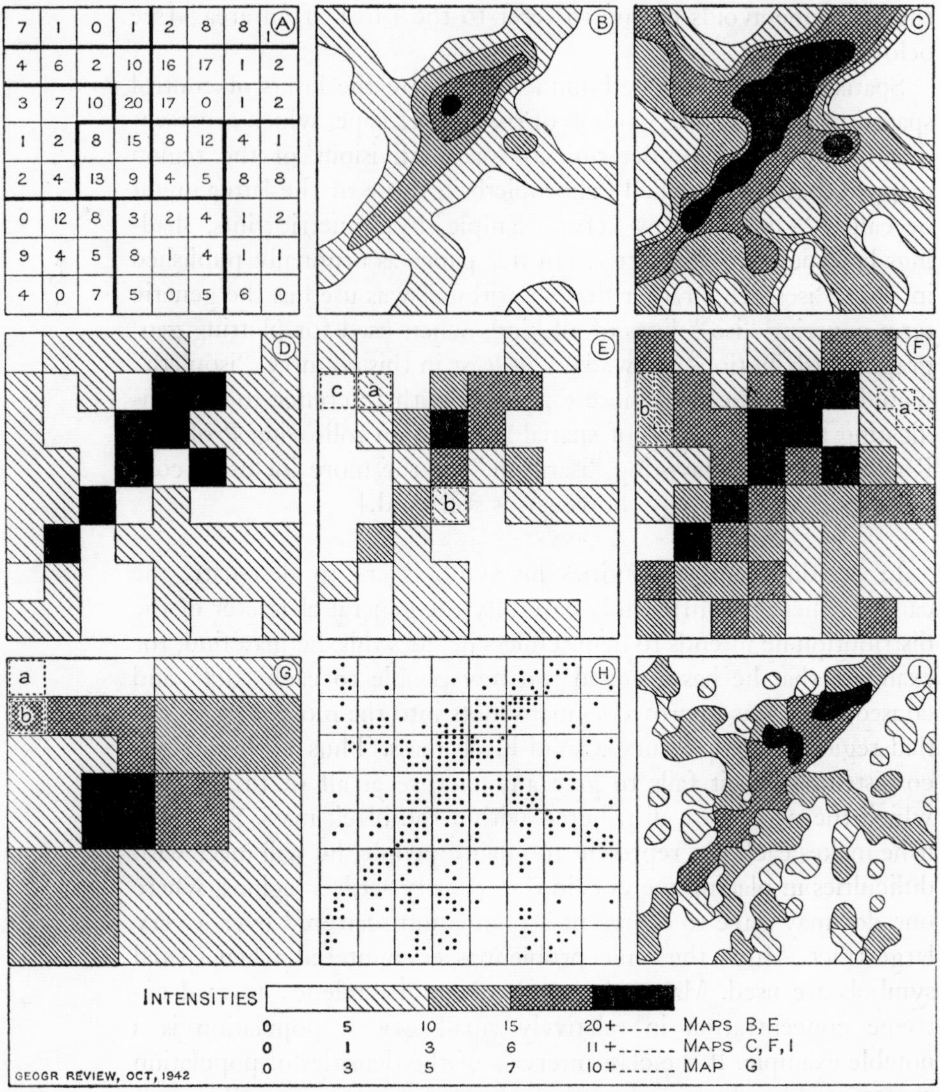

Fig. 2. These "maps" show how differently the same quantitative distribution may be mapped. The numbers in *A* are control quantities (see text). Each number might represent either a locational control quantity, such as an altitude, or a spatial control quantity, such as the density of units in the square in which the number stands. In *B* and *C* isarithms have been drawn with reference to the numbers by interpolation, and spatial symbols ("chorisarithms") have been placed on the spaces between the isarithms. In *B* the class intervals between the isarithms are equal, whereas in *C* they are narrower at the lower end of the scale, widening toward the upper end. Thus *B* presents less information with regard to the areas of lesser

Much the same principle also applies to the representation of relief on maps covering large tracts of country. As lowlands and plains are on the whole far more extensive than mountains, it is usual on such maps when relief is shown by contours and hypsometric tints to employ narrower contour intervals for the lower elevations than for the higher (Fig. 2, C). In this way more topographic detail can be shown for the lowlands, and there is less crowding on the highlands, than would be the case if the contour intervals were uniform. This, of course, has unfortunate effects when applied on maps that include extensive high plateaus such as those of the central Andes and Tibet, since the contour intervals are too wide to bring out adequately the often varied relief of the uplands.

Thus the pattern and with it the adequacy of maps showing quantities are much affected by the discrimination with which the cartographer adjusts the classification of the mapped quantities shown by the symbols to the control quantities.

As the control quantities for a density map are functions of the areas of control spaces, the information provided by them will necessarily be less detailed for the parts of the map where the control spaces are relatively large. For this reason the pattern of symbols based on this information will be inconsistent within itself, unless all the control spaces are of approximately equal size, which is almost never the case. Furthermore, and for the same reason, it

---

intensity and *C* less with regard to those of greater intensity. On *E* and *F*, where the densities are shown by "choropleths" or spatial symbols plotted with reference to the squares, the same difference appears. (*D* is merely an aid to comparison of *E* and *F:* for the areas shown in solid black on *D*, *E* provides more information than *F*, and for the areas ruled on *D* the reverse is the case.) The densities in each of ten larger spaces (bounded by heavy lines in *A*) are plotted by choropleths on *G* (which, of course, gives a far less adequate picture than any of the other maps). On *H* the distribution of units is shown by dots placed by an assumed "amplification" of the statistical information so as to bring out something of an assumed character of the distribution within the squares. *I* is a density map similarly constructed.

The following incorrect assumptions, which might easily be based on examination of the symbols on these maps, illustrate the "fallacy of over-refined inferences from statistical maps": (1) that the actual density in square *a* on *E* is more nearly like the density in *b* than like that in *c*; (2) that the densities in the spaces *a* on *F* and *G* are less than those in the spaces *b*.

is obvious that two statistical maps of the same territory showing the same distribution will present quite different patterns if based on control data for different systems of control spaces (Fig. 2, compare E and G), for example, a population map of the United States based on county statistics and one based on statistics by minor civil divisions. Likewise, two statistical maps of different regions are not comparable except in a very general way if one is based on figures for control spaces with areas that are on the average larger than those of the other.

Some of the inconsistencies due to lack of uniformity in the areas of the control spaces can be partly eliminated by amplifying the control quantities in the light of other evidence. Where point symbols are employed to show distributions, they can be arranged within each of the control spaces so as to conform to the probable character of the distribution therein. In placing them, other maps are consulted that make this probable character reasonably clear, for example, maps showing houses, villages, woodlands, mountains, and so on. A similar procedure can be followed where spatial symbols are used, by breaking down the control spaces into subdivisions, to each of which an estimated density is assigned with the aid of nonstatistical evidence, care being taken that the estimated densities are made to conform to the known total control quantity of each space.[10] Such estimated densities might be called "secondary control quantities," since it is they that actually govern the placing of the spatial symbols.

Isarithms are drawn on a map with reference to control points. Other things being equal, the smaller the interval between the mapped quantities represented by the isarithms (for example, the contour interval), the larger the amount of detail shown. The *reliability* of this detail, however, depends not only on the accuracy of the control quantities but also on the geographic density of the distribution of the control points. If the control quantities are of uniform accuracy and the isarithms are located with reference to the control points merely by interpolation, the probable reliability of the information furnished by the isarithms will vary in proportion to the density of the control points.

When the isarithms are drawn by interpolation, there is a minimum of amplification. As in the case of point symbols and spatial symbols, however, the isarithms may be drawn in such a way as further to amplify the information provided by the control quantities. Thus the suveyor further amplifies his map when he "sketches in" contours by observing the terrain in the field (or from the study of photographs) and shaping the lines accordingly. Formerly contours were nearly always "sketched in," and much margin was thus left for the exercise of judgment. To quote Professor Debenham:

> The original method of contouring used by the Ordnance Survey was to send a levelling party to fix each of the contours on the ground, and put in lines of pegs as they levelled round each slope. The pegs were afterwards surveyed by traverse by another party . . . This excellent but very expensive way is now rarely followed, and the rigorous use of the word "contour line" has largely disappeared, for the majority of maps showing contours depend on nothing more accurate than interpolation between spot heights or some such method.[11]

Today, the use of photogrammetric plotting instruments is largely eliminating this subjective factor so far as contour mapping is concerned, though it still influences the plotting of isarithms of other types—isotherms, isobars, and so forth. The map maker must first decide whether to interpolate arbitrarily or to try to draw the lines more realistically on the basis of guided conjecture. In either case the user of the map has no way of distinguishing the parts that are probably more reliable from those that are probably less so, unless the positions of the control points are shown on the map itself or on an accompanying diagram. Unfortunately this is seldom done on the isarithmic maps in most common use, though it has been frequently recommended by climatologists and others. An interesting illustration of control on isarithmic maps is given on Veatch and Smith's 1:120,000 charts of the submarine topography off the northeastern United States. No need to ask of these, "What is interpretation and what is fact?"[12]

Although some maps present no quantitative information of the types that have just been discussed, every map is quantitative in the sense that it provides information regarding distances, areas, and directions. Such quantities can be accurately determined by direct measurements on maps covering small spaces if the maps themselves are reliable. The larger the space covered, however, the less can reliance be placed on direct measurements, on account of distortions introduced by the projection. The Yankee saying, "What you make on your mackerel, you lose on your codfish," applies here. If some of your distances are in correct proportion, others will be misleading; if your relative areas are consistent, your shapes and distances will be badly out; and so forth. The plotting of a map on a particular projection is a mechanical operation with little that is subjective about it, but the selection of the most suitable projection for any given purpose is highly subjective, requiring good judgment guided by technical knowledge.

These remarks should have made it clear that the quantitative information furnished by maps is much affected by subjective influences. The most marked effects of all, perhaps, are felt on maps showing densities and other ratios. Because of the different practices employed by their makers, with different degrees of skill and judgment, such maps are seldom strictly comparable with one another or even consistent within themselves.

Synthetic Information and Generalization. Up to this point we have been dealing primarily with subjective influences on the mapping of the distribution of individual phenomena, such as coast lines, populations, and temperatures. One of the most important purposes that maps accomplish, however, is to show relations of different phenomena to one another. Such relations may be brought out by the use of different symbols on the same map to show different kinds of things: blue lines for rivers, red lines for roads, dots for cities, and so on—"mixtures" of information, in other words. Many maps present "compounds" rather than mere mixtures, that is, the symbols themselves indicate either quantitative relations between two phenomena or more or the coexistence of two phenomena or more at particular locations or within particular regions. Such "synthetic information" may be the result of the compounding of only two or three elements, as on a map showing a relation between rainfall and temperature, or between the distribution of malaria and that of *Anopheles* mosquitoes, or between the number of doctors and the total population. Or a much larger number of elements may be compounded, as on maps showing various types of "region"—climatic regions, soil regions, economic regions, land-use regions, "natural regions," "cultural regions," and so on.

The influence of subjective factors—judgment, discrimination, critical acumen—is of paramount importance in mapping of this type. Whether a particular relation between two wholly objective elements is of significance or not depends on the quality of the map maker's judgment. He may, for example, show places where, during the growing season, a rainfall between 40 and 60 inches is correlated with mean temperatures between 60° and 70°. He picks out this relation because he believes it has meaning in relation to soils, vegetation, crops, erosion, or something else. If he is mistaken in his belief, his map may mean little more than a map would that shows by counties the ratio of illiteracy to cocktail lounges, even though the objective statistical facts of this relation might be mapped with meticulous precision.

It is usually difficult if not impossible for the purposes of mapping to combine in the form of ratios, coefficients, and the like more than a limited number of statistical quantities. Consequently, the more general natural and human regions that geographers have marked out on many maps either are based on the assumption that the presence of certain individual phenomena or simple combinations of phenomena within a region is a fairly reliable indication that more complex combinations exist there, or else are arbitrarily marked out on the basis of the map maker's general fund of information. Frequently such regions are delimited to provide a framework for teaching or for the arrangement of material in textbooks—for convenience, that is, rather than to reflect absolute realities on the earth's surface. Maps of this type may be useful as pedagogic devices. They give the student at least a rough idea of how the parts of the world differ from one another in certain large respects, but next to maps of imaginary countries they represent almost the ultimate in subjective cartography. The regions they show are the equivalent of the well-defined "periods" into which history teachers divide the course of human events, and they have both as much and as little reality as such periods (see below, pp. 158–161).

HARMONY AND TASTE

We have seen how judgment and other subjective factors affect the reliability of the information presented by the symbols on a map. A symbol may, however, present a given phenomenon correctly as regards its distribution, character, and quality and yet be altogether unsuitable. Whether or not it is suitable depends on the map maker's taste and sense of harmony.

Where quantities such as altitude and density of population are shown by means of gradations of color or shading, the symbolism will not be suitable unless there is some harmonious relation between the variations in the tonal intensities of the symbols and the varying intensities of the quantities indicated by them. Although experiments have been made with a view to grading tonal intensities so that they may bear a definite mathematical relation to the mapped quantities, for the vast majority of maps the tonal grades are established by rule of thumb, just as are the gradations in the quantities that the symbols represent. In matters of color and tone the rule of thumb must be applied with the skill of the artist if the results are to be good.

Medieval maps are adorned with castles, towers, sea monsters, ships, and such things, and the employment of pictorial symbols is coming back today. It has certain advantages, especially on maps for popular use. By employing as point symbols conventionalized pictures of men, locomotives, shocks of wheat, insane asylums, whales, and the like, one can crowd a good deal of miscellaneous mixed information into a small space, but for any single distribution one cannot present as much precise and detailed distributional information as by the use of dots or other point symbols, since each little picture is likely to occupy more space than a single point symbol. Experiments have been made in the use of pictorial patterns as spatial symbols—for example, patterns made up of tiny men, or ears of corn, or cows—but such patterns tend to look fuzzy around the edges and do not lend themselves to the representation of gradations in quantity. Although in a sense they may be in better harmony with the things that they represent than flat colors or shading would be, they may also be out of harmony with the purpose of the map if that purpose is to give a clear and clean-cut conception.

The quality of a map is also in part an aesthetic matter. Maps should have harmony within themselves. An ugly map, with crude colors, careless line work, and disagreeable, poorly arranged lettering may be intrinsically as accurate as a beautiful map, but it is less likely to inspire confidence.

PROGRESSIVENESS AND CONSERVATISM

Whether a maker of maps is ever seeking and finding new things to map and developing new ways of mapping them or is a blind follower of tradition and precedent is of course partly a matter of individual character, but it is also a result of outside influences. Advances in cartography are due largely to the stimuli and opportunities that social needs give to the inventiveness of cartographers. Where a need arises, as for automobile road maps or air-navigation maps, cartographers respond. War provides a powerful stimulus, as the feverish mapping activities in and out of the government during World War II bore witness.

Conservatism in cartography is not inherently an evil when it means adherence to conventions and standards that have been tested by time and found good. The users of maps have become accustomed to certain conventions in symbolism, and too radical departures from these may be needlessly confusing. Many of the conventions have their origin in attempts to make symbolisms conform at least roughly to certain aspects of the things mapped, for example, that of showing water in blue instead of, say, pink (on medieval maps the Red Sea was shown in red), that of showing the relative importance of places by point symbols and lettering of graded sizes or of showing railroads by crosshatched lines. The convention of grading hypsometric tints more or less according to the spectrum, from greens for the lower altitudes, through yellows for intermediate levels, to reds and violets for mountain ridges, has been found by experiment and experience to give a graphic visual impression of relative altitudes. That there are different techniques altogether for representing the character of the surface configuration of the land is hardly a sufficient reason for abandoning a well-established and, on the whole, satisfactory method of representing relief.

Conservatism is an evil when it means failure to keep abreast of changing social needs and technical improvements. There is a certain danger of this in peacetime, especially in large map-making establishments where maps are produced in immense quantities according to uniform specifications. Under such conditions changes are costly and difficult, involving the training of personnel in new methods and the purchase of new equipment. This is why so many maps in atlases as well as in series follow traditional patterns or even repeat old errors long after the need for something better has made itself felt.

MAP USERS ARE HUMAN

If map makers are human, so too are map users. The qualities of integrity, judgment, critical acumen, and the like are as much required in the interpretation of maps as in their preparation. Like carpenters' tools, maps should not be misused. More should not be expected of them than they can perform. Sometimes when a critic damns a compiled map because he has found errors on it in regions that he has visited, his condemnation may reveal ignorance of the nature of cartography on his part rather than carelessness on the part of the map maker. Not all maps can be based on new surveys. Errors that originated in the sources of a compiled map frequently could have been avoided only by not making the map.

It is a misuse of maps, also, to draw unwarranted conclusions from them. To compare the area of Greenland with that of New Guinea on a map drawn on the Mercator projection is like trying to cut down an oak tree with a jig saw—it simply cannot be done; and we have seen that statistical maps are liable to similar if less obvious misinterpretation. Particularly dangerous are unwarranted conclusions as to the meaning of the facts that maps actually do disclose. Bring together all the maps there are showing languages, religions, densities of population, resources, economic regions, and the like; add to them twice as many more; compare and correlate them with all the ingenuity of which you are capable—and you will not, from such study alone, get very far toward the solution of vital international and national problems. Conditions and motives that no man can map must also be considered. Maps are indispensable tools in human affairs. That you cannot navigate a ship without charts, however, does not mean that you can navigate it by charts alone. Rudder and helmsman are also necessary.

CHAPTER 4

Human Nature in Science

HAVING been asked to emphasize in this paper "the indispensability of science for the future of civilization," I found myself thinking that this would be putting the cart before the horse. If civilization were to disappear there would be no science, and science contributes nothing to civilization when men of science fail to cultivate civilized qualities and to respond to civilized motives. Hence these remarks bear, rather, on "the indispensability of civilization for the future of science." They invite consideration of certain relations between human nature, both individual and collective, on the one hand, and science, on the other, and illustrate some of these relations with particular reference to geology and geography.

During their careers scientists acquire by bitter and sweet experience considerable information—even wisdom—concerning the influence of human nature on science. This they pass on to younger colleagues, who now and then give heed to it. Perhaps more heed would be given if the information itself were more scientific. Actually most of it is gained hit or miss. Scrappy, unorganized, and unsystematic, it breeds pet theories.

The question of how human nature affects science is surely important enough to warrant a less personal and more systematic approach. Large quantities of data on the subject are available in published and unpublished documents relating to the history of science. From analysis of these data principles could be derived and illustrative examples could be drawn that would offend no one, as the use of examples taken from contemporary observation might well do. Indeed, among the most valuable of the lessons to be learned from the history of science are those concerning the ways in which science has hitherto reflected human nature and will doubtless always continue to reflect it.

A SCIENTISTS' MACCHIAVELLI

Macchiavelli wrote a manual for aspiring rulers of men. On the basis of research in history and in the biographies of persons who had succeeded or failed in the art of government, he composed his famous *Prince*. I have in mind a sort of Macchiavelli's *Prince* for scientists—a manual that would analyze and synthesize those factors in human nature that contribute to success or to failure in the advancement of science. This manual, however, would not be Macchiavellian in its moral tone. It would deal with factors that contribute to the advancement of science rather than to the advancement of the scientist—not an unimportant distinction. Its writer, moreover, would have to pursue his biographic studies beyond the published "lives" of scientists, which deal largely with the successes of those who have been successful. Princes who fail create political havoc and hence their shortcomings are fully recorded. The shortcomings of men of science are more likely to be forgotten, though they may cause as much scientific havoc.

Let us assume that such a manual has actually been written by a Dr. Smith (a name that has no Macchiavellian connotations), that its title is *Smith's Manual for Scientists*, and that the remainder of this paper is a discussion of this imaginary work.

Like any good general introductory textbook in geology or geography, Dr. Smith's book proceeds from the elementary to the more complex. Part I analyzes the several personal qualities that influence scientific research, somewhat as minerals and rocks are considered at the beginning of a geological text; Part II surveys the motives for scientific research, along lines comparable to the treatment of geologic processes—tectonic, erosional, and so forth;

and Part III discusses scientific ideas much as formations of different periods are considered toward the close of the geological text.

PERSONAL QUALITIES INFLUENCING SCIENCE[1]

The personal qualities discussed in Part I include judgment, common sense, honesty, diligence, energy, modesty, taste, intellectual curiosity, and many others, and also their opposites, which Smith calls antiscientific qualities. Four qualities are stressed as having an especial bearing on science: originality, open-mindedness, accuracy, and scientific conscientiousness.

Originality, with which are linked imaginativeness, creativeness, and the like, shows itself in the urge and the ability to find new fields of investigation, to invent new techniques of research and exposition, and to devise new hypotheses. It provides the dynamic personal driving force in the advancement of science. The other three qualities provide governing controls that keep originality from running wild. As long as originality is held on the track, a man of science could scarcely have it in excess, whereas too much open-mindedness or too much accuracy may be as antiscientific in their effects as too little.

Open-mindedness, to which critical acumen is closely related, is the disposition to give full consideration to all the evidence and to all reasonable hypotheses that bear on any problem. Excess of open-mindedness may inhibit the scientist from adopting any hypothesis at all; it may lead him to see so many sides of a problem that he fails to espouse any one, contenting himself with "impartially presenting the facts and leaving it to the reader to interpret them." This is a not uncommon weakness, especially in the social studies today, where the facts are bafflingly complex. Dr. Smith hazards the opinion that excessive open-mindedness explains why the output of a good many geographers has been primarily descriptive rather than explanatory or interpretative, and that geologists as a group are freer from this failing. Not being a geologist I do not know whether he is correct in this. Closed-mindedness, he maintains, is a more definite antiscientific quality. It shows itself in violent denunciations of hypotheses, or sometimes even in unwillingness to recognize data, that contradict a pet theory. When combined with originality, energy, and vanity, closed-mindedness has been known to produce fanatical devotion to pet theories—especially when they are a scientist's own beloved brain children.

Accuracy, with its little sister precision, is as necessary a quality in a scientist as is sharpness in an edged tool. There is no excuse for inaccuracy that springs from sheer carelessness, and Smith advises well-meaning but "congenitally inaccurate" persons to avoid scientific pursuits. They do not always do so. There can, however, be too much precision. One does not cut down forests with razors, and, similarly, degrees of precision are often possible that may far exceed what is needed for a specific task in hand. Striving to attain such superaccuracy yields diminishing returns by reducing production without commensurate improvement in quality. Finicky men of science do less for science than those who know just where to call a halt to their perfectionist inclinations. Excessive zeal for precision has delayed the publication of the results of researches until after the date when they would have been of greatest use.

A scientific investigator might possess all the foregoing desirable qualities and yet be subject to the influence of undesirable motives. The fourth essential quality, therefore, is ability to discriminate between motives. This quality is "scientific conscientiousness," or the possession of a scientific conscience, which Dr. Smith maintains is merely a special aspect of social conscience.

MOTIVES FOR SCIENTIFIC RESEARCH

To do justice to the very broad and important subject of motives, which is taken up in Part II, Dr. Smith concedes is a task that could be adequately performed only by an exceedingly wise man—a combination of scientist, historian, and philosopher. He ventures into this field with some trepidation, which I confess to sharing with him.

He cites Webster's definition of "motive" as "that which incites to action." Every motive is either proscientific, antiscientific, or nonscientific, depending upon whether it promotes, retards, or exerts no effect on the advancement of science. Every motive, moreover, is either a personal, a group, or a disinterested motive, depending on whether it springs from a desire to serve individual interests, group interests, or the interests of no particular individuals or groups.[2] Dr. Smith illustrates these kinds of motives as they might operate concurrently in the case of a volcanologist in studying an active volcano. Curiosity as to how the volcano works and the desire to make some money in doing field work on its slopes are personal motives; an impulse to discover facts that might benefit the villagers living near the volcano and a wish to collaborate in a research program of a volcanological institution would constitute group motives; a desire to add to the general fund of a scientific knowledge of volcanoes would constitute a disinterested motive. These are all proscientific motives. Antiscientific motives would be exemplified in the volcanologist's fear of collaborating with a colleague lest the latter might steal some of his ideas, or in the cruder fear that might cause him to throw away his notebook and camera in his hurry to get away from an eruption.

PERSONAL MOTIVES

As regards personal motives, Smith does not dwell on that of making money, since this motive is in no way distinctive of scientists, and his subject is motives as they affect science itself rather than as they affect the fortunes of those who profess it. He does, however, analyze with considerable care the motives that spring from two personal desires—the desire to satisfy intellectual curiosity and the desire to be well thought of.

Cats, he says, feel an absorbing curiosity regarding mice. While they are being lured by scents and signs that indicate the proximity of the latter, they are immune to the allurement of dogs. So also, different scientists are susceptible to different allurements. Geographers are responsive to what they call the lure of place and the lure of the map.

The lure of place is an attraction that localities exert on the imagination—a curiosity concerning the nature of things as they exist in different places and regions. The lure of the map is the attraction of the shapes, forms, and arrangement of things on the earth's surface, a geometrical curiosity concerning concepts that are, or might be, shown on maps. The lure of place Smith compares with the musician's sensitivity to tone or the painter's to color; that of the map with the musician's feeling for rhythm or the painter's for design. The "born geographer," if such there be, derives personal satisfaction from responding to these lures, just as the chemist derives personal satisfaction of a different sort from responding to chemical lures to which the geographer is indifferent. Dr. Smith's analysis of the psychological nature of the various scientific allurements makes an interesting digression.

We are more concerned with his discussion of the controls that scientific conscience exerts upon the motives of satisfying intellectual curiosity. A well-developed scientific conscience, he argues, vetoes the waste of time, energy, and talent upon studies that satisfy curiosity but do little more besides. Care in the accumulation of data, brilliance in their comprehension and exposition, and ingenuity in the development of theories are to little purpose if the facts and theories are of meager concern to anybody but the man who gathers and expounds them. A geographer might be impelled by a burning desire to make an intensive study of a very small area, when someone else has already made a similar study of a similar area in the same region. If the geographer has a scientific conscience it will warn him against yielding to this temptation unless he is firmly convinced that something new and of substantial value to others than himself will come from his study. To add embroidery to concepts and principles already well established may be enjoyable, but a scientific conscience warns against doing so when there are more pressing needs for other types of geographical research. Of course, Smith admits, it sometimes happens that an investigation which seems utterly useless at the time it is made later turns out to be of far-reaching value, but he feels that the probability that this will occur must be weighed against the more obvious social needs for the services of scientists and against the fact that there are none too many scientists to meet such needs.

Scientists, he goes on to point out, are not unlike other mortals in their desire to be thought well of, and their motives are strongly influenced by the opinions of others. Scientific conscience tells the scientist which opinions to heed and which to reject. There have been times when it demands resistance to the opinions of persons who have it in their power to ruin a scientist's career or even to take his life, and science, like religion, has had its martyrs.

Smith shows that opinions or judgments of the relative worth of scientific investigations are of three main kinds: "formal" opinions, based on criteria of form and substance; "qualitative" opinions, based on criteria of scientific quality; and "pragmatic" opinions, based on criteria of effectiveness. As an example he asks us to suppose that three scientists are invited to express views as to whether or not two books should be published. The first book is a treatise on contemporary geopolitics, carelessly written, poorly arranged, and wholly unsound in its reasoning. The second book deals with the historical geography of ancient Siam; written in a beautiful style, it is based on profound erudition and is thoroughly convincing in its handling of all the available evidence. Scientist A says: "Publish the first book since geopolitics is more vital than historical geography." This would be a formal judgment. Scientist B says: "Publish the second book because it is so much better in quality," clearly a qualitative judgment. Scientist C says: "Don't publish either, since neither will exert much influence, the first because it is so bad, the second because it is so recondite." This is a pragmatic judgment.

Dr. Smith comments on the propensity of certain men of science to express sweeping judgments of whole fields and whole methods of scientific investigation. While these judgments are often penetrating in their insight and exert a beneficial effect, sometimes, on the other hand, they reveal a distinctly lower order of scientific thought than would normally be used by those who propound judgments in their own specific researches. When such opinions crystalize into widely and uncritically accepted clichés Smith believes they may be harmful, and they are likely to be particularly harmful when based primarily on formal rather than on qualitative or pragmatic criteria.

Among such clichés founded primarily on formal criteria he cites the view that experimental research and first-hand studies in the field are inherently more "scientific" than researches in libraries; also the opinion that quantitative studies yield more genuinely scientific results than those in which quantitative measurements are not feasible. Another formal cliché judgment, he says, is unconsciously expressed when certain works are damned with faint praise as "mere description," "nothing more than compilation," "simply a matter of techniques."

Science is founded on the assumption that the universe is governed by laws, and the goal of science is to discover and formulate these laws. Dr. Smith, however, with some justification, regards as a cliché the rating of those forms of scientific investigation which yield precise and reliable statements of laws as necessarily more scientific than those which do not. Some of the laws of astronomy have been formulated so accurately that eclipses may be predicted thousands of years hence. The laws of economics barely permit the prediction of what is going to happen next week. But to regard astronomy as therefore more worthy of scientific respect than economics, Smith believes, is to take a narrow, formalistic view of the scope and nature of science.

Geographers may be interested in what Dr. Smith says about certain formal clichés concerning their subject, though not all of them will agree with him. Not so long ago, he points out, geography as a whole was criticized because a number of geographers were prone to give unlimited scope to their imaginations in the matter of generalization. The whole discipline was condemned for the excesses of some of its devotees. This helped instill in certain other geographers a horror of generalization and reluctance to generalize. In other words, these latter geographers adopted a formal cliché about the nature of generalization. Rather than take risks of generalizing, they turned to "safe" quantitative studies, to microgeographical researches, and to emphasis on the accumulation rather than on the interpretation of facts. This, in turn, gave rise to yet another formal cliché on the part of nongeographers, that geographers are sterile in ideas and narrow in vision.

Since the distinctive aim of science is to establish general principles, an irrational fear of generalization per se is antiscientific. What is to be shunned is unsound generalization. Similarly microgeography per se is not necessarily narrow and without vision—only microgeography that yields little larger fruit. But when generalization is condemned for unsoundness the judgment is qualitative, not formal, and when microgeography is condemned for leading nowhere the judgment is pragmatic, not formal. Dr. Smith shows, however, that judgments in which little or no account is taken of quality or effects are easily and often made.

That qualitative judgments of the worth of scientific investigations are fairer than formal judgments, Smith holds to be obvious. A qualitative judgment takes account of the degree of good sense, originality, accuracy, open-mindedness, and so forth, to which a study bears witness. It also takes account of the suitability of the form and substance to the solution of the problem in hand. Thus field studies would be rated higher in quality than library researches only in cases where better evidence can be secured by the former than by the latter. While field observations provide the primary data for geology and geography alike, it would be ridiculous for a geologist or geographer to expect that he could solve many of the larger problems that confront him, problems involving synthesis and correlation, by investigating them in the field alone.

While the goals of science are the discovery and formulation of laws, a lot of work has to be done before laws can be formulated, and this preliminary work, in Smith's opinion, is quite as indispensable, quite as "scientific," as are the subsequent processes of interpretation to which it leads. Part of the preliminary work consists in "mere" compilation, description, and development of techniques. Qualitatively such procedures, as long as exacting, critical, and original scholarship is devoted to them, rate higher than speculative attempts to establish laws on the basis of faulty reasoning from insufficient evidence. To Dr. Smith an economic law is fully as scientific as is the law of eclipses, provided all available evidence is used in developing the economic law and used with the same degree of rationality as that attained in developing the astronomic law. That the actual probability of the economic law is less he regards as immaterial to its scientific quality.

Whole broad domains of science are cultivated not for the immediate purpose of formulating general laws but in order to understand specific conditions and processes. This is especially true of geology and geography, where the first objective is to explain the origins, nature, and relationships of particular land forms, rock formations, types of settlement, routes of trade, and what not, as they exist in particular regions. These studies prepare the way for the formulation of geological and geographical laws sometime in the future, but the way may be long. If the scientific merits of research are judged formally according to the degree to which they succeed in stating general laws, rather than according to the quality of the work devoted to such research, a large part of our two sciences of geology and geography would be denied scientific merit —something that we may unite with Dr. Smith in regarding as absurd.

Pragmatic opinions, as distinguished from formal and qualitative opinions, are those that rate scientific researches in terms of their effects. While it is usually true that the better the quality, the better are the effects, this is not invariably so. Smith shows that incomplete and even careless studies of little-understood but important phenomena may exert more far-reaching and more beneficial effects than studies of higher quality that deal either with inconsequential matters or matters already well understood in their essentials. Great works of compilation often rate extremely high from the pragmatic point of view because of the innumerable practical purposes that they serve and because they furnish the stimulus for the development of scientific theories. Even "outrageous hypotheses," as the late Professor W. M. Davis pointed out,[3] may have pragmatic worth by providing means of testing the validity of other hypotheses.

GROUP MOTIVES

The pragmatic opinions that others hold of the effectiveness of his work are largely instrumental in fashioning the nature of a scientist's response to *group* motives. This subject Smith takes up in the next chapter. Here he states that science is the product of human gregariousness and that it would be hard to conceive of a scientific investigation not motivated in part at least by a desire to serve the interests of some group of people—be it an organized group, such as a university, a society, a community, a corporation, a nation, or largely unorganized, such as the geological or the geographical professions, or the people who happen to dwell on the slopes of a volcano, be it actual or figurative. But while this desire, combined with personal curiosity, has brought science into being, certain group motives have also constituted serious obstacles to the advancement of science, and it is with these antiscientific motives that Dr. Smith is most concerned.

Antiscientific group motives spring from competition and conflict among groups—from the ambition of groups to get the better of one another and from their fear of being got the better of. These conflicts, moreover, are on many levels, ranging from quarrels between or within departments in a single university to world wars between coalitions of nations. They give rise to three types of antiscientific practice: the wilful distortion of truth in order to mislead rival groups; the suppression of the results of scientific research in order that rivals may not benefit by them; and the use of the results of scientific research to injure rival groups.

That the first of these practices—the distortion of truth—is the negation of science is self-evident, but that the second and third are antiscientific Dr. Smith believes to be less obvious. It has been argued that science is advanced whenever scientific research is conducted, whether or not the results are suppressed, and that the use of the results is of no scientific concern—in other words that these are questions of morals and not of science. Dr. Smith, however, seeks to demonstrate that ethics and science are inextricably linked and that unethical practices are not only antisocial but antiscientific.

He argues that the advancement of science demands the continuous discovery of new truths and the continuous development of new hypotheses. For this the fullest and freest possible interchange of knowledge already acquired is prerequisite and anything that hinders this interchange retards the advancement of science. Indeed, the very word "science" connotes something that cannot be hoarded in secrecy. While one says "my knowledge" or "the government's knowledge," one never says "my science" or "the government's science." Science, unlike knowledge, is indivisible, in the sense that no part of it can be the exclusive possession of an individual or a group and the whole of it is the common property of humanity. When we make a "contribution to science," we donate some of our knowledge to humanity. Knowledge may be scientific in quality, but as long as it is kept locked in files marked "secret," "confidential," or "restricted" it remains mere knowledge. It does not become science until it is made at least potentially available to anyone who wishes to use it.

Much scientific research has been conducted for the express purpose of applying its results to the injury of other groups. Wars have produced such feverish bursts of scientific activity and have so greatly accelerated certain discoveries and inventions that it has even been maintained by some that war has accomplished more than has peace to promote the advancement of science. Smith emphatically rejects this doctrine. That human enlightenment in the long run can have benefited more than it has suffered from group selfishness, conflict, and fear he regards as an "outrageous hypothesis" of no pragmatic value.

DISINTERESTED MOTIVES

The last chapter of Part II of Smith's manual deals with disinterested motives in science. These are the incentives to scientific endeavor, which, because they do not spring from the needs or interests of specific individuals or groups, are necessarily incentives to serve the general needs of humanity.

Scientific work is never motivated solely by disinterested purposes—never without any trace of the influence of desires to serve personal or group interests. Disinterested motives operate concurrently with rather than to the exclusion of group and personal motives, and one of their principal functions is to counteract or nullify the antiscientific influences of selfish and destructive group motives. By inciting men of science to resist antiscientific and antisocial practices, disinterested motives operate, Dr. Smith says, somewhat as do the preservatives put in foods to prevent their spoiling. They have, however, a more positive function—that of inspiring men of science actively to seek out those lines of endeavor that will be of service to mankind.

SCIENCE AND WAR

We come now to Part III of Smith's manual, which deals with the nature of certain larger problems in science in the light of the various qualities and motives analyzed in the two preceding parts. Space does not permit me to dwell here on more than two of these problems—those of the effects of war upon science and of the reconstruction of scientific endeavor after the war.

[The reader is reminded that this paper was written during World War II and before the Bomb had caused men of science to search their consciences more deeply than ever before; but if what follows seems somewhat dated and naïve, it may be of interest today for that very reason. (1965)]

While he regards the institution of war as detrimental to the advancement of science in the long run, he is far from regarding war as an unmitigated evil in its effects upon science. War gives complete priority to a particular group motive—that of helping one's nation to win. In wartime all other motives—personal, group, and disinterested—are overshadowed by this one, which produces an intensity of thought, a passion for hard work, and a collaborative spirit among scientists that are seldom equaled in peacetime. It also leads to rapid advances along certain specific lines of research. This is the bright side of the picture. The darker side is that scientific effort as a whole is regimented and hence distorted. In times of peace the pioneer fringe of scientific knowledge advances more or less evenly along a broad front and in the open, where all who wish to look can see its advance. In war long stringy arms shoot ahead into the unknown, while large segments of the frontier remain at a standstill and many of the most rapidly advancing arms are blotted out by clouds of censorship. The operation of disinterested motives is not only weakened but often entirely prevented.

Dr. Smith makes it clear that the problems that a nation faces in getting into and in getting out of a war are often more difficult of solution than those of actually waging the war itself and that this is true no less in the realm of science than in other realms of national life. While a war is in progress, much scientific work can proceed smoothly "according to plan." The most serious difficulties are encountered in making the transitions from peace to war and vice versa.

As a car must be refitted in the spring for warm-weather driving, so the machinery and organization of scientific investigation will require reconditioning for peacetime operation. For one thing, steps will have to be taken to reestablish the fellowship of men of science throughout the world—a fellowship in which the nationalistic impulses of aggression, fear, and suspicion can be made partly if not wholly subordinate to the disinterested service of science. Within each nation antiscientific restraints—censorship and censoriousness, in particular—must be abolished with the utmost possible dispatch. In this reconditioning of scientific endeavor, the benefits of its wartime operation must be preserved in so far as possible, especially the spirit and means of collaboration that the war has developed. The artificial academic barriers that this collaboration has done so much to break down must not be permitted to arise again. Men of science must make concerted efforts to forestall the loss or destruction of the masses of information that have been accumulated in government offices and elsewhere for wartime purposes. Peace will bring a widespread let-down, an inertia that will have to be overcome if this latent science—as it might be called—is to be saved and released promptly for use.

Imperative as will be this reconditioning, it is merely a means to an end. Just as the end in the case of the automobile is to drive it somewhere, so in the case of scientific research it is the production of needed types of science, and the larger problem is what the mechanisms of scientific research shall be made to produce when peace returns.

Some might say that it should not be *made* to produce anything in particular—that compulsion stifles originality and initiative, and that scientists should be free to study whatever they please in whatsoever manner they wish. Dr. Smith thinks there is wisdom in this view, provided one important qualification is made. While freedom of science is fundamental, like freedom of speech it is a freedom that imposes responsibilities. It is freedom to investigate whatever one wishes, subject to the dictates of scientific conscience. A developed and enlightened scientific conscience will always incite its possessor to select from among the innumerable subjects by which he is attracted in his chosen field those in which his abilities can best be employed toward meeting the greatest human needs.

Along with many other contemporary observers, Smith maintains that science has given men a mastery over nature so extensive and so skilful that, if uncontrolled, it may lead, not to making the world more civilized, but to the destruction of civilization itself. This will happen unless an equal degree of mastery can be achieved over the forces, both good and evil, in human nature. Hence Dr. Smith believes that the most urgently needed of all forms of science is that which will contribute to increasing the will and the power of human groups to collaborate with one another instead of cutting each other's throats. The development of this will and this power has traditionally been considered a task for experts in morals rather than for men of science. Good motives, however, whether on the part of an individual or a group, cannot be inculcated by moral precepts alone. A child is more likely to behave if given a rational —a scientific, in other words—explanation of the undesirability of ill behavior than if he is merely told that such and such conduct is bad. The child ordinarily misbehaves because he feels injured or thwarted, not because of the machinations of Satan or the promptings of original sin. Similarly, groups cut each other's throats because they feel, rightly or wrongly, that they are thwarted or imposed upon by other groups possessing greater advantages. Science can often disclose whether such sentiments are founded on fact or fancy, and, if they are founded on fact, science can seek for and test out measures of amelioration.

Many of the largest and toughest roots of man's inhumanity to man are embedded in the circumstance that certain groups enjoy advantages over others because they occupy or control particular areas of the earth's surface. There are interareal conflicts within every village, every state and every nation, and worst of all between nations. Neighbors quarrel over fence lines and wandering cattle; nations fight over boundaries and the control of vast territories. Hence, those branches of science that deal with areas, their occupants, and those who control them in terms of their relative advantages and disadvantages can do much to lay bare the roots of human conflict—and the laying bare of roots is a necessary preliminary to their removal.[4] Areal, or regional, research lies partly within the provinces of geology and geography. Hence Dr. Smith believes that our two sciences afford immense potentialities of service to perplexed humanity and that geologists and geographers are in a peculiarly favorable position to produce scientific fruit indispensable for the future of civilization.

CHAPTER 5

Terrae Incognitae

THE PLACE OF THE IMAGINATION IN GEOGRAPHY

TERRA INCOGNITA: these words stir the imagination. Through the ages men have been drawn to unknown regions by Siren voices, echoes of which ring in our ears today when on modern maps we see spaces labeled "unexplored," rivers shown by broken lines, islands marked "existence doubtful." In this paper I shall deal with *terrae incognitae*, both literal and as symbolizing all that is geographically unknown; I shall discuss the appeal that they make to the imaginative faculties of geographers and others and the place of the imagination in geographical studies.

THE SIRENS OF TERRAE INCOGNITAE

In earlier times literal *terra incognita* was seldom far from the hearthfires of men. To our stone-age ancestor a blue mountain range on the horizon might have marked its border. Beyond lay a country —of evil spirits, perhaps—into which he must often have wished to penetrate but dared not. If, finally, curiosity mastered his dread and with a few hardy companions he crossed the forbidden range, as like as not he found a region not so greatly different from his own. Thus the encircling border was pushed back a little way and a short step taken in a process that has not even yet quite reached its end. But although our stone-age ancestors and their descendants down until the dawn of modern times moved back the rim of *terra incognita* bit by bit, their known world was only a pool of light in the midst of a shadow—limitless, for all that was definitely understood and proved. Voyages into this shadow became a favorite theme of poets and storytellers—the theme of the Argonautic myth and the *Odyssey*, of the legends of Sinbad and Saint Brandan. Out of its darkness wild hordes poured forth from time to time to carry fire and sword across Europe—Scyths, Huns, Tartars; it was a mysterious shadow, whence came rumors of strange men and monsters, of the priestly empire of Prester John, of the Apocalyptic tribes of Gog and Magog shut behind Alexander's wall until, on the day of judgment, they shall burst out to ravage the world. *Terra incognita* was not without contact with the known world, and throughout most of history awareness of its menacing presence must have aroused an abiding wonder in all but the least imaginative.

Possibly this wonder became rooted in the inheritable subconscious of sensitive folk and was thus transmitted from generation to generation down to our day; but, whether or not so inherited, the innermost impulse that makes us take satisfaction in geographical studies seems akin to the urge that impelled our stone-age forefathers toward the lands beyond the range. In the course of field work or on a summer holiday we have all climbed a mountain and gazed over uninhabited and unfamiliar country. Behind us has lain the valley out of which we have come, the farm or ranch where we have sojourned. Before us has spread, if not a land unknown to the United States Geological Survey, at least a personal *terra incognita* of our own. In the contemplative mood that mountain tops induce, we have brooded over the view, speculated on the lay of the land, experienced a pleasurable sense of the mysterious—perhaps felt even a touch of the sinister. We have heard the Sirens' voices.

The Sirens, of course, sing of different things to different folk. Some they tempt with material rewards: gold, furs, ivory, petroleum, land to settle and exploit. Some they allure with the prospect of scientific discovery. Others they call to adventure or escape. Geographers they invite more especially to map the configuration of their domain and the distribution of the various phenomena that it contains, and set the perplexing riddle of putting together the parts to form a coherent conception of the whole. But upon all alike who hear their call they lay a poetic spell.

Nowadays geographers seldom or never have the opportunity to enter literal *terrae incognitae*—totally unexplored territories—and at first glance it may seem farfetched to compare the allurement of such unknowns with the attraction that draws us toward the region and problems with which we must actually be concerned. However, the Siren voices heard by a Columbus, a Magellan, or a Livingstone differed only in intensity but not in tone and quality from those that call us to explore our seemingly more prosaic and humdrum *terrae incognitae*. Let us, therefore, examine a little further into the nature of *terrae incognitae* of various magnitudes and types.

SOME VARIETIES OF TERRAE INCOGNITAE

Obviously, whether or not a particular area may be called "unknown" depends both on whose knowledge and on what kind of knowledge is taken into account. As used literally on the early European maps, the words *terra incognita* signified a land unknown to the map maker after he had presumably consulted all available sources of information; but if such "unknown territories" were beyond the ken of the geographers and cartographers of Western civilization, they were known to their inhabitants, if any, and frequently to peoples of other civilizations as well. China lay deep in the heart of *terra incognita* to the Romans, but the Roman Empire was equally lost in land unknown to the Chinese. We are familiar with maps depicting the extent of the "known world" at different dates, most of which illustrate, somewhat crudely, stages in the development of the geographical knowledge of a single cultural tradition, that of the West. To round out the record, similar maps would be required for other traditions, showing the progress of the regional knowledge of Chinese, Japanese, Arabs, Hindus, Mayas, and other less advanced peoples. It would also be revealing if the dynamics of this process could be illustrated cartographically, as, for example, in the sixteenth century when the establishment of contact between Europe and the Far East produced a partial coalescence of the known worlds of Occidental and of Chinese geography.

When we say "the world as known to the Greeks of the time of Eratosthenes" or "to the Americans in A.D. 1945" we mean the areas about which certain Greeks and certain Americans were in a position to ascertain something without having to conduct exploring expeditions for the purpose. The world as actually known to the great majority of Greeks or Americans was smaller. What is *terra incognita* for all practical purposes to an isolated community of hill-billies is more extensive than what is *terra incognita* to the members of the Association of American Geographers. Hence, depending on our point of view, there are personal, community, and national *terrae incognitae;* there are the *terrae incognitae* to different cultural traditions and civilizations; and there are also the *terrae incognitae* to contemporary geographical science.

The meaning of *terra incognita* depends no less on the kind of knowledge that we are considering. There are two grades of geographical knowledge: knowledge of observed facts and knowledge derived by reasonable inference from observed facts, with which we fill in the gaps between the latter. On the basis of reasonable inference, for example, I *know* that the climate in those parts of Antarctica that have never been seen by human eyes is too cold to support tropical rain forests, and that it is too warm and dry in the unexplored heart of Southern Arabia for tundras and ice fields. Thus, if *terra incognita* is conceived in an absolute sense as an area concerning which total human ignorance prevails, no *terrae incognitae* exist today on the earth's surface. At no place on this planet is the shadow so utterly dark as it was in former times. Science has reached a point where we may interpolate sound, if incomplete, geographical knowledge into every gap (see above, pp. 38–39).

I have a summer place on the Maine coast. Geographers know nothing of it except what they may reasonably infer from their general familiarity with the region in which it lies. They might infer something about its climate, and also draw some conclusions as to what it is *not*, as we do regarding the interior of Antarctica; but as to its relief, drainage, soils, vegetation, houses, roads, and other aspects of its internal geography no published information is available. One might fairly surmise that the vegetation includes firs, spruces, and tamaracks, but, for all that is really known to geographical science, my land might not have a single tree upon it. If, therefore, *terra incognita* is conceived as an area within which no observed facts are on record in scientific literature or on maps, the interior of my place in Maine, no less than the interior of Antarctica, is a *terra incognita*, even though a tiny one. Indeed, if we look closely enough—if, in other words, the cartographic scale of our examination is sufficiently large—the entire earth appears as an immense patchwork of miniature *terrae incognitae*. Even if an area were to be minutely mapped and studied by an army of micro-geographers, much about its geography would always remain unknown, and, hence, if there is no *terra incognita* today in an absolute sense, so also no *terra* is absolutely *cognita*.

THE IMAGINATION IN GEOGRAPHY

Naturally, other motives than his magnetic attraction toward the geographically unknown play their part in making a geographer and in keeping him one. Satisfaction in what he knows and in imparting it to others, as distinguished from curiosity regarding what he does *not* know, is often a powerful factor. He may relish the assimilative processes of collecting data in the field or library, or the intellectual process of thinking through complex problems, or the altruistic process of contributing something that he hopes will be of use, or at least of interest, to his fellow men. But these motives are not distinctive of geographers, since they impel others as well. What distinguishes the true geographer from the true chemist or the true dentist would seem to be the possession of an imagination peculiarly responsive to the stimulus of *terrae incognitae*, both in the literal sense and more especially in the figurative sense of all that lies hidden beyond the frontiers of geographical knowledge. Indeed, the more brightly the light of his personal knowledge shines upon a region or a problem, the more attractive to the geographer are the obscurities within it or concerning its entire extent.

Geographical research seeks to convert the *terrae incognitae* of science into *terrae cognitae* of science; geographical education to convert personal *terrae incognitae* into personal *terrae cognitae*. In both cases the unknown stimulates the imagination to conjure up mental images of what to look for within it, and the more that is found, the more the imagination suggests for further search. Thus curiosity is a product of the imagination. Now, as to curiosity, it seems a little unfortunate that this word, used to designate a nosy, impertinent characteristic of monkeys, small children, and gossips, is also applied to the loftier and more impersonal impulse that drives the astronomer to search the depths of the universe and the geographer to penetrate the mysteries of *terrae incognitae*. "Wonder" would be a preferable term for the latter could we not experience wonder in contemplating things without seeking to understand them. At all events, the less imaginative he is, the less open is a geographer to either wonder or curiosity, and geographers of weak imagination—for a few do exist—are impelled by different motives. They follow in the footsteps of others, imitating stereotyped patterns, and, if their industry and imitative ability are considerable, they may succeed in teaching and even in research, serving well to maintain geography as it is and to advance it along beaten trails, if not to mark out new ones.

The imagination not only projects itself into *terrae incognitae* and suggests routes for us to follow, but also plays upon those things that we discover and out of them makes imaginative conceptions which we seek to share with others. In the words of the late Sir Douglas Newbold: "Knowledge must pass into vision, that state of mind and heart which does not merely swallow evidence, but changes that evidence into a judgment, an appreciation, a living picture of a country."[1] Unlike the mental images that we can merely invoke from the memory—such as the remembrance of views once seen—an imaginative conception is essentially a new vision, a new creation, and consequently the less imaginative we are the less fresh and original will be our writing and teaching, and the less effective in stimulating the imaginations of others.

But a powerful imagination is a dangerous tool in geography unless it is used with care. Indeed, the imagination might better be compared to a temperamental horse than to an instrument that operates precisely and with objectivity. A highly sensitive function of the mind, it is easily swayed by subjective influences, and for this reason has come in for a share of the disrepute in which subjectivity is often held in scientific circles.

As I shall have a good deal to say about subjectivity in what follows, it may be well to stop here and analyze it. The disrepute in which it is held, I feel, is not altogether deserved and may be due to a mistaken belief that subjectivity is the antithesis of objectivity. Objectivity, we might all agree, is a mental disposition to conceive of things realistically, a disposition inherent partly in the will and partly in an ability to observe, remember, and reason correctly. The opposite of objectivity would, then, be a mental disposition to conceive of things unrealistically; but, clearly, this is not an adequate definition of subjectivity. As generally understood, subjectivity implies, rather, a mental disposition to conceive of things with reference to oneself—that is to say, either as they appear to one personally, or as they affect or may be affected by one's personal interests and desires. While such a disposition often does, in fact, lead to error, illusion, or deliberate deception, it is entirely possible to conceive of things not only with reference to oneself but also realistically. Were this not the case, the human race would long ago have become extinct. Thus we may distinguish between (1) strictly impersonal objectivity, (2) illusory subjectivity, and (3) realistic, or one might even say objective, subjectivity. To illustrate: my conception of the skunk as a small furry animal with certain distinctive abilities—not, in this case, an imaginative conception—is impersonally objective; an unobservant person's wishful conception of a particular skunk as a cat would be a product of illusory subjectivity; and a careful observer's accurate conception of a personal encounter with a particular skunk would be a product of realistic subjectivity.

PROMOTIONAL AND INTUITIVE IMAGINING

There are three imaginative processes of importance in relation to geography, in each of which subjectivity of one form or another plays a large part. These might be called promotional, intuitive, and aesthetic imagining.

The first, promotional imagining, is controlled by a desire to promote or defend any personal interest or cause other than that of seeking the objective truth for its own sake. It is subjective imagining dominated by such emotions as bias, prejudice, partiality, greed, fear, or even love, all of which may lead the imagination to produce illusory or deceptive conceptions conforming to what one would like rather than necessarily to the truth. Realistic subjectivity, however, may also influence promotional imagining. Passionate devotion to a personal or social cause may result in a no less passionate quest for realistic conceptions useful in advancing or defending that cause. Human greed for wealth and power and human partiality for particular forms of religious doctrine have yielded, as by-products, rich fruit in objective geographical knowledge.

The purpose of intuitive imagining, the second type, is objective, in that the intent here is to secure realistic conceptions. It is, nevertheless, a subjective process because it makes use of one's personal impressions of selected facts instead of impersonally considering and weighing all pertinent evidence. Much of the world's accumulated wisdom has thus been acquired, not from the rigorous application of scientific research, but through the skillful intuitive imagining —or insight—of philosophers, prophets, statesmen, artists, and scientists.

AESTHETIC IMAGINING

The third type of imagining—the type of which I should like to speak more especially—I have called "aesthetic," though I use this adjective reluctantly because of its frequent, though mistaken, mental association with the disagreeable noun "aesthete." Aesthetic imagining is merely a subspecies of promotional imagining, in which the dominant personal interest promoted is a desire to enjoy the process of imagining itself, and to give satisfaction to others by communicating the results in written or graphic form. The end purpose, therefore, is either the creation of an independent work of art or the introduction of artistry into a work of utility or of science. Much aesthetic imagining is the product of illusory subjectivity, of a disposition to create conceptions that are fictitious or fanciful, as when a painter paints a cow as she looks to no one else on God's earth. Much of it, however, is the result of realistic subjectivity, as when he paints the cow as she would look to you or me. This he can do either with or without the aid of aesthetic imagination. Not all cows are equally worthy of being painted and not all aspects of a given cow are equally worthy of emphasis. The imagination can guide the selection of a noteworthy cow to paint, or of an ordinary cow in a noteworthy setting, or of noteworthy aspects of either a noteworthy or an ordinary cow. And by the same token, a geographer may portray a place or region, either with conscientious but unimaginative attention to all details, or with aesthetic imagination in selecting and emphasizing aspects of the region that are distinctive or characteristic.

What is the attitude of geographers toward intuitive and aesthetic imagining? There are some who believe that we should explore only such *terrae incognitae* as lend themselves to exploration in accordance with rigorous scientific principles, that the purpose of such exploration should be to determine exactly what these *terrae incognitae* contain, and that in presenting the results to others we should aspire to strict, impersonal objectivity. It may be left, these say, to the artists, poets, philosophers, novelists, and politicians to develop the aesthetic and intuitive faculties of their minds; geographers should keep to a straighter and narrower path.

Others concede that many types of geographical research cannot be pursued along strictly scientific lines and that there will long remain scope in geography for skillful intuitive, if not for aesthetic, imagining. Geography deals in large measure with human beings, and the study of human affairs and motives has not yet reached a stage in which more than a small part of it can be developed as a science. Until it arrives at that stage, much geographical study will have to be considerably tinged with intuitive subjectivity. But also among those who hold this view, the prevalent attitude toward aesthetic imagining in geography is one of distrust.

Unfortunately, this deep-seated distrust of our artistic and poetic impulses too often causes us to repress them and cover them over with incrustations of prosaic matter, and thus to become crusty in our attitude toward anything in the realm of geography that savors of the aesthetic. Like the companions of Ulysses we would row along with ears stopped to the Sirens' song. If a little of its melody were to penetrate through the stopping, we would try not to let others know. Ulysses himself, however, listened to the Sirens and as a consequence, if one may interpret the matter in a fanciful vein, his whole voyage assumed to him the aura that we sense on reading the *Odyssey*. Had his companions survived, their accounts of the expedition would have been strictly objective—factual, realistic, but uninspired, and, like some of the geography of today, soon forgotten. In Homer's words (as rendered by T. E. Lawrence), Ulysses returned "spirit gladdened and riper in knowledge," and hence his account has lived forever. He was well advised to hearken to the Sirens, to allow the charm of their voices to kindle his imagination, but nevertheless to have himself bound to the mast and so pass them by. If he had paid them a visit and yielded to their allurements, and then had the fortune to escape, he would have brought back a tale so unrealistic and sensational as to repel discriminating hearers, and his account would have been forgotten even sooner, perhaps, than would the honest if prosaic stories of the members of his crew.

The legitimate and the desirable in aesthetic subjectivity. Our undue fear of hearkening to the Sirens would seem to spring from three fairly widespread notions: first, that aesthetic subjectivity is always unscientific, leading to illusion and error; second, that it is out of place in geography, serving no necessary functional purpose; and third, that geographers, by and large, are not skilled in giving expression to aesthetic sensitivity and hence should avoid trying to do so.

In considering the validity of these three notions, I shall designate as "legitimate" such practices as do not actually interfere with the advancement of scientific geography, which is and should rightly be the primary concern of the majority of geographers, though not necessarily the exclusive concern of the totality of geographers. I shall designate as "desirable" such legitimate practices as also appear to promote the advancement either of scientific geography or of geography in a broader sense.

With regard to the first notion, it is, of course, true that aesthetic subjectivity may lead to illusion and error. There is, however, a distinction between illusion and delusion. We are by no means deluded by all of our illusions. Writers frequently create illusion for the express purpose of making more effective their exposition of truths, whether they do so merely by using an occasional metaphor like "the grapes of wrath," or "the hounds of spring," or by writing whole novels or epic poems. Illusion becomes delusion only when it is either designed to deceive or unskillfully employed. Consequently, the test of the legitimacy of aesthetic subjectivity in geography is not whether or not it is illusory, but whether or not, if illusory, it leads to delusion, and it would seem entirely legitimate to enrich and add color and vividness to the style of an otherwise strictly objective geographical exposition by the use of subjective figures of speech and other aesthetic devices if they are so chosen and phrased as not to delude the reader.

Subjective elements may slip into a predominantly objective exposition in the form of words or phrases that carry emotional connotations. This also would seem legitimate provided the images that such words invoke in the reader's imagination correspond to the impressions that the majority of readers would receive in the presence of the phenomena described or exposed. We are often tempted to use such expressions as "a gloomy wood," "bitter cold," "a majestic mountain," "a menacing thunderhead," "the mysterious unknown." Budding geographers have been cautioned by their professors against employing such adjectives on the ground that they reflect the personal emotions of the writer and are not universal common denominators in the symbolism of science. A dark wood may not seem gloomy to a lumberjack, or fifty below bitter cold to an Eskimo, or the Matterhorn majestic to all the peasants of Zermatt, or the geographically unknown mysterious to some geographers. Such terms, however, are not likely to be delusive, and to cavil against their use, if it is discriminating and restrained, seems a little pedantic. Geographical works are intended to be read by persons who share a more or less common cultural heritage and whose subjective responses to like stimuli are similar. A phrase in D. G. Hogarth's book *The Nearer East* has stuck in my memory for forty years: "the awful aridity of Sinai." Few readers of that book would remain unmoved with awe upon seeing the utterly barren mountains of the Sinaitic peninsula. Surely it is legitimate in a geographical work to convey this sense of awe to the reader, even though the Bedouins of Sinai may take its dryness as a matter of course.

Naturally, imaginative fancies that stem from some special idiosyncrasy or peculiar and passing emotional state of a writer, or that are merely whimsical, have no legitimate place in geographical expositions if they create false impressions. I should be exceeding the legitimate limits of the subjective were I to describe my Maine woodlot as an abode of hobgoblins, elves, and werewolves, even though my imagination might relish so picturing it on a moonlit night.

Thus, although aesthetic subjectivity may and often does lead to delusion and error, there are ways of expressing it that do not, and hence may be regarded as at least legitimate, whether or not desirable.

The second notion, that aesthetic subjectivity is out of place in geography—that, like so much window dressing, it serves no functional purpose—brings up the question of desirability. The notion is mistaken. The functional purpose of aesthetic subjectivity is to heighten the effect by increasing the clarity and vividness of the conceptions that we seek to transmit to reader or hearer. It enables us to share with him the impressions that place or circumstance have made upon us—to bring him down to earth from the lofty observation point of the objective and make him see and feel through our eyes and feelings. Of course, there are limits beyond which this ceases to be desirable. A geographical exposition differs from a traveler's tale, in which the reader can be held at the personal level throughout. In geography the subjective should be used only to point up the objective, never permitted to crowd it out.

It is sometimes argued that the style of a scientific exposition should be as clear, simple, and concise as possible, and that more is superfluous; but it should not be forgotten that the power to arouse the imagination is also a desirable adjunct. Most of what geographers write is intended to be read by others besides a few colleagues whose initial interest in a subject is so intense that their imaginations would be fired by almost any exposition, however inartistic. Even if a geographer is not writing for the general reader, whoever that may be, he should bear in mind the possibility that his work might be used in stimulating the interest of undergraduate and graduate students in his pet subject—surely a desirable end. Hence, if he wishes his writing and also his teaching to exert their optimum influence, a certain amount of artistry—at least a touch of the aesthetically subjective—must be injected into them.

The third notion is that most geographers lack skill in giving expression to aesthetic sensitivity and hence should refrain from trying to do so. This, of course, is a *non sequitur*. There is no question but that the majority of geographers possess aesthetic sensitivity in good measure, and skill in expressing it can be developed by them once the need is admitted. A great deal has been written and more said about the nature of geography; far less about the nature of geographers. Could we subject a few representative colleagues to a geographical psychoanalysis, I feel sure that it would often disclose the geographical libido as consisting fully as much in aesthetic sensitivity to the impressions of mountain, desert, or city as in an intellectual desire to solve objectively the problems that such environments present. The Sirens, to whom I have alluded, appeal to the artistic and the poetic that lie deep beneath the surface in most of us, for Sirens themselves are artists and poets. Obviously those few who are basically deficient in aesthetic sensitivity—and thereby functionally deaf to the Sirens—will produce lamentable results when they try to express what little they may possess, and it is always preferable to avoid aesthetic subjectivity altogether rather than to give vent to it in misleading, trite, or farfetched forms. Nor is the technique of expressing it without doing violence either to scientific integrity or to good taste one that can be quickly mastered with the aid of rules and prescriptions, for taste itself is so largely subjective. But that sound geography can be written and taught with artistry has been demonstrated too often in the past to warrant the belief that it should not be attempted.

Thus, with all due respect toward those who may think differently, I do not regard the scientific and the aesthetic either as mutually exclusive or as antagonistic in geography. Repression of the poetic in our imaginative faculties may deprive us of much of the satisfaction that geographical studies could otherwise yield and render our teaching and writing less powerful than they might well be. American geography would grow rather than shrink in stature and esteem were we to give greater scope to the aesthetic operation of our own imaginations, and, when we see sparks of artistry kindling the imaginations of our graduate students and geographical colleagues, were we to resist the temptation to stamp them out.

Borrowed imaginative impressions. We are under no compulsion to rely exclusively on our own imagining or to make use solely of its original products. The imaginative perception of others, the feeling for place that many a sensitive traveler has recorded, may be keener and more accurate than ours and may often be borrowed to advantage. In interpreting the landscape of Iceland or of Arabia one might do better to quote here and there from Lord Dufferin or from Doughty than to try to give one's personal impressions. It is a standard practice in the teaching of history to cultivate the student's sense of time and contemporaneity by requiring him to read selected passages from documents written in the periods that he studies. No less valuable in the teaching of regional geography would appear to be the cultivation of the student's sense of place by requiring him to read passages from works in which the feeling for place has been most effectively expressed. Furthermore, even though we may prefer not to borrow directly from others, our own responsiveness to the Sirens song is rendered more acute by our reading the words of those who have also heard it, and the whole tone of our writing and teaching is enriched thereby.

The realm of geography—geography in the sense of all that has been written and depicted and conceived on the subject—consists of a relatively small core area (to borrow Whittlesey's phrase) and a much broader peripheral zone. The core comprises formal studies in geography as such; the periphery includes all of the informal geography contained in nonscientific works—in books of travel, in magazines and newspapers, in many a page of fiction and poetry, and on many a canvas. Although much of this informal geography offers little of value to geographers, some of it shows an insight deep into the heart of the matters with which they are most closely concerned. I venture to think that, of two geographers equally competent in all other respects, the one the better read in the imaginative passages in English literature dealing with the land of Britain could write the better regional geography of that land.

The peripheral zone also includes another even more informal type of geography; that of the subjective geographical conceptions of the world about them which exist in the minds of countless ordinary folk. In order to estimate what these are, we seldom need to go as far as the sociologists do in making ostensibly "scientific" inquiries into human attitudes. By talking sympathetically with a few intelligent folk on the ground, by consulting the files of local newspapers and other publications, and by a little adept use of intuition we may, under most circumstances, gain all that is required for our purposes. For example, the farmers of the Great Plains must look with certain sentiments on the massing of thunderheads after a long drought. Why not give life to our regional or climatologic studies of the Plains by letting the reader sense this feeling? That it combines a hopeful expectation of rain with a dread of tornadoes is a reasonable surmise, even though suggested subjectively by the imagination and only partially confirmed by conversations, rather than established rigorously on the basis of comprehensive interviews or questionnaires concerning exactly what the farmers' attitude toward the breaking of a drought may be.

KNOWLEDGE AND GEOGRAPHY

I have tried to suggest some legitimate and desirable uses of the imagination in geography. I should now like to call attention to a broad domain that lies open for much more intensive geographical investigation than it has hitherto received.

Human knowledge is generally regarded as a phenomenon of considerable importance on the face of this earth. It may be made the subject of two types of geographical research: we may either study the geography of any or all forms of knowledge or else we may study *geographical* knowledge from any or all points of view.

THE GEOGRAPHY OF KNOWLEDGE

The *geography of knowledge* is that aspect of systematic geography which deals potentially with knowledge and belief of all kinds, whether religious, scientific, philosophical, aesthetic, practical, or whatever else. The various forms and manifestations of knowledge are investigated in the light of their distribution and areal relations, precisely as landforms, cities, languages, or other categories of terrestrial phenomena are investigated in other branches of systematic geography. Human knowledge, of course, is taken into account incidentally in many of these other branches and also in regional geography. Attention, however, is there concentrated on the results that knowledge produces on the face of the earth, rather than on the geographical nature of knowledge itself.

Though closely allied to cultural geography, the geography of knowledge differs from the latter to the extent that knowledge itself differs from culture. Knowledge is more fluid than culture, often spreading rapidly from one culture area to another without fundamentally altering established patterns. The sociologists have developed the sociology of knowledge more consciously and perhaps more systematically than we have developed the geography of knowledge, and would probably regard the latter as merely a part of the former. This need not trouble us, for there are many phases of geography in which we may profit from explorations conducted by others than ourselves.

GEOSOPHY: THE STUDY OF GEOGRAPHICAL KNOWLEDGE

Though the possibilities of research into the geography of knowledge are attractive, I wish to dwell here more particularly upon the second type of investigation, the *study of geographical knowledge*. As there is no accepted term for this field comparable to "musicology" or "historiography" for the study of musical or historical knowledge respectively, I shall yield to the geographer's perennial temptation and coin one. My term is *geosophy*, compounded from *ge* meaning "earth" and *sophia* meaning "knowledge." Although this suggests theosophy, there is no connection; nor should *geosophy* be confused with *geosophistry* and *geopedantry*, both of which have been known to flourish. Also, I am not trying to introduce any of these terms into the literature of geography.

Geosophy, to repeat, is the study of geographical knowledge from any or all points of view. To geography what historiography is to history, it deals with the nature and expression of geographical knowledge both past and present—with what Whittlesey has called "man's sense of [terrestrial] space."[2] Thus it extends far beyond the core area of scientific geographical knowledge or of geographical knowledge as otherwise systematized by geographers. Taking into account the whole peripheral realm, it covers the geographical ideas, both true and false, of all manner of people—not only geographers, but farmers and fishermen, business executives and poets, novelists and painters, Bedouins and Hottentots—and for this reason it necessarily has to do in large degree with subjective conceptions.[3] Indeed, even those parts of it that deal with scientific geography must reckon with human desires, motives, and prejudices, for, unless I am mistaken, nowhere are geographers more likely to be influenced by the subjective than in their discussions of what scientific geography is and ought to be.

(Studies of geographical knowledge from the *geographic* point of view—that is, in terms of its geographic distribution, areal relations, and so forth, as suggested under the heading "Cartographic Geosophy" below—are contributions not only to geosophy but also to the geography of knowledge. This present paper is a study in geosophy but not in the geography of knowledge. Works aiming, for example, to interpret the distribution in the United States of illiterates, or of holders of Ph.D.'s, or of persons able to read Russian, would be studies in the geography of knowledge but not in geosophy. [In the present paper as first printed in 1947, I relegated to a footnote the none-too-serious suggestion that study of the geography of knowledge might be called "sophogeography," on the analogy of biogeography, zoogeography, or anthropogeography. Two years later Professor L. E. Klimm called my attention to a paper published in Great Britain in 1918 in which the term geosophy had been suggested for my sophogeography.[4]])

While it is true that subjective ideas may be studied objectively up to a certain point, geosophy certainly is not a field in which one may apply the stricter methods of analysis possible in the physical sciences and physical geography. I doubt, however, that on this account any geographer in his senses would hold geosophy to be either illegitimate or undesirable. Its value both to ourselves and to the others whom we seek to serve requires little defense. Geosophy can provide a background and a perspective indispensable to our work. It can show us where the ways in which we observe and think fit into a larger scheme. By helping us better to understand the relations of scientific geography to the historical and cultural conditions of which it is a product, it can enable us to become better-rounded scientific geographers, when that is our purpose. Recognition of its function in these respects is implied by the methodological discussions in which many American geographers take delight, and specifically by the emphasis that Sauer, Brown, Whittlesey, and others have placed of late on values to be derived from the history of geography.

There are many possible approaches to the study of geosophy. Let us consider two of these—the cartographic and the historical.

Cartographic geosophy. The cartographic approach to geosophy involves the making of maps that present information about the distribution of geographical knowledge. Obviously, every map tells us something in this regard by implicitly revealing facts concerning the extent of its maker's knowledge of its subject; a geosophic map is one designed explicitly to set forth facts concerning geography *as knowledge.*

Such maps might be grouped in two main categories. The first would comprise maps that present facts relating to what is or has been known *about* different areas. By far the most common of these are maps showing areas that have been surveyed and mapped in various ways, for various purposes, and with varying degrees of intensity and accuracy—*cartosophic* maps, in other words, because they depict cartographic knowledge. In this same category, however, would also belong maps of the world as known to the Greeks or Romans, or of the United States as supposedly conceived by Ralph Brown's friend Mr. Keystone in 1810,[5] or, perhaps, by the average contemporary Bostonian.

The second group would comprise maps that reveal facts concerning geographical knowledge, present or past, *in* different areas or at different places. This is an almost completely virgin field for ingenious experimentation. A dot map, for example, showing the distribution of members of the Association of American Geographers would disclose information of considerable interest regarding the distribution of geographical knowledge in North America, especially if each dot were colored according to the quality and made proportional in size to the quantity of geographical knowledge in the mind of each individual represented.

Whether or not this particular geosophic map would be either feasible or desirable, geosophic maps in general bring out sharply the contrast between the shadows of ignorance and the light of knowledge. *Terrae incognitae* of various forms and degrees stand forth clearly upon them to arouse our curiosity.

Historical geosophy, or the history of geography. The historical approach to geosophy implies the study of the history of geographical knowledge, or what we customarily call "the history of geography." This subject is usually understood to deal with the record of geographical knowledge as acquired through exploration and field work, and as formalized and made into a discipline, and most of the work that has actually been done in the field has been restricted to the core area of geographical knowledge to the exclusion of its peripheral zone. There is, however, merit in conceiving it more comprehensively. I have already suggested that geographical knowledge of one kind or another is universal among men, and in no sense a monopoly of geographers. All persons know some geography, and I venture to think that many of the animals do, also.

However it may be with the animals, such knowledge is acquired in the first instance through observations of many kinds—from the stone-age man's view of distant ranges to the precise geodetic measurements of today aided by the use of electronic devices. Its acquisition, in turn, is conditioned by the complex interplay of cultural and psychological factors. The data with which it deals fall within the scope of each and every one of the natural sciences, the social studies, and the humanities. Its conceptions range from the purely personal, subjective impressions of a farmer or a hunter to those gained by rigorous mathematical calculations and highly refined statistical correlations, and find expression not only in scientific forms but throughout literature and art. Indeed, nearly every important activity in which man engages, from hoeing a field or writing a book or conducting a business to spreading a gospel or waging a war, is to some extent affected by the geographical knowledge at his disposal. If, therefore, the history of geography is conceived as potentially embracing all of the geographical knowledge of the past in its various relations of cause and effect, it is an immense subject indeed. It is, however, no more immense than certain subjects of which the teaching is being promoted today—notably the history of science or of the humanities in general, or "contemporary civilization"—and has, besides, one advantage over these, in that it ties together with a unifying thread—that of geography—a record of wide and representative segments of human enterprise, thought, emotion, and techniques. For this reason, I submit that it is a subject of which the investigation and the teaching offer superb educational and cultural values.

AN ASPIRATION

I shall conclude by expressing an aspiration, quite impractical, no doubt, and not to be taken too literally. My aspiration is that there might one day be established in some of our universities or colleges chairs of geosophy and the geography of knowledge. The purpose would be to increase the effectiveness of geographical research and education by broadening their scope. One school of thought has held that the effectiveness of geography can be increased only by limiting its scope, but this school would seem to confuse the effectiveness of geography as a discipline or profession with that of the individual geographer or existing university department. The more general tendency today is to stress the need of better linkage between geography and other subjects, notably ecology, soil science, agricultural and industrial economics, and cultural anthropology, and not a few regret the loosening of ties with geology and the various branches of geophysics. To the desirability of establishing and reestablishing such contacts, I would add, as no less desirable, the reestablishment of closer connections with history and the humanities.

In the periphery that lies outside the core area of scientific geography there are alluring *terrae incognitae.* If we ourselves do not personally feel equipped or competent to conduct excursions into them, should we exclude them from the scope of our sympathies? Although most of us are committed to the advancement of scientific geography along straight and narrow paths and would do well not to deviate too far from the directions in which they lead, we may at least extend our interest and encouragement to those who daringly strike out upon other routes. There is something to be said for considering scholarship, as distinguished from science alone, as our métier. All science should be scholarly, but not all scholarship can be rigorously scientific. Scholarship, moreover, embraces not only the natural sciences and social studies but also the humanities—the arts and letters—inquiring no less into the world of subjective experience and imaginative expression than into that of external reality. The *terrae incognitae* of the periphery contain fertile ground awaiting cultivation with the tools and in the spirit of the humanities.

The professors whom I have in mind would develop their subjects along different lines according to their tastes. Some might specialize in the geosophy of scientific geography—in its history, its methods, and perhaps in comparative biographical studies of the careers of individual geographers as bearing on the larger progress of geography. Others might concern themselves with geographical conceptions, both scientific and otherwise, as influencing and influenced by particular human activities and motives, or with particular categories of geographical knowledge in relation to the changing tides of doctrine and opinion.

At least one or two should surely devote themselves to what might be called aesthetic geosophy, the study of the expression of geographical conceptions in literature and in art. Literary historians, but few geographers, have followed the Sirens' call into these *terrae incognitae.* Need we leave their exploration wholly to the literary scholars? One function of my hypothetical professors of aesthetic geosophy—though God forbid they be called by such an atrocious title—would be to prevent the oncoming generations of geographers from becoming too thickly encrusted in the prosaic and to render the study of geography more powerful than it would now seem to be in firing the artistic and poetic imaginations of students and public. These professors should be scholars in the humanistic sense—men widely read in the classics of geography and also in general literature and in literary criticism and history. Masters of a style not only clear but restrainedly artistic, they might help by their writings to raise the standards of geographical writing as a whole. Their research and teaching would be directed toward the discovery and the interpretation of geographical truth, belief, and error as these find and have found literary and artistic expression. As long as they did not come to regard themselves as the only true exponents of what geography ought to be, there would be little danger of their exerting an adverse effect upon the advancement and the prestige of scientific geography. They could do much to keep our ears open to the Sirens' song and make our voyaging into geographical unknowns a perennially satisfying venture, for, perhaps, the most fascinating *terrae incognitae* of all are those that lie within the minds and hearts of men.[6]

CHAPTER 6

The Open Polar Sea

NOT A SPECK of ice" was to be seen. "As far as I could discern, the sea was open, a swell coming in from the northward and running crosswise, as if with a small eastern set. The wind was due N.,—enough of it to make white caps,—and the surf broke in on the rocks in regular breakers." The Open Polar Sea!

The time was June 24, 1854. William Morton, one of the crew of Dr. Elisha Kent Kane's "Second Grinnell Expedition,"[1] stood on the summit of the mighty cliffs along the Greenland northwest coast to which Kane later gave the name "Cape Constitution." The latitude was 80° 30′ N, farther north than any white man had yet reached in the Western Hemisphere, and the coast line before Morton marked the northernmost known land on the face of the earth. On the other side of Kennedy Channel, Grinnell Land stretched off some 50 miles toward the North Pole.

Morton, alas, was mistaken. There is no Open Polar Sea. But during the next 25 years his report and the interpretation Dr. Kane and others put upon it gave a feverish vitality to a dying theory.

Belief in open water and a mild clime in "the realm of the Boreal Pole" had persisted for centuries. Inherently it was neither an absurd nor an unimportant theory. Hardheaded seamen and explorers and reputable geographers upheld it, and its record helps explain much that would otherwise be obscure in the history of Arctic discovery. Today it is perhaps of greatest interest as providing an illustration of the nature and growth of scientific error. Since it has been conclusively disproved, we may compare it with the facts and see clearly wherein its proponents went astray.

HOW THE THEORY GREW

During Greco-Roman and medieval times most Europeans who thought about the matter at all conceived of the region around the North Pole as a frozen, uninhabitable, and impenetrable waste. The Norsemen probably took a less uncompromising view, as Dr. Vilhjalmur Stefansson has reasonably pointed out, and there were a few theorists who argued speculatively that temperate conditions, or even great heat, might prevail in the Far North.[2] While the marked rise in temperature that has occurred in the Arctic in recent years might seem to suggest that a similar amelioration in climate in the past may have given rise to belief in an Open Polar Sea, this was not the case. The theory originated in the realm of pure speculation and at a time when the Arctic climate was more rigorous than it is today, and until the age of Columbus and Magellan the problem remained purely academic.

If, as a good many reputable scholars have believed and argued, Columbus visited Iceland in 1467 or 1477 and sailed 100 leagues beyond it to the north, encountering no ice, he might qualify as the first explorer who sought to test the navigability of the Polar Sea.[3] The evidence is none too certain, however, and a further assumption that Columbus later planned a voyage to the North Pole, and perhaps from there to the Far East,[4] seems to be based on a modern misreading of a word in one of his letters.[5]

We are on surer ground when we come to Robert Thorne. An English merchant, Thorne first set forth his ideas on this subject in 1527 when he was living in Seville.[6] The Portuguese had discovered and were developing the eastward sea route to the Spice Islands around the Cape of Good Hope, and the westward route had recently been opened up for Spain by the great voyage of Magellan and Del Cano (1519–1522). Thorne suggested that a third, much shorter route, navigable and relatively safe, might be found across the North Pole, and he recommended that the English explore and exploit it in order to find a place in the spice trade. Like much of

Fig. 3. The open water from Cape Jefferson. Reproduction in halftone of steel engraving in Elisha Kent Kane's *Arctic Explorations: The Second Grinnell Expedition . . . 1853, '54, '55, vol. I* (Philadelphia, 1856), facing p. 307.

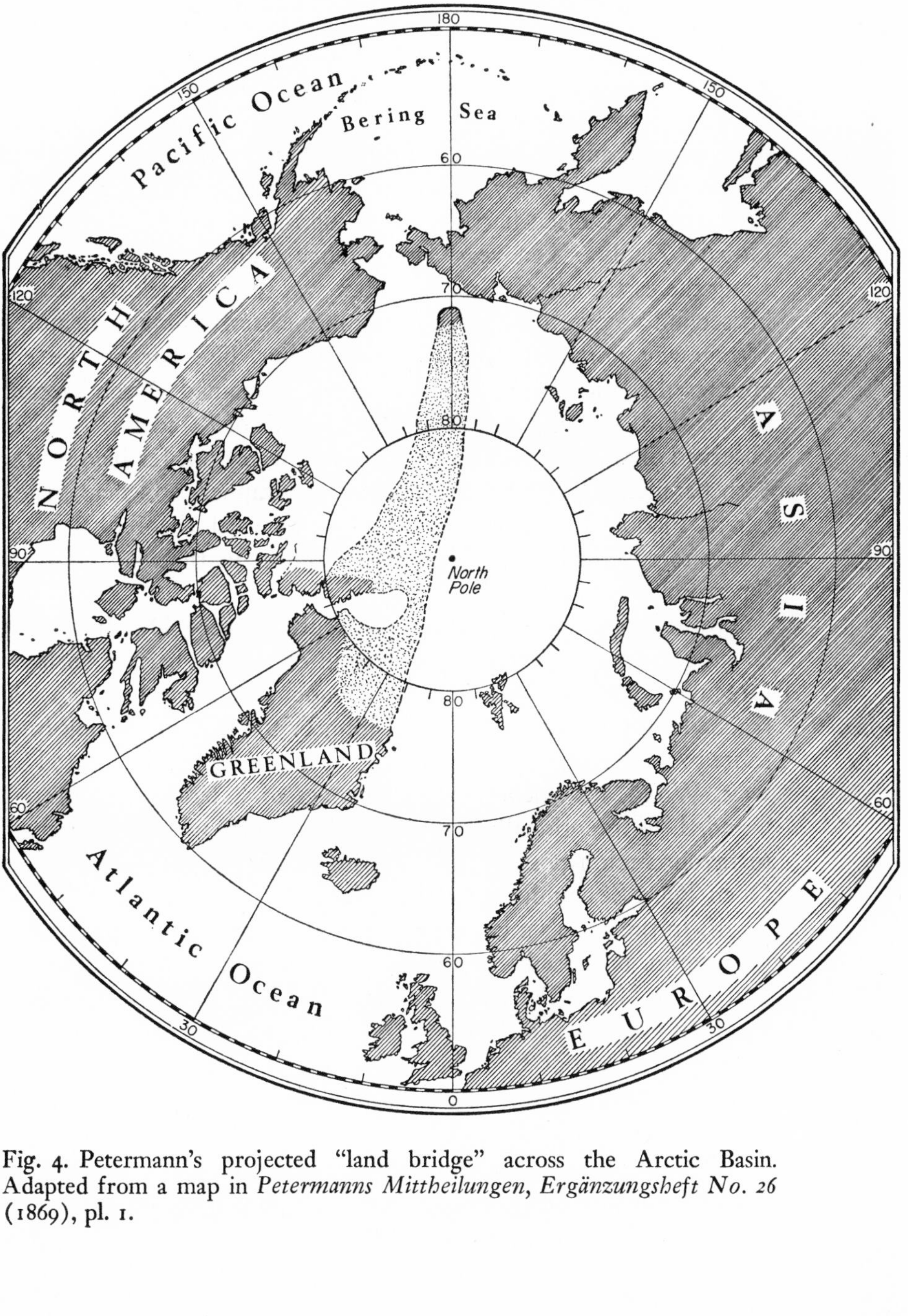

Fig. 4. Petermann's projected "land bridge" across the Arctic Basin. Adapted from a map in *Petermanns Mittheilungen, Ergänzungsheft No. 26* (1869), pl. 1.

the later discussion of an Open Polar Sea, Thorne's arguments were sired by wishful commercial thinking and born to national ambition.

About the middle of the sixteenth century the English began the search for a Northeast Passage[7] around Asia to the Pacific, and the Dutch soon followed. The earliest expeditions kept close to the continent and were blocked by ice. Willem Barents, convinced by two failures, in 1597 took a more northerly course, only to be thwarted by the ice barrier and to perish on the bleak coast of Novaya Zemlya. Ten years later Henry Hudson sailed far north along the east coast of Greenland but was turned back by ice. On his second voyage (1608) he bore more easterly, hoping to find a passageway north of Novaya Zemlya, but, like Barents, he encountered the ice front. Undiscouraged, he gained Dutch support for his third voyage (1609) on the strength of "evidence" he had observed of warmer conditions and open water not far beyond the 80th parallel. The Dutch theologian, geographer, and map maker Petrus Plancius (1552–1622) brought forward theoretical arguments to sustain both Barents and Hudson in their belief in an Open Polar Sea.[8]

Toward the middle of the seventeenth century Dutch whalers began to visit the ever-shifting margin of the ice barrier between Greenland and Novaya Zemlya. Rumors that some of them had reached the Pole inspired Captain John Wood in 1676 to look for a route to the Pacific east of Spitsbergen, but unbroken ice barred his way. Wood, his "belief in the chances of a passage" shattered, branded "as intentionally misleading the reports which had seemed to favour such a possibility,"[9] and his views put the theory of an Open Polar Sea under a partial eclipse for nearly a century and a half. In the 1770's a versatile Englishman, the Honorable Daines Barrington, proposed that the British government send an expedition to the North Pole. He presented "proofs" of the existence of an Open Polar Sea to the Royal Society, and his recommendations helped bring about Captain Constantine Phipps's expedition, on which Horatio Nelson served as midshipman. Phipps sought the Pole by way of Spitsbergen but was blocked by ice in 80° 36′ N. In the early years of the nineteenth century Russian explorers discovered tracts of open water (*polynyas*) north of the ice off the

Siberian coast, which they interpreted generously as parts of an extensive open sea, and thereafter the Open Polar Sea was universally conceived as lying within an encircling barrier of ice, through which ships might be able to make their way at favorable places;[10] Dr. Kane,[11] for example, in 1852 referred to the ice barrier as an "annulus, a ring surrounding an area of open water—the Polynya, or Iceless Sea."

The nineteenth-century advocates of the theory refurbished some of the speculative arguments dating from Barrington, Plancius, and even Thorne, and drew heavily upon the treasury of facts and hypotheses to which exploration was contributing a new quota every year. Among those whose ideas carried much weight during the mid-century were the learned geographer Dr. August Petermann in Europe and, in America, the oceanographer Matthew Fontaine Maury and the Arctic explorers Kane and Hayes.

After the Civil War a strong revival in the United States of interest in Arctic exploration bore fruit in Charles F. Hall's government-sponsored expedition of 1871–1873 in the *Polaris*, by the Smith Sound route. In Morton's "Open Polar Sea" Hall encountered ice drifting south with the current, but he succeeded in getting through nearly to the point where Robeson Channel opens out into the Arctic Basin before firm ice halted advance.

During the discussions leading up to the *Polaris* expedition, an American Navy officer, Captain Silas Bent,[12] took up the cudgels for an Open Polar Sea. His reasoning, as embroidered upon by a Professor Thompson B. Maury[13] (not to be confused with the oceanographer), brought the theory to its all-time peak of ingenuity—to put it charitably—and drew the fire of Judge Charles P. Daly,[14] President of the American Geographical and Statistical Society. In England, that able if dogmatic geographer Mr. (later Sir) Clements R. Markham, Secretary of the Royal Geographical Society, was a leading advocate of the doctrine that pack ice covers the entire inner Arctic Basin, and in his book *The Threshold of the Unknown Region* (1873) he analyzed the arguments in favor of an Open Polar Sea, pronounced them "mischievous," and asserted that they had done "much harm to the advance of discovery and the progress of sound geography."

On his expedition of 1875–1876, Sir George Strong Nares

threaded the Smith Sound passageways to the margin of the Arctic Sea, where he found that the pack ice formed a blockade so massive and formidable—the "paleocrystic ice"—as effectively to shut the door upon all hope of reaching navigable water in that direction.

Meanwhile, one after another, Swedish and Norwegian, German, Austro-Hungarian, and British expeditions were turning back. Nevertheless, an American of Dutch descent, Samuel R. van Campen, tried valiantly to reopen the European door to an Open Polar Sea in 1876 in his book *The Dutch in the Arctic Seas*. Lover of the Netherlands and student of its history, van Campen urged the Dutch to dispatch an expedition to the Pole. Like Barrington and Markham, he painstakingly reviewed the evidence and the arguments for and against a navigable route, but, unlike Markham, he was beguiled into at least a qualified belief in an Open Polar Sea.

Since the publication of van Campen's book there has been, to my knowledge, no serious discussion of the possibility of reaching the Pole by ship from the European side. For a few years, however, hope persisted that it might be done from Bering Strait. In the late sixties the French organized a polar expedition on the assumption that a continuation of the Japan Current passes through that strait to an open sea, but the Franco-Prussian War brought their plans to nought.

In the years 1879–1881 Lieutenant G. W. De Long, with the financial support of James Gordon Bennett and the geographical advice of Dr. August Petermann, attempted to reach the Pole by drifting with the supposed Japan Current and, if necessary, resorting to sledge travel. His ship, the *Jeannette*, was carried northwestward to the vicinity of the New Siberian Islands and there crushed by ice; the survivors escaped to the Siberian mainland, where De Long perished. And thus ended hope of a navigable route from Bering Strait.

After wreckage from the *Jeannette* appeared in Greenland in 1884, Fridtjof Nansen reasoned that it must have been carried, first by a current sweeping across the Pole, and thence southward by the Greenland Current; but he harbored no illusion that it might have passed through an open sea.[15] With supreme daring he designed and built the *Fram* to withstand the pressure of ice, took her to the New Siberian Islands, let her be frozen in, and committed himself and his men to the will of ice pack and current. Three years later the *Fram* emerged near Spitsbergen, having passed about halfway between Franz Josef Land and the Pole. Had there been an open sea in this vicinity, Nansen would have found it. The theory of an Open Polar Sea, as one that could be supported with any semblance of reason, was already mortally stricken before 1880; Nansen gave it the *coup de grâce.*

In 1946 the Hydrographic Office of the United States Navy published an *Ice Atlas of the Northern Hemisphere.* The atlas presents detailed information regarding the average monthly distribution of sea ice, its extreme limits, and many other things—different indeed from the scanty information available when the Open Polar Sea was still a living issue. Moreover, the atlas shows clearly "the enormous year to year variations from the conditions defined as 'average.' " Had it not been for these shifts—had the margins of the ice barrier remained fixed and mappable like a coast line—the impenetrability of the pack ice would have been established much earlier and belief that open water lay to the north might have died in infancy.

ARGUMENTS PRO AND CON AN OPEN POLAR SEA

Arguments Based on the Distribution of Land and Water. Nothing whatever was known about the distribution of land and water in the north polar region when Thorne, writing from Seville in 1527, proposed that the English seek a new navigable route to the Far East. He assumed that a Polar Sea connected the Atlantic with the ocean across which Magellan had sailed. In the back of Thorne's mind may have been a lingering memory of Amphitrites, the ocean that, according to a theory derived from the Greeks, encircled the globe from north to south, at right angles to an equatorial ocean. Professor E. G. R. Taylor[16] has also pointed out that the influence of "the excellent contemporary Spanish charts . . . can be seen in Thorne's assumption that Ocean ways lay all across the Polar regions. Academic writers all postulated a land bridge or bridges or an Arctic land that would interrupt northern navigation quite apart from ice, but the Spanish chart-makers only drew coast-lines where coast-lines had been located, and so upon their charts all the far north lay open." However derived, the assumption that the Atlantic and Pacific were connected at the north suited Thorne's purpose.

The sixteenth-century maps reveal a variety of conceptions regarding the Arctic. Some show North America as an eastward extension of Asia; others show the two continents far apart; still others, notably certain maps of Mercator and Plancius,[17] represent them as separated by a narrow "Strait of Anian" in the vicinity of Bering Strait. Mercator's delineation of the general trend of the Arctic coasts of Asia and America seems so nearly correct as to amaze one, but any illusion regarding his geographical knowledge of the region is shattered by his extraordinary symmetrical arrangement of lands in the Arctic Basin.[18] Plancius and other map makers of the period copied this formidable system of circumpolar islands and waterways on certain of their maps. Plancius, however, explained on his world map of 1594 that he included it, not because he believed in its existence, but lest someone accuse him of omitting something—a principle of cartographic compilation that fortunately has been largely abandoned!

Barrington set forth geophysical and geodetic arguments—of a sort—that there could be little or no land in the polar regions. Cook's circumnavigation of the earth during his second voyage had proved that if, as the Greeks had postulated, the continents of the Northern Hemisphere were counterbalanced by a southern continent it must be much smaller than had been imagined. Therefore, Barrington inferred, the area north of latitude 80½° N must be mostly sea, for otherwise the preponderance of land in the Northern Hemisphere would upset the equilibrium of the globe. So much for his geophysics. He borrowed his geodetic arguments from Newton's hypothesis that the earth is a spheroid flattened at the Poles, and from that inventive genius Dr. Robert Hooke, who had "brought very strong reasons to show, that there is nothing but sea at the Poles."

The progress of exploration provided better ground for theorizing. The essential questions were whether or not North America and Asia were joined, and, if not, whether there was (1) a Northeast Passage around Asia, (2) a Northwest Passage around North America, (3) a Northern Passage across the Pole, or (4) more than one of these passages. In 1648, Bering Strait was discovered, and by 1742 Russian explorations had progressed far enough along the north coast of Siberia to demonstrate that there was a Northeast Passage, though not necessarily an ice-free waterway.

This did not shed light on the question how far North America extended out into the Arctic Sea. As late as 1818, John (later Sir John) Barrow wrote that it was still unknown whether or not Greenland was connected with North America by a land bridge attached to the mainland somewhere northwest of Hudson Bay. Barrow considered it unlikely; he thought that the north coast of North America probably trended westward from the entrance to Hudson Bay, and in support he referred, among other things, to "rude charts painted on skins by the Indians, which, though without scale or compass, mark the inlets from Hudson's Bay with tolerable accuracy, and carry the coast without interruption to the Coppermine River."[19] Twenty years later enough had been learned about the waterways of the Arctic Archipelago to dispose of a land connection between North America and Greenland or any other large Polar land mass,[20] though it remained for M'Clure definitely to demonstrate, in 1851–1852, the existence of a Northwest Passage.

There was still, however, the possibility of extensive lands unconnected with the known continents, a possibility that Petermann, Kane, and others held to be a probability. Most of their arguments were developed in order to explain conditions of climate, marine currents, and ice. In partial explanation of why he had selected the Smith Sound route for his proposed expedition, Dr. Kane set forth his theory before the American Geographical and Statistical Society in December 1852. "My plan of search [for Franklin]," he said,

> is based upon the probable extension of the land masses of Greenland to the far north—a view yet to be verified by travel, but sustained by the analogies of physical geography. Greenland . . . is in fact a peninsula, and follows in its formation, the general laws which have been recognized since the days of Forster, as belonging to peninsulas with a southern trend. Its abrupt truncated termination at Staaten-Hook is as marked as that which is found at the Capes Good Hope and Horn of the two great continents, the Comorin of Peninsular India, Cape South East of Australia, or the Gibraltar of southern Spain.

This implied a substantial bulging out of Greenland or, as Kane put it, a "fan-like abutment of land"[21] on its north face. How far this reached into his "Open Polar Sea" Kane did not attempt to estimate.

Arguments Based on Reports of Open Water.

Being about 22 years ago in *Amsterdam*, I went into a Drinking-house to drink a cup of Beer for my thirst, and sitting by the publick Fire, among several People there hapned a Seaman to come in, who seeing a Friend of his there, who he knew went in the *Greenland* Voyage, wondred to see him, because it was not yet time for the *Greenland* Fleet to come home, and ask'd him what accident brought him home so soon: His Friend . . . told him that their Ship went not out to Fish that Summer, but only to take in the Lading of the whole Fleet, to bring it to an early Market, &c. But, said he, before the Fleet had caught Fish enough to lade us, we, by order of the *Greenland* Company, Sailed into the *North-Pole*, and came back again. Whereupon (his Relation being *Novel* to me) I entred discourse with him and seem'd to question the truth of what he said. But he did ensure me it was true, and that the Ship was then in *Amsterdam*, and many of the Seamen belonging to her to justifie the truth of it: And told me moreover, that they sailed 2 degrees beyond the *Pole*. I askt him, if they found no Land or Islands about the *Pole?* He told me No, there was a free and open Sea; I askt him if they did not meet with a great deal of Ice? He told me No, they saw no Ice. I askt him what Weather they had there? He told me fine warm Weather, such as was at *Amsterdam* in the Summer time, and as hot. I should have askt him more questions, but that he was ingaged in discourse with his Friend, and I could not in modesty interrupt them longer. But I believe the Steer-man spoke matter of fact and truth, for he seem'd a plain honest and unaffectatious Person, and one who could have no design upon me.

Thus wrote Joseph Moxon,[22] "Hydrographer to the King's most Excellent Majesty," in 1674. His account records one of a number of instances that Barrington cited a century later of Dutch vessels alleged to have reached or nearly reached the Pole. When Barrington appealed to Dutch skippers themselves, "he got the simple truth from them. In reply to his enquiries, they said, 'We can seldom proceed much higher than 80° 30′ N, but almost always to that latitude.' "[23] Nonetheless, Barrington was inclined to believe the rumors. In 1865 the learned Petermann also thought that they deserved credence, for the none-too-convincing reason that their originators did not report finding land, as they presumably would have done had they been indulging in pure fabrication.[24] Bent, T. B. Maury, and van Campen also accepted the rumors as satisfactory evidence of an Open Polar Sea. Judge Daly, however, subjected them to a critical analysis, which brought out their dubious character, and Markham wrote, "They are all . . . so obviously fabulous that it is astonishing how any sane man could have been found to give credit to them." Though he was referring specifically to Barrington, the implication smote Petermann and other contemporaries.

In 1810 the Russian explorer Hedenström discovered an extensive polynya off the northeast coast of Siberia, and similar reports were made by Anjou and Wrangel in 1820–1821. Wrangel wrote: "We beheld the wide, immeasurable ocean spread before our gaze, a fearful and magnificent, but to us a melancholy spectacle." Morton's "Open Polar Sea" was in reality a small polynya. Dr. Kane adds, after describing it in his book and mentioning other eyewitness accounts of open water in high latitudes since the time of Barents:

> All these illusory discoveries were no doubt chronicled with perfect integrity; and it may seem to others, as since I have left the field it sometimes does to myself, that my own, though on a larger scale, may one day pass within the same category. Unlike the others, however, that which I have ventured to call an open sea has been travelled for many miles along its coast and was viewed from an elevation of five hundred and eighty feet, still without limit, moved by heavy swell, free of ice, and dashing in surf against a rock-bound shore.

Seven years later Dr. Isaac Israel Hayes, who had been with Kane, conducted an expedition of his own to the same area. He made a sledge journey along the Grinnell Land side of Kennedy Channel and on May 18, 1861, reached a point nearly[25] as far north as the point on the Greenland side from which Morton had seen the whitecaps and breakers. Although there was ice in the channel, Hayes thought it would melt as the summer advanced and that it would be possible to take a ship into an open sea. In 1871, as we have seen, Hall in the *Polaris* pushed through the Smith Sound waterways almost to the north end. Some of the members of the expedition thought they had seen sure signs of open water ahead. The wish to believe was strong!

At a meeting of the Royal Geographical Society in 1869 caustic comments were made to that effect. Captain R. V. Hamilton read a paper controverting certain arguments of M. F. Maury in support of an Open Polar Sea. During the ensuing discussion Captain (later Rear Admiral) Sherard Osborn, a redoubtable opponent of the theory, remarked that he thought the British "ought to be very lenient with the Americans upon the question of Polynias, because he remembered the time when the English had Polynias of their own, and when we had recourse to one, in order to keep alive the search for Franklin." Captain Hamilton agreed that Maury's polynya "was placed up Smith Sound in order to work expeditions, just as the English had formerly put a Polynia up Wellington Channel in order to work an expedition there."[26]

Climatic Arguments. When we turn to the climatic, oceanographic, and biological arguments, we note a contrast between those developed before and those developed during the nineteenth century. The earlier arguments were concerned almost wholly with static regional differences in the temperatures of the atmosphere and the sea, in the manner in which ice forms, and in flora and fauna. The nineteenth-century arguments were far more likely to deal with dynamic phenomena—winds, currents, the migrations of birds and mammals—about which the progress of exploration was yielding much new information.

Robert Thorne admitted that cosmographers might object to his proposal on the ground "that passing the seuenth clyme, the sea is all ice, the colde so much that none can suffer it." The cosmographers, however, had also maintained that "vnder the lyne Equinoctiall for muche heate the land was inhabitable [that is, not habitable]. Yet since by experience is prooued no lande so much habitable nor more temperate. And to conclude, I thinke the same shoulde bee founde vnder the North if it were experimented. For as all iudge, *Nihil fit vacuum in rerum natura.*" Plancius repeated this argument from analogy and added another: because the climate is hotter at the borders of the tropics than at the equator, the cold must be greater about the 70th parallel than at the North Pole, and the higher temperature toward the Pole is probably sufficient to cause open water not far beyond the 80th parallel. A contemporary of Plancius, a German physician named Helisaeus Roeslin, concluded that it could not be as cold at the North Pole as was generally assumed. In 1610 he tried to substantiate this not only on meteorological grounds but by an appeal to astrology. The appearance of a new star in Cassiopeia was a heavenly sign that success would be achieved in finding a northern route to the Indies.[27]

Barrington reasoned much as did Thorne and Plancius. He rejected the ancient doctrine that the Poles were "sources and principles of cold," since experience had shown "that there might be a diversity of climates in the frigid as well as in the torrid zone." From a comparison of the altitude of the freezing point on Mount Cotopaxi, under the equator, with its greater altitude at the top of the Peak of Teneriffe, 5° north of the Tropic of Cancer, where no ice or snow was found, he inferred that, "as the heat varies so little between the equator and the tropical limits, it may differ as little between the Arctic Circle and the Pole."

By the mid-nineteenth century descriptive climatology had made substantial advances through the accumulation and mapping of climatic data. Von Humboldt had suggested the isotherm; Berghaus and others had used it in the preparation of temperature maps. From the imperfect knowledge thus acquired of conditions near the margins of the unknown Arctic somewhat better inferences could be drawn as to conditions within it than from analogies with the equatorial regions.

In his address before the American Geographical and Statistical Society in 1852, Dr. Kane said:

> The system of Isothermals, projected by Humboldt upon positive data, ceased at 32°; and the views of Sir John Leslie (based upon Mayer's theorem, that the north Pole was the coldest point in the Arctic regions,) have, as the members are aware, since been disproved.
>
> Sir David Brewster, by the combination of the observations of Scoresby, Gieseke, and Parry, determined the existence of two poles of cold, one for either hemisphere, and both holding a fixed relation to the Magnetic Poles.[28] These two seats of maximum cold are situated respectively in Asia and America, in longitudes 100° west and 95° east, and on the parallel of 80°. They differ about 5° in their mean annual temperature; the American, which is the lower, giving three degrees and a half below zero. The iso thermals surround these two points, in a system of returning curves, yet to be confirmed by observation; but the inference, which I present to you without comment or opinion, is, that to the north of 80°, and at any points intermediate between these American and Siberian centers of intensity, the climate must be milder, or more properly speaking, the mean annual temperature must be more elevated.

This was the old speculative belief of Plancius and Barrington supported by a flimsy framework of meteorological evidence. The facts are different. There would seem to be no pronounced "cold pole" in northern North America at any time of the year. The vicinity of the North Pole and the Greenland icecap are probably the coldest regions in the Northern Hemisphere. In winter, there is also a pronounced cold pole in the Lena Valley, but it is not there between March and October.

How did the advocates of an Open Polar Sea explain *why* there is presumably a milder climate in the vicinity of the North Pole? Before the nineteenth century they usually attributed it to the direct effects of the sun's rays. Plancius suggested that near the Pole the sun shines continuously for five months each year, and that, "although his rays are weak, yet on account of the long time they continue, they have sufficient strength to warm the ground, to render it temperate, to accommodate it for the habitation of men, and to produce grass for the nourishment of animals."

This reasoning was often repeated, in spite of two fairly obvious objections: the "weakness" of the insolation during the months when the sun shines uninterruptedly, and the fact that the sun does not shine at all during a comparable period each year. Might not these outbalance the warmth received during the months of continuous daylight? Even as late as the year 1869 Professor T. B. Maury did not think so, for he wrote: "FROM THE SUN ALONE, FOR SIX MONTHS IN THE YEAR, WE HAVE FORTY DEGREES OF HEAT AT THE POLE! Less than three fourths of this amount would liquefy and open the space around the Pole, supposing it locked in ice."

Barrington, as we have seen, brought forward Hooke's speculation that there could be no land in the vicinity of the Poles, from which he concluded that the sun's rays reflected from the flat surface of the sea must afford great heat to the air. Professor Maury went even further: he expressed the remarkable view that the flattening around the Poles brings the earth's surface there nearer to the internal reservoirs of terrestrial heat and thus enables that heat to escape more easily than elsewhere and to combine with heat received from the sun, from ocean currents, and even in substantial amounts from outer space, in maintaining warmth and an open sea.

More effective mid-nineteenth-century arguments were based, not on such fantasies as these, but on the supposition that winds and ocean currents bring warmth to the unknown regions of the Arctic. At a special meeting of the American Geographical and Statistical Society held in 1860 to discuss Dr. Hayes's plans for his polar expedition, a letter was read from Matthew Fontaine Maury. He pointed out that more than a million observations of wind directions showed a diminishing barometric pressure toward the Pole. This must be due to the liberation of latent heat, and "whence do those vapors come which liberate all this heat . . . if not from that boiling, bubbling pool of Gulf Stream water, which my observations show goes into that sea as an under current, and which we know comes out as an upper one."[29] There is a touch of rhetoric in these remarks. Maury's trust in the quantity of the observations would have rested on a firmer foundation had some of the observations come from the region where the phenomena were supposed to be. This region was completely unknown, and subsequent exploration has shown that its system of pressures, winds, and currents is not as simple as Maury pictured it.

A decade later Professor T. B. Maury ventured to write in similar vein:

> Nearly all around the hemisphere, south of and even within the Arctic regions, "*southwest and westerly winds decidedly predominate, and this statement is sustained by Dr. Kane's researches.*" (Keith Johnston's Physical Atlas, p. 48.) . . . In their course they blow directly along and over the surface of the Gulf Stream and Kuro-Siwo. From the tepid and smoking waters of these currents they take up vast quantities of heat and moisture, and bear these accumulated stores to the Pole . . .
>
> First, the heat given off by these hot streams, and otherwise wasted and lost, is stored away in the vesicles of vapor, as *latent heat*, and, by the winds, transported to the Pole, *and piled up around it*, there to be liberated at Nature's call, by condensation as *sensible* heat . . .
>
> It may be asked, but what bearing has the accumulation of vapor at the Pole upon its temperature? It is plain that, as a "*blanket*," or "*local dam*," mantling the Polar contour, it will arrest the processes of radiation and preserve to the soil there all the heat it may derive from every source . . .
>
> These facts suffice to show how potential is *aqueous vapor* in determining the temperature of the Pole.

After this warm, aqueous vapor, the following remarks of Dr. Kane's brother General Thomas L. Kane, in 1868, are like a breath of fresher air:

> Open your Physical Atlas. Look at the maps of the Isothermals, the Distribution of the Winds, the Distribution of the Rain and Snow, the Currents of the Ocean, the Tides . . . and observe where the lines all break off and abruptly come to an end . . . Here you have Hydrog-

raphy, losing all track of its tide waves and hot and cold water streams; here Thermology, vainly asking its right of way; Meteorology, youngest of the sciences, with nothing over all this space but a vacant stare . . . The votaries of all these sciences are so many hosts in arms against Ignorance, intrenched in the darkness surrounding the North Pole.[30]

OCEANOGRAPHIC ARGUMENTS

In the domain of oceanography, one of the stock arguments employed in support of an Open Polar Sea until the nineteenth century was that ice can form along coasts only. Barents reported that ice near the shores was "the onely and most hinderaunce to our voyage . . . whereby it appeareth, that not the nearenesse of the North Pole, but the ice that commeth in and out from the Tartarian Sea, about Noua Zembla, caused vs to feele the greatest cold." Hudson explained that, instead of penetrating into the open sea, which is never frozen on account of its depth and the great force of its currents and waves, he kept near the coast, but there he found the sea frozen.

Barrington developed a complex line of reasoning, based in part on dubious experiments, to show that "floating ice, which is observed both in high Southern and Northern latitudes, cannot be probably formed from sea water . . . Hence it seems to be almost demonstration, that the floating ice met with by navigators, being both solid and sweet to the taste after dissolution, cannot be produced from the water of the ocean." Moreover, the turbulence of the open sea would destroy ice. The Black and Baltic Seas freeze because of the absence of tides and the great influx of fresh water into them. Since ice forms along shore lines only, and since, as Dr. Hooke has shown, there is little or no land in the vicinity of the Pole, therefore, said Barrington, there can be little or no ice in the Polar Sea.

Professor Maury outdid Barrington in the matter of turbulence. He described "the *friction* of vast masses of watery matter meeting and clashing at the Pole. Here, vertically, horizontally, obliquely, the mighty currents and counter-currents underrun, overlap, and rub against each other in their fierce and ardent struggle to preserve oceanic equilibrium and circulation. As in the beehive, 'fervet opus.' Such tremendous attrition must excite warmth far from inappreciable." Scoresby, however, had long since "found that ice was formed in the Spitzbergen seas during nine months of the year; and that neither calm weather nor the proximity of land were essential for its formation." He "often saw ice grow to a consistence capable of stopping the progress of the ship, with a brisk wind, even when exposed to the waves of the Atlantic."[31]

During the nineteenth century ocean currents, actual and alleged, also were often cited as evidence in support of an Open Polar Sea. Three matters were considered: (1) cold currents flowing outward from the Polar Basin, (2) warm currents flowing inward into it, and (3) the distribution of land and water in relation to these currents.

Markham thus stated and commented upon the argument relating to outflowing cold currents:

> It is that the enormous fields and floes of ice which drift away to the south during the summer, leave a wide space of open sea round the North Pole. By way of proof, it is urged that in the Antarctic regions Sir James Ross pushed through 800 miles of pack ice, and reached an open sea to the south of it; being the space whence it had drifted.

Dr. Petermann reasoned along these lines and, in this country, Dr. Kane and M. F. Maury. Markham explained that the analogy with the Antarctic was false,

> as Admiral Collinson well pointed out . . . in 1865. The Antarctic pack was drifting away from a solid line of immovable grounded ice cliffs, and of course left open water in its rear, because there was no moving ice further south to take its place. Unless there is a continent or a similar immovable line of ice cliffs at the North Pole, the North Polar pack does nothing of the kind. The exact analogy to the voyage of Sir James Ross is that of Scoresby. The Antarctic pack, in latitude 75° S., is analogous to the ice met by whalers in the early spring in 75° to 76° N., through which they can usually pass. The open water north of Spitzbergen is analogous to the open sea found by Ross in the south; and the Polar pack which Scoresby found bounding that open water to the north, from whence the ice he had passed through had drifted, is analogous to Ross's line of impenetrable ice barrier.

Markham gave two reasons why there was probably no land or grounded ice barrier north of Spitsbergen, from which the ice could become detached. "One is that the masses of Siberian drift-wood on the Spitzbergen Islands and elsewhere would be intercepted if there was an extensive continent in their way; the other is that, as Parry advanced to his extreme point in 82° 45′ N., he found the water north of Spitzbergen rapidly becoming of very great depth."

M. F. Maury attributed the Open Polar Sea to mighty undercurrents flowing *into* the Polar Basin from Bering Strait, Baffin Bay, and the North Atlantic. As proof he cited Arctic voyagers in the vicinity of Davis Strait who had described "huge icebergs, with tops high up in the air, and of course the bases of which extend far down into the depths of the ocean, ripping and tearing their way with terrific force and awful violence through the surface ice or against a surface current, on their way into the polar basin."[32] Maury was dramatizing, though it is true that icebergs in coastal waters are occasionally swept back against a surface current by vagaries of local tides at greater depths. Captain Hamilton, in his critique of Maury, pointed out that the general movement of the icebergs in Baffin Bay and Davis Strait was from north to south.

In a letter to the Secretary of the Navy written on November 8, 1856,[33] and also in the 1857 edition of his *Physical Geography of the Sea*, Maury called attention to the fact that in 1855 Commander John Rodgers in the U.S.S. *Vincennes* had taken temperatures in the Arctic north of Bering Strait. "Warm and light water [was found] on the top, cool in the middle, 'hot and heavy' at the bottom." Maury conjectured that the lowest layer was heavier because of its increased specific gravity owing to evaporation when the water composing it must have been at the surface. Its warmth showed that it must have come from the tropics; for observations had demonstrated that undercurrents can transport water great distances without change in temperature. This was in line with Maury's general conception of ocean currents, as due, not to the winds primarily, but to differences in salinity and specific gravity produced by the effects of climate upon the surface of the sea. Numerous cold currents were known to flow out from the North Polar Basin, and, as they were all salt, "we cannot look for their genesis to the rivers of hyperborean America, Europe and Asia, and the precipitation of the Polar basin." Presumably, therefore, they were southward-returning undercurrents that had upwelled to the surface and been cooled. Rodgers's observations had disclosed temperatures of 40.2° at 25–28 fathoms, close to the bottom. "An extensive layer of water at the temperature of 40°

would, when brought to the surface in those hyperborean regions, tend mightily to mitigate and soften climates there." "How did this hot and heavy water that was found at the bottom get there?" Maury queried. "Did it come through Behring's Strait with the warm water of the surface? or did the Gulf Stream pour it into the Polar basin?" At all events, Rodgers's discovery furnished the only link missing "to complete from known facts the theory of an open water in the Arctic Ocean."

Unlike Maury, Captain Silas Bent held that both the Gulf Stream and the Japan Current (or Kuro Siwo) remained surface currents throughout their entire course from the tropics to the vicinity of the Pole. In 1856, Bent, then a lieutenant, read a paper before the American Geographical and Statistical Society on the "Japanese Gulf Stream," based on observations he had made while serving on Perry's expedition to Japan supplemented by data compiled from the writings of Cook, Krusenstern, and other earlier navigators. He emphasized the striking likeness between this great current in the Pacific and its counterpart in the Atlantic. Twelve years later he stated the gist of his theory in a letter to Judge Daly:

> Now, since these streams [the Gulf Stream and the Japan Current] possess such a wonderful power of retaining their heat, so long as they do not touch the land, as to raise the climatic temperature 30 or 40 degrees over half a continent lying eight thousand miles distant from the points in the Tropics from whence they spring, and from which they derive their heat, it does not seem unreasonable to believe that those portions of the streams which pursue their courses direct into the Arctic Ocean, carry with them warmth enough not only to dissolve the ice they encounter, and keep their pathways open all through the year, but also, to raise the temperature permanently above the freezing point of a large area of the sea around the Pole, and thus prevent this extremity of the earth becoming locked in eternal ice, and overburdened, in the lapse of ages, with the accumulations of snow precipitated from the winds, loaded with moisture taken up by evaporation, and carried thence [thither?] from more southern and warmer regions of the earth's surface.

Judge Daly published Bent's letter in part in the *Journal of the American Geographical and Statistical Society* but for a while kept to himself his views regarding it. In 1870 the publicity that the theory had received, especially as enlarged upon by Professor Maury, led the Judge to speak out in no uncertain terms in one of his annual addresses before the Society. He felt that, "in behalf of the Society, and in the general interest of geographical science in this country," he should "examine calmly what foundation there is for so confident an assumption." He deemed it unnecessary, however, to go into the deductions that Captain Bent had drawn, but devoted himself to a discussion of the more direct "evidence": (1) the assumption that the Gulf Stream and the Kuro Siwo "are each prolonged to the vicinity of the Pole" and (2) reports that various navigators had actually seen the Open Polar Sea. Daly denied that there was any positive information regarding the course of the Gulf Stream beyond the coast of Norway and quoted a letter from the hydrographer George W. Blunt. That current, Blunt had written, had been much misrepresented by "the inventions of Maury, the stupidities of weather predictors, and the assumptions of meteorologists—enough, either of them, to crush out the vitality of any thing which had not so perfect an organization as the Gulf Stream has." After analyzing some of the alleged reports of Dutch and other visits to an Open Polar Sea, Daly concluded by expressing assurance that the Pole would ultimately be reached but grave doubt whether an open passage to it exclusively by water would ever be found.

Dr. Petermann believed that the Gulf Stream entered the Arctic Basin between Spitsbergen and Novaya Zemlya as a surface current that flowed northeast along the Siberian coast. Unlike Bent, he did not think it maintained an open passageway through the ice. In 1852 he urged the British Admiralty to send out an expedition by this route in search of Franklin, whose ships he surmised would be found somewhere northeast of Bering Strait.[34] After pushing through the ice barrier the rescue vessel would take advantage of the open sea, of which Barents had found "unimpeachable" evidence, and of the great polynya discovered by the Russians. Petermann recommended that the voyage be made in winter, during which season, he thought, ice conditions would be preferable. The "general opinion" was that Franklin had passed through Wellington Channel. "If so, it is beyond doubt that he must have penetrated to a considerable distance further, so as to have rendered it exceedingly difficult, if not impossible, to retrace his steps, should he have found it impracticable to proceed in any other direction." Petermann gave no clear explanation of how Franklin might have reached the vicinity of the Siberian coast. The Admiralty did not act on these suggestions.

In later years, in the face of increasingly convincing evidence of the impenetrability of the ice barrier east of Spitsbergen, Petermann concluded that the ice must there be pressed back and compacted by the Gulf Stream, which consequently cannot offer a practicable sailing route. The Greenland Current presumably does offer such a route because it brings the ice *out* from the Polar Basin and hence must furnish at least a partly ice-free passageway. Petermann, accordingly, argued that this was the best route, even though it would necessitate sailing against an adverse current.[35]

As explorations yielded new evidence concerning the set of currents and the nature of the ice in the Arctic, it was interpreted as shedding light on the distribution of land and water there. Barrow in 1818 had argued against the supposition that Greenland was a part of North America. Strong currents ran south through Roe's Welcome and along the Greenland coast, bringing "logs of mahogany and the remains of the North American ox" from the northwest. A whale struck in the sea of Spitsbergen had been taken the same year in Davis Strait. Why the whale might not have come round the south end of Greenland, Barrow did not explain.

Dr. Petermann divided the Arctic into two major geographical regions by a line running from northeast Greenland past the Pole to the vicinity of Bering Strait.[36] The region on the European and Siberian side he believed to consist mostly of sea; the region toward North America, to consist more largely of land and to have a colder climate. This was a projection into the unknown of conditions observed to the south. Petermann was convinced that Greenland extended either as a continent or as a chain of islands almost to the northeast Asiatic coast. He envisaged the waters of the Gulf Stream as turning to the left off this coast and returning to the Atlantic as the Greenland Current. Without a land barrier to prevent them, these waters would continue into the American half of the Arctic and produce a warmer climate and less massive sea ice than seemed to prevail there. Immense quantities of driftwood brought down by the Siberian rivers were found along the eastern shores of Spitsbergen and northeast Greenland, but little or no driftwood had been reported from the waterways of Smith Sound. If the current making south through Smith Sound were a branch of the current east of Greenland, it, too, would bring driftwood with it. That it did not, Petermann held to be further evidence of a land barrier.[37] Petermann's theory confined the Open

Polar Sea chiefly to the Eurasian side of the Arctic, in contrast with the much more extensive sea that Russian and American enthusiasts imagined as occupying most of the Arctic Basin. Moreover, again unlike the Russians and Americans, Petermann did not regard it as an iceless sea but as "open" only in that it was clear enough to permit ships to navigate it among the ice floes at certain times of the year.[38]

Like Dr. Petermann, Admiral Sherard Osborn and others postulated land at various places in the Arctic to explain currents and ice. Their arguments were complex and need not be considered. But the words of an American sailor, Captain George Tyson,[39] a hero of the *Polaris* expedition, deserve quotation:

> In describing the hydrography of Smith Sound I free myself entirely of the geographers, chart-makers, and romancers, and relate only what I observed with my own eyes . . .
>
> I do not think there is enough current in Robeson Channel to warrant the theory of an open Polar Sea. Neither such a sea, nor a portion of such a sea, could empty itself through such a narrow channel as Robeson. Any large sea to the northward coursing through such a contracted passage would cause such a powerful current as to make it unsafe for navigation by any but the most powerful steamers, yet the *Polaris* overcame it without difficulty.

Tyson believed there was an archipelago north of Robeson Channel. With regard to the nonexistence of an Open Polar Sea he was right, of course, if by this he meant an iceless sea, and one may applaud his honest aim to relate only what he "observed with his own eyes." Otherwise his reasoning seems none too coherent.

Biological Arguments. Advocates of the Open Polar Sea also brought biological evidence to its support. Henry Hudson reported "that the more northwards he went the less cold it became; and that whilst in Nova Zembla, the land was barren, and there were none but carnivorous animals of prey, like bears, foxes, and the like, he had found under the eighty first degree [in Greenland and Spitsbergen] grass on the ground and animals that lived on it." He failed to tell us what the carnivorous animals of Novaya Zemlya lived on. Dr. Kane in his address of 1852 said:

> In West Lapland, as high as 70°, barley has been and I believe is still grown; though here is its highest northern limit. If 80° be our center

of maximum cold, the Pole at 90° is at the same distance from it as this West Lapland limit of the growth of barley.

Kane also spoke of

the migrations of animal life. At the utmost limits of northern travel attained by man, hordes of animals of various kinds have been observed to be traveling still further.

The Arctic zone, though not rich in species, is teeming with individual life.

He stressed the abundance of birds, whales, and walruses. The polar bear

is further to the north than we have yet reached; and this powerful beast informs us of the character of the accompanying life, upon which he preys.

The ruminating animals, whose food must be a vegetation, obey the same impulse or instinct of far northern travel. The reindeer (*Cervus Tarandus*), although proved by my friend Lieut. McClintock to winter sometimes in the Parry group, outside of the zone of woods, comes down from the north in herds as startling as those described by the Siberian travelers, a "moving forest of antlers."

In describing the arrival of Morton and the Eskimo Hans at the "Open Polar Sea," Kane wrote:

Animal life, which had so long been a stranger to us to the south, now burst upon them. At Rensselaer Harbor, except the Netsik seal or a rarely-encountered Harelda, we had no life available for the hunt. But here the Brent goose, (*Anas bernicla*,) the eider, and the king duck, were so crowded together that our Esquimaux killed two at a shot with a single rifle-ball.

There were also many gulls, sea swallows, and kittiwakes. Flocks of brent geese, whose "habits may be regarded as singularly indicative of open water," which had not been seen since Kane had entered Smith Sound, were flying off to the north and east.

Professor Maury made much of this.

As to the birds seen in such numbers by the Grinnell explorers, we may ask, Where would they find supplies of food *nearest?*

Better geographers than we or Dr. Kane, they know that, south of them, there was no food short of the coasts of Labrador, distant fifteen hundred miles; while, if they flew poleward, in less than five hundred miles . . . they would soon enter the balmy air of the Open Sea and refresh themselves with the generous spoils the Gulf-Stream transports . . . in its capacious volume to the Pole.

Van Campen also commented on Morton's observations of these flights of birds:

Here, surely, are the truest guides—dumb though they be—to the region we have been wont to imagine so hemmed in by eternal barriers of ice that life was impossible within its circle. Man may err, but Nature does not lie, and in following closely upon the heels of Nature it is impossible to mistake the clues that are available for the solution of Polar problems.

Though right as to Nature's truthfulness, van Campen was too optimistic regarding the impossibility of man's mistaking her clues. "Dr. Kane," wrote Markham, "mentions that great numbers of seals and sea-fowl were seen by Morton, and adduces this as a proof of an open Polar sea; but Rink remarks, on the contrary, that the flocking together of sea-animals and birds is a sign of a single opening in a sea, the rest of which was covered with ice." And Captain Hamilton denied that climate was a cause of birds' migrating and cited evidence of the wintering of certain birds in high latitudes.

In the discussions of the Open Polar Sea the whale figures prominently. Barrington contended that

if the ice . . . extends from North Latitude 80½° to the Pole, all the intermediate space is denied to the Spitzbergen whales, as well perhaps as to other fish . . . If this tract of sea also is thus rendered improper for the support of whales, these enormous fish, which require so much room, will be confined to two or three degrees of latitude in the neighbourhood of Spitzbergen; for all the Greenland Masters agree, that the best fishing stations are from 79° to 80°, and that they do not often catch them to the Southward.

In his *Physical Geography of the Sea* M. F. Maury wrote

of whales that have been taken near the Behring's Strait side with harpoons in them bearing the stamp of ships that were known to cruise on the Baffin's Bay side of the American continent . . . it was argued therefore that there was a northwest passage by which the whales passed from one side to the other, since the stricken animal could not have had the harpoon in him long enough to admit of a passage around either Cape Horn or the Cape of Good Hope.

Maury's own study of the logbooks of whalers had

led to the discovery that the tropical regions of the ocean are to the right whale as a sea of fire, through which he can not pass, and into which he never enters. The fact was also brought out that the same kind of whale that is found off the shores of Greenland, in Baffin's Bay, &c., is found also in the North Pacific, and about Behring's Strait, and that the right whale of the northern hemisphere is a different animal from that of the southern . . .

In this way we were furnished with circumstantial evidence affording the most irrefragable proof that there is, at times at least, open water communication through the Arctic Sea from one side of the continent to the other, for it is known that the whales can not travel under the ice for such a great distance as is that from one side of this continent to the other.

Maury admitted that "this did not prove the existence of an open sea there; it only established the existence—the occasional existence, if you please—of a channel through which whales had passed." He did not explain what kind of channel he had in mind, whether continuous or broken by belts of ice under which the whales could pass. Professor T. B. Maury, though he enlarged his namesake's "channel" into an "occasional open avenue through the Arctic waters," also admitted that the information concerning the whales did not alone suffice as proof of the existence of a navigable highway to the Pole.

M. F. Maury pointed out that

navigators have often met with vast numbers of young sea-nettles (medusæ) drifting along with the Gulf Stream. They are known to constitute the principal food for the whale; but whither bound by this route has caused much curious speculation . . . Now . . . at first there is something curious to us in the idea that the Gulf of Mexico is the harvest field, and the Gulf Stream the gleaner which collects the fruitage planted there, and conveys it thousands of miles off to the hungry whale at sea. But how perfectly in unison is it with the kind and providential care of that great and good Being which feeds the young ravens when they cry, and caters for the sparrow!

In *The Sea Around Us* (1951) Rachel Carson has made it clear that, although there is a greater variety of species in the warm, tropical waters than in the polar seas, the lack is compensated by "the enormous abundance of the forms that inhabit" the latter. Until the immense productivity of the polar waters had been demonstrated, it was natural to assume, on the analogy of the barrenness of the polar lands, that the waters also must be barren except where ocean currents carried provisions to them from warmer climes. Taking a cue, perhaps, from his namesake, Professor Maury enlarged lavishly upon this theme:

Midway in the Atlantic is found . . . the famous *Sargasso Sea* . . . This great, *living* bank of marine growth is computed, by Humboldt, to cover with sea-weed . . . an area *almost six times as large as Germany* . . .

The mighty equatorial current or drift of waters, from the east, rushes through this enormous deposit and crowds the channel of the Gulf-Stream with the vegetable matter transported from the "Weedy Sea" . . . The Gulf-Stream bears it as part of its freight, in measureless quantities, to the Polar Sea.

A similar freight is borne in from the Sargasso Sea of the Pacific by the Kuro Siwo, and together they provide "*food* for myriads of marine inhabitants . . . The rations daily transported and served out to these swarming hosts, no mind may rightly compute . . . Could so much life exist *beneath a frozen ocean?* Could it exist anywhere *save in an open and tepid sea?* The infusoria of the sea are known to wear away the ice which fringes Arctic shores."

MOTIVES FOR, AND NATURE OF, THE DISCUSSIONS

In the discussions of an Open Polar Sea the wish was usually father to the thought. Until the time of Barrington the predominant "wish" was for the discovery and development of a navigable commercial passageway to the Pacific; in the nineteenth century it was for the discovery of a navigable passageway to the North Pole. Nearly all of those who took part in the discussion did so in the interest of a particular route, which either they themselves or their compatriots proposed to follow. Judge Daly's arguments, however, seem to have been free of special pleading and in keeping with the spirit of scientific inquiry.

No matter how modestly and tentatively they may be first suggested, some scientific theories spawn a progeny of subsidiary speculations in the minds of the more ingenious and credulous wishful thinkers to whom they appeal. This was true of the theory of an Open Polar Sea. The hazards of Arctic travel have been feared since early times, and the Open Polar Sea was conceived partly as a palliative of such fears. Thus Thorne affirmed that the polar seas were not dangerous, having "in them perpetuall cleerenesse of the day without any darknesse of the night [!]: which thing is a great commoditie for the navigants to see at all times rounde about them, as well the safegardes as daungers." Barrington pointed to fear of perishing from excessive cold as a stock objection to further attempts at polar voyaging and sought to counter this, as we have seen, by offering "proofs" of increasing warmth north of the 80th parallel.

Captain Bent said that

> other and higher objects than the mere accuracy of my theory—something more elevated than the just and honorable feeling of satisfaction that would, were it to prove correct, certainly belong to him who could claim priority in such an important discovery—has actuated me at the present time . . . It is the actual saving of human life—the benefits that will accrue to many departments of science, and the solving of a geographical problem which is now, for the most part, conjectural.

Professor Maury thought that, if proved correct, Captain Bent's theory would not only prevent "a recurrence of those disastrous expeditions, which have already cost the world a frightful amount of human life," but might well lead to even greater benefits to mankind. He tried to revive the old hope of a commercial route to the Far East and visualized a flourishing transpolar trade with China. The goods that the Chinese would receive in exchange for their glut of surplus products might do away with the scourge of famine in that unhappy land.

Furthermore,

> who shall say that, within the Arctic circle, dwelling upon some of the islands or shores of that sea Dr. Kane [actually it was Morton] saw rolling and beating at his feet, there may not yet be found—
>
> "One touch of nature makes the whole world kin!"
>
> some vestige of humanity—some fragment of our race, wafted thither by these mighty currents we have heard of, whose cry of welcome is yet to greet the mariner who finds them, and amongst whom there may, at least, be found some one of God's elect?
>
> Did ever wider or whiter field expand before the world's restless philanthropy?

There is another "law" relating to the development of scientific theories, the "law of personal pride." Once committed to advocacy of a theory, all but the most judicial-minded tend to promote and defend it through thick and thin. Dr. Petermann is an illustration. Undoubtedly he first worked out his conception of the geography of the Arctic as a disinterested scientific theory, but he overelaborated it and personally sponsored the German Arctic expedition of

Koldewey to prove its correctness.[40] Sometimes personal pride takes the form of personal pique. Judge Daly's criticism of Bent's theory stung Bent to a rejoinder in which he imputed to Daly not only ignorance but bad faith. He complained that Daly had been misled through partiality to Dr. Hayes, who favored the Smith Sound route for the proposed government expedition as against the Gulf Stream route that Bent himself advocated. Daly's reasoning, however, also tended to remove the props from Dr. Hayes's own cherished belief in an Open Polar Sea.

In an address before the American Geographical and Statistical Society in 1868, Dr. Hayes sought unsuccessfully to enlist support for a new expedition. He proposed to establish a colony of hunters and natives as a base on the west coast of Smith Sound, and from there to push north, first across the ice barrier and then by boat over the open water. As to the existence of an open sea, he admitted no shadow of doubt.

> That such an extensive area of water should be frozen would appear to be impossible, even without other evidence than an ordinary acquaintance with the phenomena of nature. Water is a restless object. Lightsome and limpid, it treads the earth or mounts to heaven, and its natural state is one of incessant change. When the evening shades are coming on it leaves the floating air and nestles through the night upon the gladdened leaf; with the early flush of morning it melts away upon the first returning sunbeam, or steals to earth and seeks the fountain and the brook; and when the winter frosts threaten it, then it flees before the danger in the rivulet and rolling wave.

Even the rhapsodic Professor Maury could not have done better. Here, in harmonious combination, are high-sounding words and reasoning from analogy. Though suitable in poetic fancy, the use of analogies—or similes, or metaphors—can be dangerous as a method of scientific argument. Resemblance of things in one respect is not necessarily proof of resemblance in other respects; usually it merely signifies a higher or a lower degree of statistical probability that other similarities exist.

Professor Maury commented thus on Captain Bent's theory:

> This profound and beautiful hypothesis may boast no sanction of high authority, nor count as its advocate any Arctic explorer. For a while, it may have to rest its claims on deductions of science, and be ushered into notice on the quiet authority of mathematical calculation. Was it not so with the theory of Columbus? What of this?

Maury's implication—and van Campen employed the same analogy—was that, because Columbus' theory was right (which was not the case), therefore Bent's might well also prove to be right. The theories were analogous in that they both were geographical and both were held to be of a "crackpot" nature when first put forward. The probability that any such "crackpot" theory will ultimately be vindicated is not great. But it would not have served Maury's or van Campen's purposes to point that out.

The arguments regarding the Open Polar Sea also reveal other forms of defective reasoning, which, like the stretching of analogies, were due to failure to think the matter through. After the Open Polar Sea had ceased to be a living issue, Professors T. C. Chamberlin[41] and W. M. Davis formulated principles applicable to the "thinking through" of theories, especially in the fields of geology and physical geography. Had Davis developed the theory of an Open Polar Sea as he developed his own theory of the formation of coral reefs,[42] he would, no doubt, have made a thorough study of all the hypotheses that others had proposed, together with new ones of his own devising—hypotheses, for example, postulating in the unknown Arctic much land, no land, land distributed in various places, ice-free waters, partly ice-free waters, icebound and ice-free waters, all in various relation to one another, to the postulated land masses, and to currents, winds, and so forth. From the different conditions conceived according to these hypotheses he would have deduced as many necessary or probable consequences as he could and would have looked for evidence of these consequences as tests of the validity of each hypothesis. After doing all this, he would have given his support to the hypothesis that best met the tests. He might still have been wrong, but his reasoning would at least have rested on a more solid foundation than most of the reasoning we have been considering in this study.

It is hardly fair to the advocates of a discarded theory to cast aspersions on their thinking in the light of facts and methods that were not known to them. The quality of their thought should be judged according to what they had to work with.[43] If so judged, the proponents of an Open Polar Sea might not appear conspicuously less rational than its opponents. Science would not progress without the aid of new hypotheses based on evidence that scientists of later times regard in retrospect as scanty and inconclusive, and, as Davis made clear, even "outrageous" hypotheses have their value, if they arouse intelligent interest and invite attack.[44] In itself, the theory of an Open Polar Sea was not "outrageous" during its lifetime. Some of its more credulous and rhetorical proponents, however, embroidered upon it in a manner that was "outrageous" as measured by the scientific standards even of their own day. But this at least lent the theory a special charm.

CHAPTER 7

From *Kubla Khan* to Florida

In Xanadu did Kubla Khan
A stately pleasure-dome decree:
Where Alph, the sacred river, ran
Through caverns measureless to man
Down to a sunless sea . . .
And here were forests ancient as the hills,
Enfolding sunny spots of greenery . . .
And from this chasm, with ceaseless turmoil seething . . .
A mighty fountain momently was forced: . . .
Huge fragments vaulted like rebounding hail . . .
It flung up momently the sacred river.
Five miles meandering with a mazy motion
Through wood and dale the sacred river ran,
Then reached the caverns measureless to man,
And sank in tumult to a lifeless ocean.

SURPRISINGLY ENOUGH, these visions came partly from the Blue Sink, the Manatee Spring, and Salt Springs Run in Florida.[1]

In *The Road to Xanadu*, on the basis of much study of Coleridge's notes and correspondence and of works which the poet is known or believed to have read, the late Professor John Livingston Lowes traced far back toward their origins most of the images in *Kubla Khan* and in *The Rime of the Ancient Mariner*. He showed[2] that the lines just quoted were inspired in large part by passages in William Bartram's *Travels*,[3] where the Philadelphia botanist wrote of three places that he had visited in north-central Florida in 1774. Two of these Bartram identified by their then current names: "Alligator Hole" and "Manate[e] Spring." The other he did not name but described thus:

In front, just under my feet, was the inchanting and amazing chrystal fountain, which incessantly threw up, from dark, rocky caverns below, tons of water every minute, forming a bason, capacious enough for shallops to ride in, and a creek of four or five feet depth of water, and near twenty yards over, which meanders six miles through green meadows, pouring its limpid waters into the great Lake George.[4]

With a view to locating the three springs I first tried to compare the text of the *Travels* with modern maps, but soon found that this led nowhere. Bartram's text is much too vague in its topographic detail to make identifications possible in this way for any but a few of the places mentioned. The key to the problem was found in Dr. Francis Harper's annotated edition of Bartram's more factual *Travels in Georgia and Florida, 1773–74: A Report to Dr. John Fothergill* (published from a unique manuscript in 1943[5] and hereinafter referred to as *Report*). This work, which Dr. Harper thinks was presumably compiled from Bartram's long-lost original field journals,[6] supplies topographic and other details not presented in the earlier-published *Travels* (first edition, 1791). By comparing it with other descriptions and maps and by tracing Bartram's routes in the field, Dr. Harper has been able to locate with varying degrees of assurance most of the places that Bartram mentions both in the *Report* and in the corresponding parts of the *Travels.* Thus Harper explains in his commentary that the Alligator Hole is probably "Blue Sink, a mile north of Newberry in western Alachua County"; that the "Manate[e] Spring," which still bears the same name, "is approximately 7 miles west of Chiefland in Levy County"; and that the "inchanting and amazing chrystal fountain" is undoubtedly Salt Springs, at the head of Salt Springs Run, which flows into the northwest corner of Lake George.[7] In the *Report* Bartram's account of this last locality already dimly but unmistakably foreshadows what was to come in *Kubla Khan:*

We put into a large swift running Creek, that come from a vast Spring five or 6 miles up to it . . . when we came towards the head The Creek widened . . . at last [we] came to the head of the Creek, an immense fountain four or 500 Yards over where were, a great number of boiling holes throwing the water up in prodigious ebulitions . . . I continued somewhat higher up to the principle Fountain which, boiled up in an incredible maner out of the Chasms of deep Rock between tow [two] steep high hills.[8]

In his editions of the *Report* and *Travels* Dr. Harper includes photographs of this spring and also of the Alligator Hole and Manatee Spring. In the pictures these springs are hardly as impressive as one might expect from Bartram's and Coleridge's words. The Salt Springs, it would seem, though they have gained in poetic grandeur, have lost in actual physical force. Dr. Harper writes me that there has been a "progressive diminution in the height of the ebullition, as recorded by Bartram (1774), Michaux (1787), Le Conte (1822), and myself. Possibly the limestone has dissolved away to such an extent as to result in a larger passage for the water, with less forceful ebullition."

Dr. Harper's commentary on the *Report* gives many page references to corresponding passages in the *Travels*. His concern, however, was with Bartram as a naturalist and traveler and hence he did not point to the links with Coleridge, Wordsworth, and other poets.[9] Conversely, the literary historians who have approached the *Travels* from the poet's domain[10] have had little interest in the exact localities that Bartram visited and described, and, even had they sought to identify them, it would have been next to impossible to do so until after the publication of Harper's studies. Thus it has come about, as we have seen, that Lowes associated Coleridge's meandering river with Bartram's meandering creek, and Harper identified the latter with the Salt Springs Run, but hitherto no one seems to have thought of crediting the Salt Springs Run,[11] of little fame, as well as Alpheus of classic fame, with a share in the ancestry of "Alph, the sacred river" (see above, p. 20). Bartram explored Salt Springs Run up to the Salt Springs on the last of four journeys that he made between April and September 1774, from Spalding's Lower Store, near the present town of Palatka. On the first of these trips he traveled westward to the Alachua Savanna (modern Payne's Prairie), visiting the Alachua Sink ("a most remarkable Place for Allegatores"),[12] which he named and described in the *Travels* as the "Great Sink." Coleridge had this in mind when he wrote the first draft of *The Wanderings of Cain.*[13]

The second (May–June) and fourth (August–September) trips both took Bartram up the Saint John's River beyond Lake George. While southbound on the first of these he had an adventure with alligators at "Battle Lagoon," as he named it, a spot that Harper identifies with Mud Lake, near the entrance of the Saint John's into

Lake Dexter. In the *Travels* the story of this encounter is followed by a vivid account of the breeding habits of the alligators, and these passages made a strong impression upon Coleridge's imagination.[14] Here also (or near here) Bartram tells us that he got "some fine trout for supper," and he describes glowingly the trout's "orient attendants . . . the yellow bream, or sunfish." Some of Bartram's very words—"blue," "green," "velvet black"—describe the water snakes that "coiled and swam" within the shadow of the becalmed ship of the Ancient Mariner. Returning northward, Bartram "launch[ed] again into the little ocean of Lake George, meaning now . . . to coast his Western shores in search of new beauties in the bounteous kingdom of Flora. I was however induced to deviate a little from my intended course and touch at the inchanting little Isle of Palms." I find no reference in *The Road to Xanadu* to the simile of "the little ocean," although Lowes quotes the passage from the *Travels* describing Lake George as the outlet of the meandering creek[15] and discusses other co-ancestors of Coleridge's "sunless sea" and "lifeless ocean." Lowes makes much, however, of Bartram's magical description of the "inchanting little Isle of Palms," with which he associates the "forests ancient as the hills, Enfolding sunny spots of greenery" of *Kubla Khan.*[16] Dr. Harper has written me: "The 'Isle of Palms' at the head of Lake George does not appear (at least as an island) on current maps. I can only guess that it has been engulfed by the delta; or perhaps it never was, even in Bartram's time, more than an island in a marsh rather than in the lake."

(In 1775 Bartram made a journey into the Cherokee country, during which he climbed alone to the top of the "Jore mountains," whence he "beheld with rapture and astonishment a sublimely awful scene of power and magnificence, a world of mountains piled upon mountains." On the way down he passed through "spacious high forests and flowery lawns." These, according to Professor Ernest Earnest, "became Coleridge's 'forests ancient as the hills, Enfolding sunny spots of greenery.' "[17]

Bartram made the ascent from and returned to the Indian village of Cowee, the site of which has been located at Wests Mill, on the Little Tennessee River, in western North Carolina. Dr. Harper, who has been over the ground, is certain that "Bartram's Jore Mountains were the Nantahalas and that his Jore River was Burningtown Creek.")

Bartram's third Florida journey of 1774 (June–July) brought him west as far as the Suwanee River. While on this trip he visited the "admirable Manate Spring" and the Alligator Hole. Lowes believed that the former helped create not only the image of the "mighty Fountain" at the head of the "sacred river," but also that of the shadow of Kubla Khan's pleasure dome, which "floated midway on the waves."[18] As to the latter, we read in the *Report* that "this Sink is called by the Indians & Traders the Alegator Hole from a Prodigious large Alegator that has lived here from time immerorial."[19]

William Bartram's *Travels* are an "inchanting and amazing" forest, explored from one side by lovers of poetry and from another by lovers of nature, without much meeting or recognition of each others' knowledge of the trees and glades. This note may suggest one of many possible pathways along which these forest devotees from different sides might meet and exchange their lore with mutual advantage and delight.

CHAPTER 8

What's "American" about American Geography?

WHAT'S "AMERICAN" about American geography?—and what could possibly be more ambiguous than this question?[1]

I shall not try to answer the question in comprehensive terms, but shall comment, rather, on what has seemed "American" about the work of three former American geographers and about a symposium brought out in 1954 by a representative group of geographers.

This paper is part of a discussion that has just kept rolling along like Ol' Man River since the seventeenth century, producing a huge "literature."[2] With so much smoke one might expect a little scientific fire, but the study of what is and has been American about this or that—let us call it *americanistics*—would seem to be still in virginal innocence of scientific sophistication. My first purpose, therefore, is to set up a rickety frame of reference into which answers to the theme question might be fitted in an embryonically scientific manner. This involves terms and definitions.

DEFINITION OF TERMS

By "American geography" is meant here geographical studies as pursued by Americans of the United States and the antecedent colonies, *not* American geography in the sense of the geographic actualities of America as "given" by God and Nature and as altered by man, although it would be interesting to try to figure out what's American, and more especially what's *not* American, about these (for instance, my mother used to say the landscape in the part of Ohio where she grew up was "English"). As to what's "American" about American geographical studies, two overlapping but divergent varieties of "Americanness" must be considered. These may be designated the *americanistic* and the *geoamericanistic*, respectively.

As pertaining to the United States, the adjective "American" may mean anything from merely "existing in" (Mount Washington is an "American" mountain), "coming from" (an "American" automobile in Tibet), "descriptive of" ("American" history), or "belonging to" this country (the "American" navy), through "loyal to the Stars and Stripes" or "free from the taint of 'un-Americanism,' " to "characteristic of the civilization of the United States, or of some aspect of that civilization." The americanistic attributes of American geography are those generally characteristic of the civilization of the United States as distinguished from other cultures; the geoamericanistic attributes are those generally characteristic of American geography as distinguished from British, French, and other geographies. Hence the meaning of "generally characteristic of" is the crux of the matter (and please note that, paradoxically enough, "a characteristic" of something is not necessarily "characteristic of" that thing in the sense of "characterizing it" or distinguishing it from other things).

To be "characteristic of" the subject to which it pertains, an attribute or element must be neither too generally prevalent elsewhere nor too localized within its subject. Thus it is characteristic of American culture—therefore americanistic—to play or watch baseball because a relatively large proportion of all Americans do so as contrasted with the relatively small proportion of all non-Americans who do. Although an even larger proportion of Americans play and watch games, this is not americanistic because of the high percentage of non-Americans who do likewise; nor is it americanistic to drink mint juleps, because so relatively few Americans drink them. Similarly, it was once geoamericanistic for American map makers to base their longitudes on the prime meridians of Philadelphia or Washington, but not geoamericanistic for American geographers either to make maps or teach geography as they do at the University of California. The former, like playing and watching games, is too widespread outside; the latter, like drinking mint juleps, too restricted inside the United States.

Or, perhaps I am wrong about the mint juleps and the California geography. Both might be americanistic, and the California geography, if not the juleps, might also be geoamericanistic. I introduce this note of doubt because no clean-cut criteria have yet been established for identifying either americanistic or geoamericanistic traits, not to mention criteria for drawing nice distinctions among them and between each of them and its opposite—nor have statistical techniques been applied to their measurement. Until we have such criteria, can draw such distinctions, and apply such techniques, we shall be in no position to treat the Americanness of American geography quantitatively and thus achieve scientific sanctification in inquiries like this. So, prior to entering that millennium, the best we can do is to muddle along without benefit of quantitative clergy. One kind of muddling is to take account of what people have said both explicitly (without, of course, using my terms) and by implication, to be either americanistic or geoamericanistic characteristics of or elements in American geography, some of which, naturally, seems rather mythological. Another kind of muddling is to try to perceive for ourselves what such traits are and have been. In this paper there is muddling of both kinds. Were americanistics already a science, I could quote an orthodox universally accepted definition of "generally characteristic" and therefore of "americanistic," but as it is not we shall have to use intuition. Be it said, though, that matters sometimes described as "quintessentially,"[3] or "uniquely," or "distinctively" American may or may not be characteristically American; also that, while much that is geoamericanistic is also americanistic—characteristic alike of American geography and of American civilization in the large—the two qualities are far from identical. There are aspects of American geography which, though geoamericanistic, are not americanistic, and vice versa. (Many things that are americanistic are also generally characteristic of one or more [but not too many] non-American civilizations—for example, baseball, which nowadays is as *japanistic* as it is americanistic—and the same principle no doubt applies to certain things that are geoamericanistic.)

The term *nonamericanistic* may help to solve the problem of what to call those elements in or aspects of American civilization that are *not* characteristically American. Obviously *un-American*, with its disagreeable implications, won't do, and *non-American* too strongly suggests what is exclusively foreign or otherwise outside the pale of American civilization.

One might, accordingly, arbitrarily class the qualities that constitute American civilization in six principal categories, the americanistic and five categories of nonamericanistic qualities. The latter are the *universal*, the *Occidental*, the *exotic*, the *subamericanistic*, and the *individualistic*.[4] The universal elements, of course, are those that American civilization shares with all civilizations, as, for example, the use of writing; the Occidental are those generally characteristic of the West European cultural tradition, such as the use of the Latin alphabet; the exotic are imported elements that have not become assimilated; the subamericanistic are elements characteristic of a small part only of American civilization, such as the culture of a section or city or social group (for example, the Proper Bostonians);[5] and the individualistic elements reflect personal idiosyncrasies. The elements that make up any important component of American civilization, whether American philosophy, art, science, literature, or geography, could, of course, be classified according to this scheme.

Regarded, however, as parts of the world's body of philosophy, art, science, literature, geography, and so on, the same elements could, also, in each case be grouped in six parallel categories. On this basis, American geography would consist of *geouniversal* elements, exemplified in the use of maps; *geo-occidental*, exemplified in the use of the Mercator projection; *geoexotic*, exemplified in the use of British or German or other foreign geographical techniques; *geoamericanistic*, or those generally characteristic of American geography; *geosubamericanistic*, exemplified in a point of view regarding geography characteristic solely of the University of Kanbraska; and *geoindividualistic*, exemplified in the brilliantly original geographical ideas of Professor Jedidiah Morris Bowman.

(These *geo-* terms will shock linguistic purists easily horrified by the marriage of Greek with anything other than Greek. To add *geo-* to this or that is a great convenience, however, and seems harmless, if care is taken not to confuse *geo-* as referring to the earth [as in geodiversity, geophysics, or geochemistry] with *geo-*

as referring to geographical knowledge or belief [as in geo-american, geoexotic, geouniversal].)

FOUR EXAMPLES

With this pedantry behind us, let us turn to the three men and the symposium. The men—Jedidiah Morse (1761–1826), William Morris Davis (1850–1934), and Isaiah Bowman (1878–1950)—were among the five or six outstanding geographers that this country has produced to date. The symposium was published in 1954 under the title *American Geography: Inventory and Prospect.*[6]

Jedidiah Morse. Before the Revolution, American geography, in the main, was geo-occidental, or, indeed, geoexotic. Most of our geographical books, maps, methods, and ideas came from England. The winning of independence stimulated a lively interest in geography among the American people and we began for the first time to write and publish geographical works in considerable quantities.[7] In these, americanistic influences were felt. Some of these works, such as Thomas Jefferson's *Notes on the State of Virginia* (1785), dealt with individual states. The most ambitious, however, was the Reverend Dr. Jedidiah Morse's *American Universal Geography*, which covered the whole world in two substantial volumes.[8]

Dr. Morse explained his purpose in the preface:

> So imperfect are all accounts of America hitherto published, even by those who once exclusively possessed the best means of information, that from them very little knowledge of this country can be acquired. Europeans have been the sole writers of American Geography [by which he meant, here, "geography concerning the United States"] and have too often suffered fancy to supply the place of facts, and thus have led their readers into errors, while they professed to aim at removing their ignorance. But since the United States have become an independent nation and have risen to Empire, it would be reproachful for them to suffer this ignorance to continue; and the rest of the world have a right to expect authentic information. To furnish this, has been the design of the Author of the following.

Accuracy was not one of Dr. Morse's paramount virtues. In reality, much "knowledge of this country" was then available, and a goodly proportion of it had been made available in published works by Americans, as Morse himself revealed by quoting at length from Jefferson, Lewis Evans, Jeremy Belknap, and others.

His comment that "Europeans" had "been the sole writers of American Geography" was incorrect.

The late Ralph H. Brown's understanding studies of Morse do show,[9] though, that the appellation "Father of American Geography" is deserved if we construe "geography" as embracing geographical manuals, textbooks, schoolteaching, and little else. Brown has demonstrated that Morse zealously compiled his chapters on the United States from printed and manuscript sources, interviews, answers to questionnaires, and to some small degree from personal observations in the field. Hence, these chapters may be deemed American with regard to their authorship, subject matter, and sources; as americanistic where they reflect an ardent and characteristically American patriotism in passages extolling our republican institutions and defending America against European disparagement; and as subamericanistic where they reveal strong sectional or denominational prejudices. They follow British models, however, in their scope and general arrangement, and the other chapters—that is to say, all of those concerning geographical generalities and the world outside of the United States—are American solely in the sense that an American citizen compiled them and they were published in the United States. Otherwise, they are geo-occidental, or, more specifically, British.[10]

William Morris Davis. Let us jump ahead a hundred years. In the late eighteen nineties and early nineteen hundreds, geography was coming to be taught increasingly in our colleges and universities and an American geographical profession was coming into being under the leadership, more especially, of William Morris Davis, professor of geology at Harvard. The ranks of the new profession were recruited largely from geologists, like Davis, who found the physiographic aspects of geology of greatest interest. As a consequence, from the late eighties until the First World War physiography was strongly emphasized in American academic geography. As the profession grew, which it rapidly did after World War I, a change set in and the pendulum swung far over toward the "human" side of geography. Even so, in the administrative setup of a good many of our colleges and universities geography is still subordinate to geology, a reminder of what was once an even more markedly geoamericanistic state of affairs.[11]

Why did physiographers play such a prominent part in the

founding and early development of our geographical profession? Was there anything either geoamericanistic or americanistic about this? Davis, himself, has given partial answers to these questions. Physiography in the Davisian sense, or geomorphology as it is now more often called, is the study of landforms, especially as related to the underlying rock structure.[12] Such relations had been examined with care and brilliantly interpreted on the basis of surveys carried out in the mid-nineteenth century in two regions of the United States where the relations are readily and impressively apparent—the folded Appalachian mountains of east-central Pennsylvania and the arid West.

Davis quoted J. P. Lesley, of the Pennsylvania Geological Survey, as follows: "We . . . became, not mineralogists, not miners, not learned in fossils, not geologists in the full sense of the term, but topographers, and topography became a science and was returned to Europe and presented to geology there as an American invention." By "topography" Lesley meant geomorphology, and the form in which it "was returned to Europe" might be classed as geoamericanistic. Davis went on to explain that Lesley and his colleagues saw in the zigzag ridges of Pennsylvania the surface expression of structures (and again he quoted Lesley) "so vast that to our eyes, familiar with rock curves ten and twenty miles in radius, underground or in the air, all that customary European local research which filled the proceedings of societies in lieu of new and larger matter seemed tedious and puerile,"[13] a comment that is surely americanistic.

The geological and geographical surveys of the sixties and seventies in the West offered even broader scope for the observation of "new and larger matter." That J. W. Powell and Grove Karl Gilbert could make momentous contributions to the interpretation of landforms, Davis, in a paper published in 1924,[14] attributed to three circumstances. To paraphrase, the first circumstance was that neither Powell nor Gilbert had had much formal training in geology; that both men were "little trammeld by the conventions of geological science" and, hence, were freer than they might otherwise have been to develop their own points of view and their lines of thought. The second circumstance was that their surveys were carried out in the "Great American Desert," where, owing to the absence of a cloak of vegetation, the relation of understructure to surface form was so manifest that it was forced upon their attention. And the third was that the principles of land sculpture as worked out by Powell and Gilbert "were announced at a time when the violent old doctrine of catastrophism was rapidly losing ground" before the gentler doctrine of uniformitarianism, and hence the geological world was favorably disposed toward a rational treatment of such matters.

This third circumstance, of course, was in no way characteristically American, but, rather, geo-occidental. Whether the first circumstance—that Powell and Gilbert were self-taught as scientists—is a sound explanation of their successes may be questioned, and Davis's assertion to that effect may well reveal as much about Davis as about the men to whom he applied it. Although Davis himself was not a self-taught scientist like Powell and Gilbert, he was not unduly trammeled by the conventions of geological science —or of English spelling, for that matter. ("Geomorphology should ever be thankful to Davis for having freed it from the bondage of classical geology," wrote Henri Baulig, a French disciple of Davis, in 1950.)[15] But do we not detect in Davis's paper a good old americanistic confidence in the virtues and advantages of the self-made man? Elsewhere in the same paper Davis commented upon the meteorological studies of four mid-century Americans: "The labors of these four men were in every respect distinctivly American, and they were performd at a time when next to no training in meteorological investigation was provided in our colleges; hence each of the four . . . was essentially self-developt."[16]

In ascribing Powell's and Gilbert's successes in part to the second circumstance—the absence of vegetation in the arid country—Davis was on firmer ground. It has often been said that "one can study geology from the train window" in the Far West and it seems reasonable to argue that the development of modern American physiography was largely inspired and greatly facilitated by the barren character of the western half of the country. This could hardly be called an americanistic influence, since I have defined "americanistic" as applying to American civilization rather than to the physical environment (although it is true that the Grand Canyon of the Colorado has been called "perhaps the most American, the most self-explanatory, and the most self-confident of natural prodigies").[17] It did, however, give rise to an important geoamericanistic trend in American geographical studies which prevailed for a good many years.

Davis worked out an exquisitely logical system of geomorphological exposition which was adopted by many geographers in this country and had a large vogue abroad. He also had considerable artistic skill, which he employed to advantage in illustrating both the principles of his system and detailed examples of its application. His ingenious line drawings or block diagrams portray blocks sliced, as it were, out of the earth's crust. The underlying structure appears on the vertical faces, the landforms themselves are shown in perspective relief on the top surfaces, and, in some instances, different hypothetical stages of cycles of erosion and deposition are suggested by the juxtaposition of different blocks. Davis did not invent this technique. His inspiration came from similar diagrams, notably by W. H. Holmes and Gilbert, which had appeared in the reports of the Hayden surveys.[18] But Davis improved upon and elaborated the technique so highly that it has charmed geologists and geographers all over the world into using it to illustrate countless books and articles. The Davisian block diagrams have also had a vigorous progeny in the shape of the comprehensive landform maps of such geographer-draftsmen as A. K. Lobeck, Erwin Raisz, and R. E. Harrison. They are superb examples of the geoamericanistic. It must be said, however, that the very clarity and relative simplicity of Davis's exposition of geomorphic processes and the cycle of erosion have been to some degree its undoing in the minds of a later and, naturally, far wiser and more sophisticated generation of geographers and geomorphologists. It has suffered the frequent fate of that which is relatively simple and clean-cut.

Isaiah Bowman. Isaiah Bowman studied under Davis, and until his appointment as Director of the American Geographical Society in 1915 hewed close to the Davisian line of geomorphology in his most productive original research. Thereafter he shifted the center of his interests to the human side of geography, as did the American geographical profession as a whole at about the same time, for reasons that have never been adequately explained.

(Bowman in 1934 discussed the "distinction . . . between a geography which has been chiefly physical in America and which has been and is so largely cultural or social in Europe," attributing the difference partly to our "wilderness-conquest history" and to the circumstance that "we have been bred in a land of pioneering traditions." He added, "Today the European and American schools are not so far apart. Our wilderness period is past. Altogether revolutionary is the change in point of view that has taken place in recent years among American geographers," that is, the shift in emphasis to the "human" side of the field.)[19]

From among the varied interests of this dynamic and versatile man I shall call attention to two.

Bowman attributed his interest in pioneers and pioneering partly to "life as I lived it as a boy in Michigan."[20] Actually the frontier had long since moved west, but Bowman, like Frederick Jackson Turner, drew an americanistic inspiration for his pioneer studies from the memories and myths of pioneer days still fresh in the Middle West during his youth. In 1907, 1911, and 1913 he studied frontier communities in the Andean valleys; in 1925 he launched a long-range collaborative program of pioneer-belts studies in different parts of the world;[21] and in 1931 his book, *The Pioneer Fringe*, was published.[22] In the preface to the book he explained that the purpose was "to sketch the outlines of a 'science of settlement' " and that such a science was desirable "not merely to provide means by which to attract men to a new land" but also to furnish guidance for "the ultimate withdrawal of the borders of settlement in the least favorable situations." Bowman's conception was both americanistic and geoamericanistic. He envisaged an applied science concerned with the control of the settling or desettling of specific tracts of land. Recently in the study of settlements American geographers, under European influence, have shown what would seem to be a geoexotic preference for the examination of existing settlements in the light of their historical origins and changing structures and patterns.[23] Bowman, like the teacher of creative writing, strove to establish principles for directing and guiding a *process;* the more recent studies, like those of the literary critics, have sought, rather, to interpret the forms resulting from processes—but it is of interest in this connection to compare the following remarks by two American scholars in the fields of geography and of English literature, Professors Preston E. James and John A. Kouwenhoven, respectively:

Geographers need to know more about the processes as they are measured and described in the cognate systematic fields . . . They are charged with the study of the operation of processes in the total environments of particular places [James].[24]

"America" is not a synonym for the United States. It is not an artifact . . . It has not order or proportion, but neither is it chaos except as that is chaotic whose components no single mind can comprehend or control. America is process. And in so far as Americans have been "American"—as distinguished from being (as most of us, in at least some of our activities, have been) mere carriers of transplanted cultural traditions—the concern with process has been reflected in the work of our heads and hearts and hands [Kouwenhoven].[25]

When the United States entered the First World War Bowman placed his services and the resources of the American Geographical Society at the disposal of the Government. This led to extensive use of the Society by the experts whom Colonel House brought together to gather and prepare geographical and other data for the coming peace conference. Bowman played a responsible and influential part in these studies and later at the Paris Conference, to which thousands of books and maps from the Society's collections were sent.[26] This experience gave him the incentive and background for his best-known book, *The New World*, which first appeared in 1921.[27] A broad survey of the political geography of the world at the time, the book had a remarkable success and was published in several editions both in English and in translation.

These events could, of course, be cited as supreme examples of the impact of exotic factors upon American geography. But they also had an americanistic aspect. That our government had maintained no agency of its own capable of supplying the information it needed in meeting the diplomatic crisis and was obliged to turn to a private society was, of course, due partly to the political isolationism and aloofness of the United States—to our freedom from diplomatic entanglements and from the fear of wars and invasions, and our consequent neglect of preparations for meeting them. This same international aloofness also goes far toward explaining the fact that nearly all of our basic governmental mapping, until recently, was done by civilian agencies—the Geological Survey, the Coast Survey, the General Land Office, and so forth—whereas in European countries the fundamental mapping has been done mostly by the military.

The success of *The New World*, likewise, may have sprung in part from this aloofness. Even though we had entered the war, we were less emotionally involved than were the European belligerents. Could a British, a French, a German, an Italian geographer in 1921 have surveyed the politico-geographical problems of the world in so unimpassioned a manner as Bowman did? Even if he could, would not his nationality have been prejudicial to the widespread acceptance of his book outside his own country as American nationality might, unfortunately, be comparably prejudicial today?

American Geography: Inventory and Prospect. In 1954 the Association of American Geographers brought out a book under this title.[28] In the preface the claim is made that the book "represents the combined thoughts of hundreds of professional geographers." Besides a general introduction, there are twenty-five chapters, each dealing with either a branch of geography or a geographical technique as then currently being developed in the United States.

An elaborate organization of committees was set up, and, with one or two exceptions, each chapter reflects the presumed consensus of one or more committees. Some of the committees met frequently, and there was much passing of the manuscripts back and forth for criticism, revision, and rewriting. In the fall of 1953, a general committee went over the entire text paragraph by paragraph during a four- or five-day series of sessions around a great open fire in an Adirondack lodge.

Most of the participants found all this very agreeable indeed. We Americans like to get together and talk,[29] and the Depression and the Second World War had given a mighty impetus to the doctrine that scholarly work can be better done at committee meetings than by lone scholars burning the midnight oil. Indeed, the lone scholar was coming to be regarded as ineffectual if not precisely un-American.[30] *American Geography, Inventory and Prospect* is thus a product of americanistic gregariousness—or, if you prefer, of our characteristically American genius for working together harmoniously, or more or less so. I cannot prove the proposition that Americans work together in groups any better than Englishmen, or Frenchmen, or Siamese do, but it *is* hard to conceive of any but an American group getting out such a book.

At all events, whoever may wish to make a study of the americanistic and the geoamericanistic in contemporary American geography will find this book useful. For example:

> Marketing geography is concerned with the delimitation and measurement of markets and with channels of distribution through which goods move from producer to consumer . . .
>
> The best place to develop the field of marketing geography is in business. The marketing geographer should either work for business full time or should seek to conduct research for business on a part-time consultant basis. Many problems in marketing geography cannot be studied without the private information and facilities possessed by business. Similarly the application and evaluation of ideas and methods needs the laboratory of actual business operations.[31]

Though americanistic in its emphasis on salesmanship, this does not strike me as particularly geoamericanistic—characteristic, that is, of American geography in the large—except perhaps as revealing a concern for the practical; but, whatever else it may be, it is certainly geoindividualistic, reflecting, as it does, the special views and interests of one geographer.

In other passages views are explicitly expressed as to characteristically American or European elements and trends in American geography, and in several instances branches of geography as developed in this country are compared with the same branches as developed abroad. Although the distinction that I have made between the americanistic and the geoamericanistic is nowhere drawn, as such, it is recognized implicitly here and there—for example, in the two following passages relating to the geographical study of cities. The first, from the section on urban geography, reads thus:

> Studies of American cities generally place less emphasis on the contrasts in street layout and in architectural types than do studies of European cities. This may be due to the fact that most cities in the United States lack long histories involving changes through a succession of strongly differentiated cultures. American urban studies have to deal with another kind of complexity, however. This is the rapidity of the changes taking place because of the development of the technology of transport.[32]

In the other passage, from the chapter on political geography, the authors explain that European geographers have given much attention to the study of the location of capital cities, but that American geographers have largely neglected this matter.

Perhaps the latter have felt that the location of capitals in the United States . . . was geographically less interesting than in countries where capitals appear to have become established because of intrinsic advantages of location rather than by legislative fiat . . . Or, perhaps, they have had *less interest in a topic which appears to be unrelated to current practical problems and the study of which requires intensive historical research* [italics mine].[33]

A SWEEPING GENERALIZATION

This last comment gives the clue for a generalization, albeit one of no great originality, namely, that the most americanistic as well as the most geoamericanistic, and also the most largely developed, phases of American geography are now and long have been those that would seem to bear the most obviously and directly upon practical and contemporary economic problems; and that, in so far as American geographical studies have been concerned with historical, theoretical, cultural, methodological, and other matters of which the practical bearings are obscure or indirect, the inspiration has come mostly from abroad.[34] In other words, our "applied" geography has tended to be americanistic and geoamericanistic, our "pure" geography predominantly nonamericanistic and geoexotic.[35]

Commenting on this generalization, a reader for the publisher of this book wrote, "European geographers have done far more with applied geography than have American geographers (this includes climatology, land use, and more recently planning). Perhaps marketing geography is being confused with applied geography." With the first of these comments I agree, although it does not wholly invalidate my observation; as to marketing geography, while no doubt it is capable of purity, when actually pursued is it not more frequently "applied" than "pure"? But whether or not the generalization is sound, it exemplifies what has been said to be a markedly geoamericanistic trait—addiction to sweeping generalizations, an addiction that dates from the days of Jedidiah Morse or perhaps even from those of Cotton Mather.

Reviewing fifteen American college textbooks in geography in 1937, a much-honored British geographer also made some generalizations—and be it noted carefully that sweeping generalizations, when based upon a certain amount of evidence, are not *necessarily misleading.* It would appear that in the English university a subject, if taken at all, must be studied systematically branch by branch, according to the technique laid down by specialists. Hence textbooks are rather narrowly specialized, and wide, sweeping views are condemned: the student must work toward his own conclusions, and he is encouraged to be cautious, critical, and hesitant in doing so. Nor must a further fundamental difference of outlook be forgotten. England is a small country, highly aware of the diversity that exists not only within herself but among a score or so of small or medium-sized countries that are her neighbors; the United States bestrides a continent and so uses a larger measuring stick. It is highly significant that not one of the books under review takes less than a world view. Moreover, most of them are written from a subjective, rather than (as in England) from a purely objective viewpoint; in other words, they seek to establish particular theses or to present particular points of view rather than to provide a detached record of observed facts.[36]

To conclude on a different note. Actually what has interested me the most in considering this elusive question of "What's 'American' about American geography?" is not so much the answers, as such, as the problem of how to go about finding answers. Concerning the answers themselves I retain an uncomfortable sensation of having been chasing will-o'-the-wisps in a fog of ambiguities. To me, the challenge is how to get rid of the fog—how to define and delimit the categories for which names have been suggested, how to classify specific items according to these or similar categories, how to assess the relative degrees of *americanism* and of *geoamericanism* in different works, how to tell the genuinely from the spuriously americanistic, and how, having got everything neatly classified, to draw sound conclusions as to the "hows" and "whys". If all this could be done, the outlines of a science of americanistics might (or might *not*) emerge out of the dissolving fog. (And perhaps, indeed, they are already so emerging, if we are to believe Michael McGiffert, whose interesting collection of readings from varied sources on *The Character of Americans* was published late in 1964. Professor McGiffert holds that "from the disciplines of social psychology, psychoanalysis, and cultural anthropology, linked in the study of 'culture and personality,' national character has acquired formulas which, though not altogether free of imperfection, have greatly reduced the liabilities under which the concept formerly labored . . . When the findings of the behavioral scientists are illuminated by a knowledge of history it becomes evident that never before has there been better hope and more adequate intellectual equipment for discovering what the American is, has been, and may become.")[37]

In any event, the question "what's *not* 'American' about American geography?" is as inseparable from "what's 'American' about American geography?" as the back from the front of a coin, and together they would seem to constitute one of three important pairs of questions regarding American geography as it has been in the past and as it now is today. The other pairs consist of the questions "what's modern" and "what's unmodern" and of "what's true" and "what's false" about American geography?" A good way to organize a truly comprehensive study of the history of American geography might be in terms of these three pairs of questions.

CHAPTER 9

The Heights of Mountains "An Historical Notice"

As the perpendicular elevation of mountain summits above the level of the sea . . . is still, like all that is difficult of attainment, an object of popular curiosity, the following historical notice of the gradual progress of hypsometric knowledge may here find a suitable place.

—Von Humboldt.[1]

SOME MOTIVES FOR MOUNTAIN MEASURING

THE ELEVATION of mountains continues to be "an object of popular curiosity." This springs partly from the human love of superlatives and records.[2] It is pleasurable to know the elevation of the highest point in one's town or state or country and compare it with that of Mount Everest. Mountain climbers, in particular, find solace (or pain) in comparing the heights to which they have attained with those to which others claim to have attained. On June 23, 1802, Humboldt made his famous ascent on the slopes of Chimborazo (*20,577*) in Ecuador, to an altitude that he estimated at 19,286 feet. (Recent estimates in English feet are shown in italics and, unless otherwise indicated, are derived from *The Columbia Lippincott Gazetteer of the World* [New York: Columbia University Press, 1952].) Twenty-six years later he wrote his friend Heinrich Berghaus: "All my life I prided myself on the fact that of all mortals I had reached the highest point on earth, I mean on the slopes of Chimborazo. It was therefore with a certain feeling of envy that I learned of the accomplishments of Webb and his companions in the mountains of India."[3] But Webb's accomplishments did not necessarily deprive Humboldt of his supposed climbers' altitude record. They merely showed that Chimborazo is *not* the world's highest peak. In 1880, however, Edward Whymper, who made the first ascent to the summit of Chimborazo, was led to harbor doubts that Humboldt could have gone even as high as certain ledges that he, Whymper, photographed at 18,528 feet,[4] and A. H. Bent in 1913 pointed out that if Humboldt had not exceeded that elevation his boast of having climbed higher than any other mortal would be invalid:[5] in 1677 a party from Arequipa had reached the summit of El Misti (*19,166*), and, what is more, they had been fairly energetic while there, having "exorcised the crater, cast in holy relics, celebrated Mass, and set up a great cross on the highest place."[6] In 1907 Mrs. Workman was troubled by Miss Peck's claim to have reached 24,000 feet on Huascarán (*22,205*) in Peru, a record that would have surpassed her own of 23,263 feet in the Himalayas. Accordingly she dispatched an expedition of three French topographical engineers to triangulate Huascarán. To her gratification they reported it to be only 22,187 feet high.[7]

Though it had a competitive, sporting component, Humboldt's interest in mountain heights was prompted mainly by scientific motives.[8] It was a by-product of his researches concerning variations in the general elevation of land masses and of the snow line and life zones—studies that went far toward putting the geography of plants and animals on a three-dimensional basis. Indeed, the principal incentives for the costly surveys to measure the heights of mountains that have been in progress since the early eighteenth century have been scientific, the leading motive being, perhaps, the advancement of geodesy and through it of mapping. The altitude markers embedded in the rock of many a mountain summit were put there to aid surveyors, not to amuse tourists and climbers. "Accurate triangulation over large areas made it necessary to erect signals on hills and mountain tops which could be seen many miles. The accurate computation of a meridian line made it necessary to reduce distances to sea level. This could not be done unless the heights of mountain stations above sea level were known."[9]

Despite such popular and scientific interest, the history and the bibliography of mountain measurements have been neglected; and hence this "historical notice," in which three matters will be touched upon lightly (but, perhaps, with sufficient bibliographic information in the notes to start an inquirer on his way): (1) the development of methods of measuring mountain heights; (2) the question of the world's highest known mountain; and (3) certain early mountain measurements and estimates in the Americas.

THE DEVELOPMENT OF METHODS

Peschel in his now old but continuingly useful *Geschichte der Erdkunde*[10] traced the development of the theory and practice of mountain measuring down to about 1850. More recently the late Professor Florian Cajori[9] (1929) and Professor Knut Lundmark[11] (1942) have dealt comprehensively in periodical articles with the history of methods of mountain hypsometry. Cajori distinguishes three chief ways in which the ancient Greeks sought to determine altitudes: measurement of angles, observation of the distances from which mountains may be seen at sea, and precise leveling. Of these, the last was by far the most accurate, but, though used effectively in engineering operations (the construction of canals, aqueducts, and the like), it does not seem to have been applied to mountains until modern times. Of about a dozen specific mountain altitudes on record as having been determined by ancient Greeks, even the most accurate, that for Mount Cyllene, may have been an overestimate of as much as 17 percent of the actual height (*7792*).[12] Probably the first occasion upon which the height of a mountain was measured with any close approach to accuracy was during the Cassini survey of an arc of meridian in France in 1700–1701, when the altitude of Mont Canigou, in the eastern Pyrenees within sight of the Mediterranean, was determined by triangulation with an overestimate of less than 1 percent of the present figure (*9137*).[13]

What might be called "atmospheric" methods of hypsometry—methods involving measurements either of barometric pressure or of the temperature at which water boils—were not developed until modern times.[14] Experiments were begun in the measuring of heights by means of the barometer soon after Torricelli's discovery of the principle of that instrument. But, although before 1686 John Caswell had calculated the height of Snowden (*3560*) at 3720 feet "by levelling and by means of a mercury barometer" and J. J. Scheuchzer had estimated the altitudes of certain Alpine passes and peaks barometrically in 1706–1707, barometers were not sufficiently perfected for general hypsometric use until the 1770's.[15] The Rev. Dr. Samuel Williams of Vermont wrote in 1794: "The altitude of mountains has been one of the curious inquiries which the philosophers of this century have been solicitous to determine." The most common method had "been by the Barometer," though in its application that instrument had "generally given very different altitudes to the same mountain." "Geometrical mensurations" were of "greater certainty and simplicity," but their "difficulty and expense" had "prevented any great progress being made in this part of the natural history of the earth."[16] Indeed, the heights of few mountains were actually determined by any instrumental procedures before the nineteenth century. Thus in 1807 Humboldt could count only 62 (122?) measured mountains throughout the world: 30 in the Americas, of which he himself had measured 24; 28 in Europe (including Iceland and Spitsbergen); 2 in Africa, and 2 in Asia.[17] By 1829, however, some 4000 measured heights were listed in a German encyclopedia.[18]

THE WORLD'S HIGHEST KNOWN MOUNTAIN

In former times men were prone to conjure up visions of mountains of enormous height in unexplored parts of the world. Such was the mountain that Cosmas Indicopleustes (sixth century after Christ) located in the far north to explain the succession of night and day; such was Dante's Mount of Purgatory, the supposed precise altitude of which has been learnedly discussed,[19] and also "Mt. Hercules, 32,783 ft. above the sea," which appears on a map in, and is pictured in the frontispiece of, a book of faked explorations in the interior of New Guinea published in 1875.[20]

By contrast, the actual measurements of mountain heights undertaken in ancient Greek times gave rise to a long-lasting belief among the more scientifically minded that the world's loftiest mountains are only about the height of Mount Washington, N. H. (*6288*), or perhaps half again as high. With the aid of a dioptra, a primitive form of theodolite, first the distance of a mountain from the observer and then its altitude could be determined by simply comparing similar triangles. This is believed to have been the method employed by Dicaearchus (*c.* 350–290 B.C.) in estimating the altitude of Mount Pelion (*5252*) in Thessaly at 1250 paces (6062 feet), of Mount Cyllene (*7792*) in the northern Peloponnesus at "less than 15 stadia" (9101 feet), and of Mount Atabyrius (*3986*) in Rhodes at "less than 10 stadia" (6068 feet), and probably also by a certain Xenagoras, whom Plutarch records as having measured the height of Mount Olympus (*9570*) at "10 stadia and a plethrum less four feet," or about 6163 ft. (1 *stadium* = 6 *plethra* = 600 Greek feet = 606.75 English feet).

It will be seen that Dicaearchus' figures are overestimates, the most accurate being the one for Cyllene, whereas Xenagoras' figure for Olympus does scant justice to the abode of the Gods. Dicaearchus and Eratosthenes (*c.* 276–*c.* 194 B.C.) are both said to have maintained that the highest mountains do not exceed 10 stadia (6068 feet) in height, although this does not accord with Dicaearchus' figure for Cyllene. Cleomedes (second century after Christ) wrote as follows: "Those who say that the earth cannot be spherical because of the depressions occurring in the ocean and the elevations of the mountains, speak without reason. The height of mountains does not exceed 15 stadia (9101 feet) and the depth of the ocean is not more. But 30 stadia have a vanishingly small ratio to 80,000 stadia (the earth's diameter). It is like dust upon a ball."[21]

Aristotle, in a frequently quoted passage in his *Meteorology*, gave as evidence of the supposed immense height of the Caucasus the fact that it can be seen from certain points at great distances from its base and also the purported observation that "the sun shines on its peaks for a third part of the night before sunrise and also after sunset."[22] From the latter observation later students, unlike Aristotle, drew inferences as to the specific height of the Caucasus. Thus, Scipione Chiaramonti (1565–1652) calculated that the altitude might be anywhere between 115 and 1539 miles, extremes dependent upon whether the height is reckoned with reference to a nocturnal illumination occurring when the sun is at the summer solstice or at the winter solstice, respectively.[23] A contemporary of Chiaramonti, Father J. B. Riccioli, S. J. (1598–1671), author of a learned treatise on mathematical geography and its history,[24] felt that Chiaramonti's figures were "absurd," but that if refraction is considered Aristotle's statement would imply that the Caucasus is not less than 47,000 nor more than 57,000 paces high (43–52 miles).[25] Riccioli believed it not impossible that the world's highest mountain—some peak unknown to him—might reach to an altitude of 64,000 paces (59 miles).[26]

In a table Riccioli gives 125,000 and 1,715,000 paces, respectively, as Chiaramonti's minimum when the sun is in Cancer and maximum when it is in Capricorn. According to Riccioli, 8 stadia = 1000 paces = 1 Italian mile. I have converted his data into English miles on the basis of 1 stadium = 606.75 feet. My friend Mr. O. M. Miller of the American Geographical Society, New York, has been kind enough to calculate that at the summer solstice the sun actually sets only about 15 minutes later at the summit of Mount Elbrus (*18,481*), the highest peak in the Caucasus, than it does at a sea-level horizon in the same latitude; also that a mountain in the latitude of the Caucasus would have to be about 290 miles high for the sun to shine directly on its summit at the summer solstice for two-thirds of the total period between sunset and sunrise (or about 101 miles high if what Aristotle meant was one-third of the total period). These calculations took no account of refraction.[27]

Peschel shows that between the early sixteenth and early eighteenth centuries the world's highest mountains were located by different authorities in the Urals, in Novaya Zemlya, in the Canaries (Peak of Teide in Teneriffe), and in the Alps (the mountains near the Saint Gotthard Pass, the Titlis, and finally Mont Blanc).[28] Meanwhile, Acosta's experiences with mountain sickness (*soroche*) in the Andes had induced him to believe that the South American mountains were even higher. In his *Natural and Moral History of the Indies* (1590) he wrote: "The ayre is subtle and piercing, going into the entrailes, and not onely men feele this alteration, but also beasts, that sometime stay there, so as there is no spurre can make them goe forward." The assumed absence of such effects in Europe led Acosta to "holde this place [the Andes] to be one of the highest parts of land in the worlde, for we mount a wonderfull space. And in my opinion, the mountaine of Nevada of Spaine, the Pirenees, and the Alpes of Italie, are as ordinarie houses in regard to hie Towers."[29]

In 1738, during the surveys in which he was collaborating with Charles Marie de la Condamine to measure an arc of meridian at the equator, Pierre Bouguer, by a combination of barometer readings and triangulation, measured the altitude of Chimborazo (*20,-*

577) at 3220 toises (20,592 feet).[30] Thenceforth Chimborazo was generally considered the world's highest mountain until the results of Lieutenant W. S. Webb's Himalayan surveys of 1816 became known.[31] These gave to Dhaulagiri (*26,810*) an altitude of approximately 27,000 feet. (This mighty peak in Nepal again achieved publicity when the French expedition of Maurice Herzog in 1950 sought in vain to climb it before turning to Annapurna.)[32] In 1852 the palm passed to Mount Everest, after a few years of doubt as to whether Kanchenjunga (*28,146*), which had been discovered in 1847, or Dhaulagiri were the higher.[33]

In 1852 Mount Everest was computed to be 29,002 feet high, the most famous single value in the annals of mountain hypsometry. The computation was based upon angle measurements made during the years 1849–1850 at six stations on the plains at an average distance of about 110 miles from the mountain. The work was conducted under the direction of the Surveyor General of India, Lieutenant Colonel A. S. Waugh, who later proposed, in a letter to the Royal Geographical Society in London, that the mountain be named for his predecessor, Colonel George Everest. The letter was read in 1857 at a meeting of the Society at which Everest was present. Called upon for comments, he remarked that, while appreciative of the honor, he found the proposal objectionable lest the natives of India, unable to pronounce "Everest," might confuse the name with "O'Brien."[34] "Everest," however, has stuck, and the altitude 29,002 feet remained authoritative, though sometimes challenged,[35] for just over a hundred years. In 1952–1954, however, the Survey of India carried out comprehensive geodetic operations in Nepal, much nearer the mountain than those of 1849–1850 had been. From these a new value of *29,028* feet was computed, "which, it is hoped, is not likely to be in error by more than 10 feet." Mr. Gulatee of the Survey holds that "it will serve no useful purpose to push the accuracy further by more observations, as the seasonal fluctuations of snow on the summit could well be of this order—10 feet. The older value of 29,002 feet was vague and was computed in a most incomplete manner."[36]

From time to time since 1852 there have been rumors of yet higher mountains than Everest. Especially during the 1930's and during the Second World War they were reported looming in remote Chinese-Tibetan borderlands,[37] or not far from the Indo-

Chinese "Hump," or in Greenland, or elsewhere, but all have failed of confirmation, and it is now most unlikely that Mount Everest's supremacy will again be seriously questioned.

SOME EARLY AMERICAN MOUNTAIN MEASUREMENTS

The Highest American Mountain. Although Chimborazo lost the "world's championship" in 1817, it continued to be honored as the highest known peak in the Americas until the late 1820's, when J. B. Pentland measured the Nevado de Sorata (also known as Illampu, *21,490*) and Illimani (*21,185*) in Bolivia at 25,250 and 24,000 feet respectively. During the voyage of the *Beagle* the height of Aconcagua (*22,835*) was measured at 23,200 feet. After Pentland in 1848 had scaled down his estimated altitudes of Sorata and Illimani to less than 21,300 feet, Aconcagua was recognized as the highest known mountain in the Western Hemisphere and is still so regarded.[38] Mount McKinley was first acclaimed as North America's "champion" early in 1897. William A. Dickey, a prospector, had observed it with a transit the year before from the Chulitna Valley and had estimated the height at over 20,000 feet. Triangulations by U.S. Geological Survey parties in 1898 and 1902 yielded a rounded average of 20,300 feet, which remained the official value until the 1950's, when the Geological Survey and the U.S. Coast and Geodetic Survey jointly agreed upon *20,320*.[39] Before 1897 the North American palm had been variously awarded to Mount Saint Elias, Popocatepetl, Orizaba,[40] and a goodly number of peaks in the western parts of the United States and Canada, and at an earlier period to New Hampshire's Mount Washington, North Carolina's Mount Mitchell, and even Virginia's Peaks of Otter.

Some Early Measurements in the United States. The earliest instrumental mountain measurement in the United States of which I have found a record (though without having undertaken any extensive search) is mentioned by the Rev. Dr. Jedidiah Morse, the "Father of American Geography." It was a barometric estimate made in 1777 by an unnamed observer of the height of Mount Wachusett (*2006*, per U.S.G.S.) in central Massachusetts at 2989 feet.[41] Three years later James Winthrop, Esq., measured the height of the Grand Monadnock (*3165*) in southwestern New Hampshire at 3254 feet, an overestimate of 2.8 percent. He used "a barometer and the table of corresponding heights in Martin's *Philosophia Britanica*,"[42] and his result was the most accurate North American mountain measurement (at least to my knowledge) dating from before 1816 (see Fig. 6).

One of the earliest attempts in this country to determine a mountain's altitude by "geometrical mensuration" was made by Dr. Samuel Williams in 1792. From a base line which he had measured on the "plain" near Rutland, Vermont, he triangulated the altitude of Killington Peak (*4241*) and found it to be 2813 feet above his base. By totaling measurements and estimates of the heights of falls and descents of rivers between Rutland and sea level at Quebec, Williams figured the elevation of the "plain" at 641 feet (actually about 600), thus giving the mountain a total height of 3454 feet (roughly 18 percent too low).[43] Governor John Drayton of South Carolina records a similar procedure carried out not later than 1802 whereby the altitude of Table Mountain, now called Table Rock (*3157*), in western South Carolina was reckoned at 3168 feet above its base and "not less than" 4800 feet above the sea (an overestimate of some 36 percent).[44] In 1785 Thomas Jefferson wrote that the Blue Ridge and especially the Peaks of Otter "are thought to be of greater height, measured from their base, than any others in our country, and perhaps in North America" and that the highest peak was "about 4000 feet perpendicular."[45] And in 1815 Jefferson measured the heights of the Peaks of Otter by triangulation. In a manuscript note in his own copy of the *Notes on the State of Virginia*, he described this operation thus:

> In Nov. 1815, with a Ramsden's theodolite of 3½ I, radius, with nonius divisions to 3′, and a base of 1¼ mile on the low grounds of Otter River, distant 4 miles from the summits of the two peaks of Otter, I measured geometrically their heights above the water of the river at it's base, and found that of the sharp or S. Peak, 2,946½, of the flat or N. peak . . . 3,103½.

Estimating that the base was 100 feet above tide water, he gave the higher peak (*4001*) an altitude of 3203½ feet (about 20 percent too low).[46] French surveyors had triangulated the height of Mont Canigou in the eastern Pyrenees in 1700–1701 (as already mentioned) and also that of the Peak of Teide in Teneriffe in 1771 with plus and minus errors of less than 75 and 15 feet respectively.[47] If these results make those of the Early American triangulations look amateurish (a minus error of 887 feet in the case of Killington Peak and of 798 feet in that of the higher of the two Peaks of Otter, and a plus error of 1643 feet in that of Table Rock), it should be remembered that the American measurements were of mountains in wooded country at considerable distances from the sea, whereas the French surveyors were working in more open country within sight of the sea.

The height of Mount Washington (*6288*) has probably been more frequently computed than that of any other mountain in the United States. A note published in 1867 lists some two dozen values dating from between 1784 and 1863 and ranging from 6013 up to 12,729 feet.[48] The first ascent of Mount Washington was made in 1642, and the mountain is known to have been climbed several times between then and 1784.[49] Even so, Robert Rogers, of Rogers' Rangers fame, wrote in 1765 that he had never heard of the White Mountains' having been climbed, and that he was inclined to share the Indians' belief that they were inaccessible because of some different quality of air over them.[50]

In 1784 the Rev. Dr. Jeremy Belknap, with his friend the Rev. Dr. Manasseh Cutler and several others, made a trip to the White Mountains to carry out scientific observations on the highest peak, which had not, as yet, been named Mount Washington. Belknap thus describes the difficulties encountered:

It happened, unfortunately, that thick clouds covered the mountains almost the whole time, so that some of the instruments, which, with much labour, they had carried up, were rendered useless. These were a sextant, a telescope, an instrument for ascertaining the bearings of distant objects, a barometer, a thermometer and several others for different purposes. In the barometer, the mercury ranged at 22,6, and the thermometer stood at 44 degrees. It was their intention to have placed one of each at the foot of the mountain, at the same time that the others were carried to the top, for the purpose of making corresponding observations; but they were unhappily broken in the course of the journey, through the rugged roads and thick woods; and the barometer, which was carried to the summit, had suffered so much agitation, that an allowance was necessary to be made, in calculating the height of the mountain, which was computed in round numbers at five thousand and five hundred feet above the meadow, in the valley below, and nearly ten thousand feet above the level of the sea.* They intended to have made a geometrical mensuration of the altitude: but in the meadow, they could not obtain a base of sufficient length, nor see the summit of the sugar loaf; and in another place, where these inconveniences were removed, they were prevented by the almost continual obscuration of the mountains, by clouds.

Their exercise, in ascending the mountain, was so violent, that when Doctor Cutler, who carried the thermometer, took it out of his bosom, the mercury stood at fever heat, but it soon fell to 44°, and by the time he had adjusted his barometer and thermometer, the cold had nearly deprived him of the use of his fingers . . . The sun shone clear while they were passing over the plain [probably the shoulders of the mountain at an altitude of about 5000 feet] but immediately after their arrival at the highest summit, they had the mortification to be inveloped in a dense cloud, which came up the opposite side of the mountain. This unfortunate circumstance, prevented their making any farther use of their instruments.[51]

The asterisk in the original refers the reader to a footnote in which Belknap points out that the estimate of the altitude of the mountain at 10,000 feet was made by Cutler. Belknap's own view was that his "ingenious friend" was "too moderate" and that the figure would ultimately be found to be greater. Since the altitude of Mount Washington is 6290 feet, 10,000 feet would represent an overestimate of 3710 feet, or nearly 60 percent. The measurement of the height above the base at the "meadow" (probably the Glen, near Pinkham Notch), however, was only about 1000 feet, or 22 percent, in excess. Where Cutler went completely off the track was in giving the base an altitude of 4300 instead of only 1800 feet.

Commenting on these figures, Williams expressed a more "moderate" view.[52] He estimated the altitude at not more than 7800 feet. Various reports of snow and ice on the mountain during the summer months suggested that it rose nearly but not quite to "the line of perpetual congelation." Observations in Europe had shown that in the latitude of Mount Washington this was at an elevation of 7872 feet. Owing, however, to the "greater coldness of the American climate," it must be somewhat less than this "in a similar American latitude." Therefore, Dr. Williams felt that 7800 feet was a probable maximum.

Morse in his earlier editions quoted Belknap's figure of more than 10,000 feet, but in the later editions upped this to 11,000 feet,[53] possibly influenced by the Rev. Dr. Timothy Dwight's calculation of the height at 12,729 feet on the basis of the distance from which the mountain could be seen from ships at sea.[54]

The first reasonable approximation of the height of Mount Washington (6225 feet) was made in 1816 by a party of botanists from Boston.[55] In 1851 Professor Arnold Guyot, as part of a program of barometric observations on our eastern mountains which occupied him for several summers, estimated the altitude of Mount Washington at 6291 feet (a figure that he reduced to 6288 in 1861), and closely comparable results were reached by spirit leveling carried out in 1853 by Captain T. J. Cram, assistant in the Coast Survey.[56] Thus Mount Washington's stature was determined more or less definitively within a year or so of the time when Mount Everest was first recognized as the world's loftiest eminence.

Outside the White Mountains of New Hampshire there are three other localities in the eastern United States where mountains rise to above 5000 feet: central Maine, with its lone summit, Katahdin (*5268*); the Adirondacks, culminating in Mount Marcy (*5344*); and the Southern Appalachians, culminating in Mount Mitchell (*6684*).

Morse (1796) mentioned Katahdin as "a remarkable high mountain" near the forks of the Penobscot, but he did not name it.[57] The first recorded ascent was made in 1804 by Charles Turner, Jr., a surveyor.[58] Katahdin was rumored to be 6000 to 6400, and 10,000,[59] feet high, until 1820, when the height was estimated barometrically at 5335 feet by Messrs. Loring and Odell, surveyors representing the United States and Great Britain, respectively, "under the Treaty of Ghent."[60]

In 1838 William C. Redfield (a decade later to serve as the first president of the American Association for the Advancement of Science) published an account of the initial exploration of the fastnesses of the higher Adirondacks, which makes good reading, albeit not quite on a par with Thoreau's narrative of his journey to Katahdin. "The mountains of this region," Redfield wrote, "appear to have almost escaped the notice of geographical writers, and in one of our best gazetteers, that of Darby and Dwight, published in 1833, the elevation of the mountains of Essex county, wherein rises Mt. Marcy, is stated at one thousand two hundred feet."[61] In August 1836 a party, of which Redfield was a member, carried out a reconnaissance near the headwaters of the Hudson. They took the compass bearings of a mountain seen towering above its neighbors, on which Redfield bestowed the name "High Peak of

Essex." Later in the year Professor Ebenezer Emmons of the New York State Geological Survey observed the same mountain from the opposite direction (that is, from the summit of Whiteface, some 15 miles to the northeast). During the following summer Emmons, Redfield, and others made the first ascent of the "High Peak," and measured its altitude barometrically at 5467 feet. In his official report Professor Emmons proposed that the mountain be named for the then Governor of New York, W. L. Marcy,[62] and, as the latter seems to have offered no objection, the name has persisted.

At about the same time Professor Elisha Mitchell (1793–1857) of Chapel Hill, North Carolina, made barometric observations of certain of the mountains of the western part of his state. "The Black Mountain," he wrote, "cost me nearly a week's labor fixing upon the peak to be measured and the measurement." This was determined at 6476 feet,[63] later found to be too low when Guyot estimated it barometrically at 6707 feet in 1856 and Major J. C. Turner, the next year, by spirit leveling arrived at the figure 6711 feet.[64] The highest peak was appropriately named for Professor Mitchell, who wrote in 1837 that the Black Mountain "has some peaks of greater elevation than any point that has hitherto been measured in North America, and is believed to be the highest mountain in the United States." (I first learned from a historic marker on the Blue Ridge Parkway while contemplating Mount Mitchell's cloud-capped majesty in the cool autumnal dusk, 1963, that Professor Mitchell was killed accidentally while carrying out further explorations in the Black Mountains twenty years after he had measured their highest summit.)

Between 1830 and 1864 the Wind River Mountains in Wyoming (Gannett Peak, *13,787*), certain peaks in central Colorado, Mount Hood (*11,245*), and Mount Shasta (*14,162*) were all by one writer or another reported as the highest in the United States.[65] In this reconnaissance period many mountains—for example, Pike's Peak (14,110), Fremont's Peak (13,730), Mount Lincoln (14,284), Mount Hood,[66] and Mount Shasta—were arbitrarily assumed to be 18,000 feet high, which caused Professor Brewer to comment: "Why this 18,000 feet should be applied to all mountains I don't know."[67] Mount Whitney (*14,495*) has been recognized as the highest mountain in the main part of the United States since

Clarence King in 1864 climbed to what he estimated was an altitude of 14,740 feet on its slopes; he thought that the summit was "just over 15,000."[68]

Except as regards Mount Everest and Mount McKinley I have not tried to carry the record of even the few mountains that have been considered much beyond the mid-nineteenth century, when the presently accepted heights were first closely approximated. The later progress of mountain hypsometry all over the world, with new techniques and ever-increasing precision, is another story. And, obviously, vastly more could be assembled, digested, and written about all phases of this subject—not only about the hard mathematical and physical core of the history of mountain measuring but also about the enticing record of fantasies, exploration, adventure, and experimentation that surrounds the core. But this paper is not intended as a meal—merely as a plate of *hors d'oeuvres* for those who love both books and mountains.

CHAPTER 10

On Medievalism and Watersheds in the History of American Geography

[Like the Siamese twins, this paper consists of two more or less independent parts between which there is a vital link. In its original form the paper was read on October 18, 1960, at the opening meeting during that academic year of the Columbia University Seminar on American Civilization. The Seminar had chosen "continuity versus change in American history" as its theme for the year, and, in preparation, the members had been asked to read a chapter in H. S. Commager's book, *The American Mind*, which opens with the words: "The decade of the nineties is the watershed of American history."[1]]

THE FIRST part of this paper consists of comments on long-lasting characteristics that might be called "medieval" in Early American geographical studies, as bearing on the question of historical continuity; the second part, of a discussion of the proposition that Commager's statement pertaining to the historical change in the nineties may be a trifle too categorical.

MEDIEVALISM IN EARLY AMERICAN GEOGRAPHY

Some years ago, while reading certain passages in American geographical writings of the period before 1800, I was struck by their resemblance to similar passages in geographical writings of the Middle Ages that I had run across more than forty years previously.[2] This led to attempts to assess various representative early American geographical writings (or parts thereof) in terms of their

Fig. 5. Mountains of the world, showing Dhaulagiri as the highest. From an engraving by W. and D. Lizar, Edinburgh (*c.* 1850).

3
4
2
8
15
7
6
A
1
2
3
4
6
B
C
D
E
5
F
6a
G
7
8
H
10
9
J.K.W. 1965
10
9

Fig. 6. Certain mountains mentioned in this book (redrawn from field sketches by the author). (A) Mount Washington and the White Mountains from Old Speck Mountain (*4180*), from a sketch made from the top of a spruce tree on the heavily wooded summit, September 21, 1911; the Mahoosuc Range ("*terra incognita*," see above, p. 2) appears in the foreground. (B) Monadnock from Wachusett, June 17, 1965. (C) An impression of Canigou from near Perpignan, April 26, 1949. (D) Katahdin from the east, August 16, 1953. (E) Wachusett from the north, April 1965 (hotel on summit omitted). (F) Killington Peak and Green Mountains from Acorn Hill ("The Pinnacle," *1371*), Lyme, New Hampshire, spring 1965. (G) The Black Mountains from near Bear Den (*3360*), Blue Ridge Parkway, North Carolina, September 1964; Mount Mitchell in the clouds. (H) The Peaks of Otter from the "low grounds of Otter River" whence Jefferson measured their altitudes by triangulation, 1815; April 1965. By intersections of vertical and horizontal lines through marginal reference numbers the following points may be identified: (1) Mount Washington; (2) Carter Dome (The Glen, whence Belknap, Cutler, and party climbed Mount Washington, lies between the latter and Carter Dome); (3) Chocorua Peak; (4) Goose Eye (in the Mahoosuc Range); (5) Killington Peak; (6) Mount Hayes; (6, 6a) Pico Peak; (7) Connecticut River; (8) village of Lyme Plain, New Hampshire (see above, p. xix); (9) Flat Top Mountain; (10) Sharp Top Mountain; (9) and (10) are Jefferson's "North Peak" and "South Peak," respectively.

medievalism as contrasted with their modernity, or, more broadly, in terms of their unmodernity as contrasted with their contemporaneity.

We have seen (above, p. 125) that from its literal application to a geographic region as variously delimited, the term "American" has acquired a variety of meanings, both more extended and more limited. In much the same way, from its literal application to a period of time in the history of a geographic region, also as variously delimited, the no less evocative term "medieval" has likewise acquired divers other meanings. Literally, "medieval" has reference to (1) anything that existed or happened during the Middle Ages (as, for example, a volcanic eruption or the discovery of the Azores). More specifically, it has reference to (2) circumstances that not only existed or occurred during the Middle Ages but were also characteristic of Occidental civilization in those times (for example, the building of Gothic cathedrals). These two meanings of "medieval" are inapplicable to Early American times, since the latter did not begin until after the close of the Middle Ages. Both of the other meanings, however, may be applied to divers ingredients in or attributes of Early American geography. The third has reference to (3) anything that originated in the Middle Ages and has persisted into later times (a "medieval" wall, a "medieval" town), regardless of whether or not it was characteristic of the civilization of the Middle Ages; whereas the fourth has reference to (4) anything that has existed or happened either before or since the Middle Ages that *would have been* characteristically "medieval" in sense (2) had it existed or happened in the Occident during that era. On this basis, certain aspects of ancient Egyptian and ancient Chinese civilization have been called "medieval." Such circumstances in modern times fall into two groups—those that *did* (4a) and those that *did not* (4b) originate in the Middle Ages, (4a) being, of course, a particular variety of (3).

In recent years historians such as S. E. Morison, Perry Miller, and Stow Persons have shown clearly that characteristically medieval ingredients and attributes persisted in American thought and education until late in colonial times, notably in science and theology.[3] Their writings shed light on how conceptions in the domains of physical and biological geography were determined or colored by Aristotelian scholasticism, but to an extent and in a manner that has never been specifically investigated, so far as I know. The particular earmarks of "medievalism," however, that have most attracted my attention are not so much these direct survivals as the presence of characteristically medieval states of knowledge, habits of thought, and attitudes of mind. Those of a theological nature are illustrated below, Chapter 14. Others will be considered very briefly here under three headings. (In a personal letter Professor Wilbur Zelinsky of the Pennsylvania State University suggests another: "The absence of the notion of progress and the perfectibility of knowledge that is implicit in just about all current geographical endeavor.")

The Immensity of Terrae Incognitae. Early American geography was closer akin to the geography of the Middle Ages than to that of today in the vast extent of its literal *terrae incognitae* (see above, pp. 68–70). Some of the unknown tracts lay right near at hand, as in the Southern Appalachians, the Adirondacks, and northern Maine. Of the Adirondack wilderness, which remained largely unexplored until the 1830's, Lewis Evans wrote in the 1750's, "This country by reason of Mountains, Swamps, and drowned Land is impassable and uninhabited."[4] Jedidiah Morse's description of northern Maine also conveys a sense of mystery; he refers to a "remarkable high mountain" near the forks of the Penobscot.[5] Thus Katahdin looms in anonymous majesty.

As nature abhors a vacuum, so the cartographer and geographer of the Middle Ages sought to fill the *terrae incognitae* of those times with the fruits of speculation and fancy, and much the same spirit persisted as long as there were extensive unexplored regions. Morse, in the 1796 edition of his *American Universal Geography*, copied from Guthrie (1795)[6] an account of a vast inland sea, comparable to the Baltic or Mediterranean, in northwestern Canada, the lineal descendant of the many inland seas in North America that sailors like Verrazano and Captain John Smith, map makers like Sanson and Delisle, and sinners like Francis Billington (the "*Mayflower's* bad boy") had imagined.[7] Indeed, until toward the early years of the twentieth century—earlier or later in different countries—serious interest in geography as a field of scientific and scholarly investigation was centered mainly upon *terrae incognitae* and what they contained, and upon getting rid of them. Narratives of exploring expeditions filled most of the geographical periodicals and the issues that aroused the liveliest controversies at geographical-society meetings had to do with the things that might exist in *terrae incognitae* and with technical problems of pioneer exploration and cartography. Today, instead of quarreling over theories about imaginary lakes in North America and Africa, the Great Southern Continent, and open polar seas,[8] geographers wrangle about more subtle and purportedly more sophisticated matters of definition, methodology, and interpretation with regard to the geographical study of the so-called known world. This change has occurred simultaneously with the professionalization of geography, and is one of the most radical changes in the nature of Occidental geography that has occurred since the Middle Ages. Beazley wrote a great book in three volumes on the history of Occidental geography from A.D. 300 to 1420 and entitled it *The Dawn of Modern Geography*.[9] A history of the final elimination of *terrae incognitae* during the last 1100 years might, with equal propriety, be entitled *The Sunset of Medieval Geography*. Everything depends on the point of view.

Lack of Specialization and of Sophistication. The second category of "medieval" characteristics in Early American geography may be ascribed to the unspecialized nature of that body of knowledge. As in the Middle Ages, there was no organized profession of geographers and, correspondingly, almost no attempt to define geography, write its history, develop its methodology, or give serious consideration to problems relating to the organization and presentation of geographical information and ideas—problems of the kind now so frequently, volubly, and sometimes tiresomely discussed. The components of an unspecialized body of knowledge are likely to be regarded as the common property of all interested parties. Hence the typical treatise on geography published in the Middle Ages was a mosaic of excerpts from earlier authorities, strung together without acknowledgment. Today, this "isn't done"; specialization has developed among scholars a proprietary interest in their own inimitability, which puts such easygoing predativity outside the pale. Scholarly predativity is still practiced, of course, but with greater delicacy and finesse. In this respect, the mores of Early American writers on geographical subjects seem nearer to those of today than to those of the twelfth century; but, even so, the "Father of American Geography" in the second edition of his volume on the Eastern Hemisphere cribbed long passages, footnotes and all, from British works without letting the reader know that they were not his own.[10]

A spirit of scientific adventure, with its antithesis, a spirit of scientific restraint, animated the geographical writers of the Middle Ages as it still does the geographical writers of the present, but these spirits did not operate in those days quite as they do now. During the Middle Ages and long thereafter there was a great dearth of reliable quantitative data on which to base and against which to verify geographical theories. Hence the spirit of scientific adventure prompted the use of the relatively few available data in the elaboration of theories that often strike us as venturesome in the extreme, as, for example, in an estimate of the altitude of the Caucasus at anywhere from 115 to 1539 miles. It will be seen later in this book (Chapter 13) that this sort of "drawing a longbow" with mathematics cropped up in Early American geography.

During the Middle Ages the spirit of scientific restraint was much less likely to challenge a geographical theory because it was founded on inadequate evidence or faulty reasoning than because it contradicted an established and authoritative doctrine, particularly in the domain of theology, and this was also true in the Early American period (see Chapter 14). Few American geographers today would hesitate to accept or propound a theory simply because it was theologically unorthodox. However, it is another question whether the spirit of scientific restraint operates less powerfully today than it did in the Middle Ages in curbing theorizing that is economically or socially or politically or academically unorthodox, or is opposed to the mores and conventions of institutionalized scholarship. Maybe we differ from our predecessors of the Middle Ages chiefly in being afraid of different things, and the seemingly ultramodern may often be merely an old medieval acquaintance in modern dress. But certainly the spirit of scientific restraint inhibits contemporary geographers from taking liberties with evidence that they would not have hesitated to take had they lived in the Middle Ages or the Early American period.

Interest in Curiosities and Wonders. The last kind of medieval characteristic of Early American geography to which I should like to call attention was a frank and refreshing interest in curiosities and wonders for their own sake. In his chapter on Norway Morse described, under the heading "Uncommon Animals, Fowls, and Fishes," the kraken, a sea monster "said to be a mile and a half in circumference," and added "the existence of which fish being proved, accounts for many of those phenomena of floating islands and transitory appearances at sea, that have hitherto been held as fabulous by the learned."[11] Like the mighty, carnivorous mammoth, which Jefferson, crediting Indian rumors, believed might still be found in the forests to the northwest of the upper Missouri,[12] the kraken was a belated cousin of the roc, the unicorn, the mantichora, and the umbrella-footed men of the medieval maps. However allured by the unusual he might be, the American geographer of today would consider it naïve to devote a section of his textbook or regional study to "Curiosities," as did our Early Americans. In Morse's opinion, "Mecca and Medina are curiosities only through the superstition of the Mahometans."[13] Caves and caverns were among the most popular curiosities; Governor John Drayton described one in "an extremely rocky and romantic country in South Carolina said to have been the hiding place of some tories" during the Revolution "as it still is of wildcats, wolves, and other vermin."[14] Jeremy Belknap, Jedidiah Morse, and others described an "active volcano" in Chesterfield, New Hampshire, not far from the future site of the world's largest basket store (see below); but President Dwight, who visited the place in 1798, called it a "very humble" volcano.[15] Today, American geographers occupy themselves with curiosities of a different kind, such as the distribution of mountain moonshine stills or of farmhouses with connecting barns or of covered bridges or of the use of manure, and they call it "cultural geography" or "cultural animal geography," as the case may be.[16]

In Early American geography there was much that could be regarded as medieval. There might even be some in American geography today.

HISTORICAL WATERSHEDS IN AMERICAN GEOGRAPHY

There is a gigantic sign near Putney, Vermont. It says, "The world's largest basket store." Such a categorical, superlative, unproven, and perhaps improbable assertion might be called a *categorilla*, a term that suggests both the categorical and the superlative—for the gorilla is superlative among the apes just as we scholars are among humans.

(I am collecting scholarly categorillas, such, for example, as the following, in which the italics are mine: "The spring below the mound of Jericho is *literally* the source of human civilization."[17] It is a magnificent specimen of *Categorilla historica gigantica*, or Toynbee's Categorilla. The next one belongs to a different species: "The history of science is the *only* history which can illustrate the progress of mankind."[18] Students of the language of the categorillas have observed that these creatures are prone to introduce their warcries by clucking sounds such as: "logically," "obviously," "palpably," "it is self-evident that," "it is a universal rule of scholarship that." The effect is to soothe the hearer into unquestioned and uncritical acceptance of what follows.)

Out of context, Professor Commager's assertion that "the decade of the [eighteen] nineties is *the* watershed in American history" (italics mine) looks like a rather formidable categorilla. Read where it belongs in his book, however, it tames down, as other ferocious animals do when restored to their native haunts. At all events, if the statement is a categorilla, it is a sincere, worthy, and hard-working one, useful as a clarifier of thought, systematizer of data, and instigator of discussion. But did the nineties constitute *the* watershed, or only one among other watersheds, in American history?

On many occasions watersheds have been perceived or imagined by geographers in looking back over the development of geographical knowledge and teaching. Here are some examples:

In the preface to the second edition (1793) of his *magnum opus* Morse asserted that

> before the Revolution . . . we humbly received from Great Britain our laws, our manners, our books, and our modes of thinking . . . To . . . receive the knowledge of the Geography and internal state of our own country, from a kingdom three thousand miles distant from us . . . would certainly be a disgraceful blot upon our literary and national character.[19]

So Morse had set about to correct this, and, like other reformers, yielded to the temptation to magnify the significance of his reforms by minimizing the merits of the matters reformed.

The American Geographical Society, the first institution of its kind in this country, was founded in 1852. Addressing its members in 1859, the Rev. Dr. J. P. Thompson said: "This science of geography, once regarded as a mere matter of dry but necessary information, is now seen to have vital relations to man in his physical, mental, social, and moral development."[20]

About thirty years later Professor Libbey noted that Guyot's textbooks had helped convert geography from "a dry collection of disjointed facts" into "an exact science . . . anything but the superficial and fragmentary summary of natural and political divisions"[21] that it had been 25 or 30 years previously.

Of a Conference on Geography held in Chicago in 1892 Professor Dryer said that it had "affected the teaching of geography as profoundly as the Pleistocene ice age affected the distribution and evolution of plants, animals and men."[22] And with regard to the report of that Conference we read that "a new element has been introduced into the science by the modern school. Rational geography has supplanted mere description."[23] There are few more damnatory words than "mere" and "description."

In 1899 Professor Dodge stated that, as a result of Frye's textbooks, "geography has largely ceased to be the driest and most uninteresting subject in the school curriculum."[24]

In 1910 Professor Tower explained that geography had recently been "given the necessary qualities of a coherent science, with a perfectly distinct field not covered by any other existing science" and was no longer open to criticism as being merely "a heterogeneous agglomeration of dissociated facts."[25]

In 1952 the British geographer J. N. L. Baker spoke of British geography in terms applicable, as well, to an attitude not unknown on this side of the Atlantic:

> One of the most remarkable facts of the last fifty years is the growth of academic geography in this country [Great Britain], and many of the geographers who are reaping the fruits of that growth display two characteristics—arrogance and despondency. Looking back on the past they thank God that they are not as those so-called geographers were. They know what geography is: their ancestors did not.[26]

Though it is not quite historiographically *comme il faut* to speak of "laws of history," I shall label what I have been talking about as the "law of the disparagement of the past," of which the motto might be *de mortuis nil nisi malum*.[27] Disparagement may be positive, as in the foregoing examples, or negative, as when past developments pertinent to and necessary for an understanding of a subject are disregarded. In the words of T. S. Kuhn, "The temptation to write history backward is both omnipresent and perennial," but "the depreciation of historical fact is deeply, and probably functionally, ingrained in the ideology of the scientific profession."[28]

In 1939 an elaborate methodological study entitled *The Nature of Geography* by Professor Richard Hartshorne was published, a work that has had a considerable influence on the thought of a good many American geographers, including myself. In the Introduction we read: "Geographers are wont to boast of their subject as a very old one . . . But often when geographers in this country discuss the nature of their subject . . . one has the impression that geography was founded by a group of American scholars at the beginning of the twentieth century."[29] The book did much to correct this misapprehension, but neither in it nor in the author's smaller book on the same subject published in 1959[30] is the probability considered in any detail that American geography in the twentieth century may have had important roots in American geography in the nineteenth.

Somewhat the same observation may be made regarding a symposium entitled *American Geography: Inventory and Prospect*, which was published in 1954 in commemoration of the semicentennial of the Association of American Geographers and in the editing of which I had a hand (see above, pp. 135–137). It contains 26 chapters, each prepared collaboratively by a committee (though in most cases one individual set the tone and did most of the work). Of some 1550 references given, only 4.1 percent were published in the United States before 1900.

The "Normal Period of Living Memories." That a large change in the direction, scope, and magnitude of American geography as a branch of scholarship set in during the nineties and nineteen hundreds seems incontestable, but the patent fallacy of the implication that it marked the emergence of something unprecedented led to the inclusion of the following sentences in the Foreword to *American Geography: Inventory and Prospect:*

> The fifty years of the Association of American Geographers coincide with the normal period of "living memories," that is, of personal

experience added to what was learned at first hand from teachers and older associates. This easily results in magnifying the significance of developments in the last half-century, or at least in over-emphasizing the gap between this last period and the preceding period. While there is no doubt that the founding of the Association signalized a great change and advance in American geography, it did not make a beginning out of nothing.[31]

The theory that the "normal period of living memories" extends back about fifty years might be supported by some such reasoning as the following. The actual duration of the period of living memories of any given individual depends both on personal variables—age, retentiveness of memory, past associations with older people, and so on—and on our definition of the term "living memories." Let us define it so as to include, at any particular time, what one has learned from early childhood through experience (as distinguished from study or research) and what one has learned of earlier or contemporary times by listening to others talk about them from personal experience. On this basis, most of us have scrappy and inexact memories of family history, politics, wars, calamities, and so forth, going back to the childhoods of our parents, uncles, aunts, and even grandparents. As far as one's profession or special field is concerned, however, it may be assumed that, on the average, the period of living memories opens more or less simultaneously with one's birth—but not, of course, because one's conscious personal memories begin to accumulate that soon. After starting graduate work in their early twenties, most scholars begin to store away living memories concerning their fields, partly through firsthand observation, but partly also through partaking in the memories of reminiscently inclined teachers and older associates. If these are assumed very arbitrarily to embrace on the average a preceding period of twenty years, they would date back to about the time of birth of the listener. The median age of the membership of the Association of American Geographers is approximately 43,[32] but, because the memories of the older members tend to carry somewhat disproportionate weight in shaping widely prevalent conceptions, attitudes, and misconceptions regarding the past development of the field, it is perhaps not out of line to place the opening of the normal or composite period of effective living memories for the profession as a whole at somewhere around 1916, or about fifty years ago.

Some Questions about Historic Watersheds. Accordingly, four or five questions may be posed:

(1) Is there evidence in fields other than that of American geography that scholars have tended to regard the nineties, or some other period, as marking a major watershed in the development of their subjects? My son, who has written a book on the American short story, gives what looks to me like excellent evidence that there *was* such a watershed about 1920 separating the short stories of the period 1890–1920 from those of the last forty years.[33]

(2) Might the view of the nineties as "*the* watershed of American history" be colored by the circumstance that for many Americans whose writings have tended to mold our views of the recent past, the nineties mark the last years of the pre-memory eons—the great dark domain of time recoverable only through deliberate, conscious efforts of historical study?

(3) As successive new generations of scholars grow in maturity, do they tend, in looking back over their fields, to think they see nonexistent watersheds, or to exaggerate the height of existent ones, at about the time of their birth or shortly before? Do such watersheds move down the stream of time, much as actual waterfalls move upstream?[34]

(4) Are scholars beset by a temptation to overemphasize not only their own periods of living memories but other periods that they have studied? If so, might not this temptation be regarded as one manifestation of a universal "law" of human behavior that calls for magnification of the relative importance of one's own times and of everything else with which one is especially familiar—city, village, state, country, religion, irreligion, denomination, school, family, class, club, hobby, profession, lack of profession? Might not this "law of proprietary magnification" operate with respect not only to a scholar's discipline but to all of his lesser specialties and pet theories, and, naturally, to the periods with which he is most familiar? And might it not also operate in a negative sense by inducing minimization of periods with which he is unfamiliar, and therefore exaggeration of the heights of the watersheds between the familiar and the unfamiliar periods?

Here I should like to throw in an observation for what it is worth. In discussing the question of the relative significance of continuity versus change in history, historians are considering a problem comparable to one that geographers wrangle about with respect to space. The inadequacy for many purposes of geographical interpretation of the obvious and fairly clean-cut subdivisions of the earth's surface as delimited by coastlines and political boundaries has led to countless ingenious attempts to differentiate regions on the basis of other criteria—in particular, regions assumed to be coherent and homogeneous in terms of whole combinations of "natural" or "human" circumstances, or of both. Correspondingly, the unsuitability for many purposes of historical interpretation of the clean-cut subdivisions of time as defined by the durations of dynasties, reigns, administrations, wars, eras of peace, and so forth, has led to attempts to differentiate periods on the basis of other criteria—in particular, periods assumed to have been relatively stable in terms of large combinations of human circumstances.[35] Thus the "regional concept" in geography[36] has an analogue in the "period concept" in history, with an implication of relatively uniform and homogeneous periods separated by "watersheds," and geographers and historians alike argue over the reality and utility of such regions and periods, and over the nature, locations, and sizes of the zones and "watersheds" between them. In both fields, moreover, the "law of proprietary magnification" has often induced intemperate enthusiasm for specific systems and methods of subdivision and corresponding adverse reactions, in which the emphasis has been laid upon the superior merits of concepts of spatial uniformity and temporal continuity.

I joined the staff of the American Geographical Society in 1920 and from then until I retired in 1956 was fortunate in being able not only to devote a good deal of time to studying the history of geography but also, through firsthand experience, to learn not a little about the nature both of geographers and of geographical knowledge and how it develops. Fieldwork on the part of the student of geography-as-actuality is considered indispensable in order that his studies may have an air of authenticity. For a student of the history of geography, service on the staff of the Geographical Society was the equivalent of the fieldwork in "unadulterated" geography, if such there be. In the late forties and early fifties, I wrote a history of the Society, which was published in 1952 in commemoration of the institution's centenary. Before beginning work on it, I knew very little about the Society during its 68 years prior to 1920; what little knowledge I had gathered from conversations with older staff members was to the effect that until the First World War the Society had been a weak, pedagogical, insignificant, and uninfluential institution. Subsequent documentary study led me largely to revise this view. While retaining the concept of a watershed in the Society's history at about the time of the War when Dr. Isaiah Bowman became the Society's first Director, I came to perceive that the watershed was not nearly as high as I had thought, that the Society had played a continuingly influential role in the advancement of geography, and that its founders and early leaders deserved much credit for substantial achievements. In my book I tried to bring this out. We have in this, therefore, a little case history illustrating some of the points that I have tried to make in this paper. My personal attitudes regarding the history of the Society had been a sort of microcosm of much more widespread attitudes affecting the study of much vaster domains of history. My own ignorance of what had happened at the Society before the dawn of my personal memories concerning it combined with what I had been told about it in those days to induce a picture in my mind of a much greater change than had actually occurred. Both of the so-called "laws"—"the law of proprietary magnification" and its converse, "the law of the disparagement of the past"—had operated to exaggerate the altitude of the watershed between the historic past and the period of living memories.

CONCLUDING QUALIFICATIONS

In conclusion, be it said emphatically that these remarks are not to be construed as a criticism of Commager's splendid book. As regards that particular work they may well be irrelevant. Thus, Professor Zelinsky has written me as follows:

> I firmly believe that there has been a sharp, irreversible, and rather sudden transition in all the arts and sciences from medievalism, classicism, romanticism, or whatever the most convenient catchword may be to the present mode of thought and expression, and that it occurred somewhere in the years from 1890 to 1920. In some instances we can pinpoint the event with great precision—certain critical art exhibits, literary manifestos, a couple of notorious musical premieres, the fateful days that the quantum and relativity theories were published, or the astonishingly simultaneous rediscovery of Mendel's theory by three different investigators. There were premonitory rumbles in many cases,

to be sure, such painters as Bosch or El Greco who were several centuries too soon, or in music and literature such lonely pioneers as Berlioz, Stendhal, and Melville. Whatever the reasons, there has been a large measure of simultaneity in the transition to uniquely modern forms . . . The timing of all this was peculiarly significant to the United States, since this transitional period corresponded so perfectly to the era in which the nation came of age as a political, economic, and cultural power. For American geographers, the period is triply meaningful since their profession first attained professional status at that time. The delimitation of historical periods is as perilous as drawing regional boundaries, but we do have our Himalayas, our Sierras, our sharp lines between Maghreb or the Nile valley and Sahara, or between the Laurentian lowlands and uplands. Let us grant the historians their 1890 watershed.

These thoughtful remarks, however, would not seem to invalidate either the general thesis that there is a tendency in scholarship to oversimplify complex conditions and events or the proposition that, whatever actually happened in the 1890's, Professor Commager's suggestive watershed metaphor has provided a concise text for a not-very-concise sermon on habits of thought that unquestionably have at times affected the interpretation of both historic and geographic actualities.

CHAPTER 11

Daniel Coit Gilman
Geographer and Historian

ALTHOUGH geography of some sort had been taught in American colleges and universities since the mid-seventeenth century,[1] probably not until Arnold Guyot (1807–1884) was appointed professor of geology and physical geography at Princeton in 1854 was it taught to any substantial degree in a modern spirit and more or less independently. Daniel Coit Gilman, however, appears to have been the first person in the United States to hold a professorship devoted, nominally at least, to geography unlinked with any other subject.[2]

This fact alone is sufficient reason why he should arouse the curiosity of an American geographer a hundred years later. But Gilman has other claims on our attention. Of three American university presidents of the middle and late nineteenth century who were profoundly interested in history and geography, Gilman was the most active as a geographer and did the most for the later advancement of geographical scholarship.

(The others were Jared Sparks and Andrew Dickson White. Of Sparks, president of Harvard, 1849–1853, Ralph H. Brown wrote: "One may gather from his writings that he maintained a deep interest in geography throughout his long and useful life.[3] White, first president of Cornell, 1867–1885, was a lifelong friend of

Gilman's from their college days. He wrote Gilman in 1859: "You have chosen a noble field—one of whose existence few among us have any inkling, one which Ritter first showed me and toward which I have looked with longing eyes ever since. And I have to do a little at it myself, for there is ever present to me Dr. Arnold's dictum that to teach History without Geography is impossible. Some of my students do work which would delight you."[4])

Gilman was born at Norwich, Connecticut, in 1831, and graduated from Yale in 1852, the year of the founding of what is now the American Geographical Society. From 1856 to 1865 he served as assistant librarian and librarian of Yale College, from 1863 to 1872 as professor of physical and political geography in the Sheffield Scientific School, and for a few months in 1860 as general secretary of the American Geographical and Statistical Society of New York (as the Society was called until 1871). In 1872 he accepted an appointment as the second president of the University of California, where he remained for three years, until called to the presidency of the Johns Hopkins University in Baltimore, then being founded. After a quarter of a century of influential leadership in this position, he retired in 1901, only to be summoned once again to the task of launching a great organization for the advancement of learning, the Carnegie Institution of Washington, of which he was president from 1902 to 1904. Meanwhile, in 1896–1897 he participated in the labors of the United States commission concerned with the Venezuela–British Guiana boundary dispute.

At memorial services held at the Johns Hopkins University in 1908, soon after Gilman's death, James Bryce commented on his special qualifications "for the work that fell to him of determining the lines upon which this new seat of learning ought to be developed," especially his ability to keep "in close touch with very different lines of study and inquiry. He was in touch with what we call the sciences of Nature . . . And . . . equally in touch with what we call the Human studies—literature, history, political science, economic science."

Although Bryce did not mention it, the study of geography calls on its devotees to be ever in touch with both "the sciences of Nature" and "the Human studies," and it may not have been wholly fortuitous that Gilman, who began his career as a geographer and always remained one at heart, should have become the builder of a great university. At any rate, it suggests a parallel between his life and that of Isaiah Bowman (1878–1950),[5] who also taught geography at Yale as a young man, supervised the work of the American Geographical Society, advised the government with regard to international boundary disputes, and served as president of Johns Hopkins. The parallel should not be stretched too far, however. Until he went to Baltimore at the age of fifty-six, Bowman's foremost interest was in geography, and he attained to the highest eminence in that field. Gilman was first and foremost an educational planner and executive. According to Dr. Abraham Flexner:

> Our latter-day geographers, while not doubting his command of texts available at the time, point out that he made no original contributions to his subject. He was, as a matter of fact, throughout his life, not an investigator, but a great educational executive—probably the greatest we have as yet developed. His ideas were not original; he sought them here, there, and everywhere, combining and adapting them to American needs and conditions.

Although this may help to explain, it does not wholly justify, the scant attention that "our latter-day geographers" have given to Gilman. Not only does the record of his interest in geography shed light on the nature and growth of geographical studies during his lifetime, but his constructive work as an educational planner and builder has affected the condition of geography today in ways of which most American geographers are perhaps unaware.

GILMAN'S EARLY GEOGRAPHICAL ACTIVITIES

After receiving his bachelor's degree at Yale, Gilman enrolled as a graduate student at Harvard in order to study lexicography, and while in Cambridge he lived in the home of Arnold Guyot. His association at this time with Agassiz and Guyot no doubt inspired him to abandon lexicography for geography and "shaped, to no small extent, his views on education." During a two-year sojourn in Europe (1853–1855) he studied for a while in Berlin under the African explorer Heinrich Barth and made "lasting friendship" with Carl Ritter and Georg Heinrich Pertz, two of Germany's most distinguished scholars of all time in the subjects that interested Gilman the most, geography and history. This journey to Europe, the first of ten, broadened his familiarity with current European studies in geography, history, and many other subjects. Among the early fruits of these contacts was a series of "Geographical Notices" that Gilman contributed to the *American Journal of Science and Arts* (1858–1864, 1869).

Gilman's teaching at Yale (and presumably also at the University of California, for he did some teaching there while serving as president[6]) embraced not only geography but history and economics as well. Half a century later, when Bowman was professor of geography at Yale, his subject had become somewhat better established, and Bowman was neither obliged nor tempted to range outside its limits. Gilman, besides his regular classroom instruction, did a good deal of special lecturing. In 1859–1860 he gave six talks on geography in the Normal School at New Britain, Connecticut, and in 1867 he held a Sunday-evening course for Yale students on Biblical geography (like Guyot, he had once seriously considered making the ministry his career). In 1871 he pinch-hit for Professor Guyot, who was ill, by delivering twelve lectures before the senior and junior classes at Princeton, and in that year and the next he addressed the American Geographical Society on current geographical work in the United States.[7] For six months in 1860 he had commuted from New Haven to New York to administer the affairs of the Society as its general secretary, a fact apparently overlooked by all his biographers. (Possibly the Reverend Joseph P. Thompson (1819–1879), one of the most active members of the Society's Council from 1857 to 1871, got this job for Gilman, who was his brother-in-law. In December 1860, when Gilman gave up the position, "it was ordered that there be added to the items of debt due by the Society a claim of $70, amount advanced by Mr. Russell, for six months commutation of D.C. Gilman Esqr. General Secretary on the New Haven Rail Road Company."[8])

Lecturing in the Congregational Church of Oakland, California, on "Berkeley: The Bishop and the Site of the University," President Gilman in 1873 discussed

> the proper laying out of the college city and . . . advised a proper regard for the topographical features of the landscape, preserving and utilizing the irregularities of the surface. He would have carriage ways, roads for equestrians, and broad areas of approach . . . [also] a commodious hotel, with restaurant attached that would provide meals for families.

GILMAN ON THE NATURE OF GEOGRAPHY

"Geography is one of the plainest and most straightforward of sciences . . . There is nothing of theory or hypothesis about it." So declared Professor J. D. Whitney[9] in 1875. Gilman would not have subscribed to this. His thoughts often turned to the human and historical aspects of geography, where theories and hypotheses abound. Nevertheless, there is evidence to show that fact-gathering and fact-describing appeared to him as the most urgently needed, and in many ways the most interesting and satisfying, forms of geographical enterprise. Thus his "Geographical Notices" in the *American Journal of Science* dealt mainly with discovery and exploration and in particular with scientific expeditions. They consisted of some 110 notes, about half of which had to do with expeditions in unexplored or little-known areas and a third with topographic, geologic, and hydrographic surveys, chiefly in the United States. The others reported on such miscellaneous topics as the publications of the Royal and American Geographical Societies, the life of Carl Ritter, Guyot's wall maps of the continents, current estimates of the world's population, and current works on the history of geographical discovery. In the first "Notice" (1858) Gilman called attention to Dr. Petermann's geographical *Mittheilungen*, which had been started three years earlier but was little read in this country at the time (the American Geographical Society did not acquire a set until 1873). He made use of this journal in compiling his notes, and Petermann valued Gilman's comments on geographical work in the United States as the most convenient source of information on the subject.

Some ninety-three of the notes dealt with specific regions, and the relative quantity of notes devoted to the different large parts of the world accords so well with the corresponding distribution of papers published 1852–1865 by the American Geographical and Statistical Society in its periodical[10] as to suggest that the accordance may reflect a common broad range of interest on the part of geographically minded Americans rather than merely Gilman's personal interests and the institutional interests of the Society. In both cases the United States received the most attention, followed in order by Asia, Africa, and the Arctic, where explorations were then much in the public eye. The poor showing of Europe implies that the complex geography of that continent was of relatively little concern to Americans at this time, or, perhaps, that geographical interest everywhere tended to center on exploration (see above, pp. 156–157).

Livingstone, Burton, Speke, and Barth in Africa, Schlagintweit in Central Asia, McClintock, Hayes, and Hall in the Arctic, Leichhardt in Australia, Page on the Plata, were among the explorers mentioned. A goodly number of the notes dealt with explorations in the African lake region and in search of routes for a ship canal across the Central American isthmuses, for a transatlantic cable, and for a transcontinental railroad across North America. Gilman called attention to the measurement of the height of Mount Everest at "over 29,000 feet above the level of the sea"; to the discovery of K2, the mighty peak in the Karakoram, second loftiest in the world; to Professor Whitney's mountain measurements in the western United States (Mount Shasta was then believed to be the highest in the country and Popocatepetl the highest in North America); and to Guyot's surveys in the Appalachians from Georgia to Maine.

In his address in 1871 before the American Geographical and Statistical Society Gilman emphasized the need of a "National Topographical and Hydrographical Survey," a matter much on his mind at this period. After attending the International Geographical Congress in Paris, he wrote the trustees of the Johns Hopkins University on August 23, 1875, that he had spent three days examining the map exhibit,

> which was vast and comprehensive, and well arranged in one of the remaining wings of the Palace of the Tuileries. It has often seemed to me desirable that one of the specialties of the Johns Hopkins University should be the training up of young men to be the surveyors and engineers by whose skill our interior country will be mapped—in its topographical, geological, agricultural and economical aspects.

He was impressed by the "great topographical maps of England, France, Switzerland, Austria, Prussia, Russia, the Scandinavian peninsulas—and the remote countries tributary to or explored by these powers. Our own country appeared to great disadvantage." Honors, however, had been "awarded to the Coast Survey, Dr. Hayden, the Census, Gen. Walker (for his statistical atlas) and to some other American works." Professor Fulmer Mood has shown that the series of census maps of density of population, which partly inspired Turner's frontier hypothesis, was initiated as a result of pressure that Gilman and other Sheffield professors at Yale brought to bear on the Census Bureau in 1872, stimulated by the example furnished in August Meitzen's atlas of Prussia (1869). Mood maintained that Turner misinterpreted the symbols on the census maps as signifying limits of settlement and therefore frontier lines.[11]

A printed synopsis of Gilman's course of lectures at Princeton in 1871[12] presents the salient features of the panorama of geography as he viewed it when at the height of the geographical part of his career. He designed the course "to incite young men to geographical inquiries and to give them a glimpse of modern geographical progress." Although tantalizingly brief, the synopsis gives enough details to make reasonably clear the range and nature of Gilman's conception of geography.

Like Guyot, he urged the study of geography in the field and from large-scale maps, photographs, and other firsthand documents. He spoke of "the value of training the eye to observe and the hand to delineate characteristic natural features." Maps were par excellence "the language" of geography, and in the introductory lecture he described projections, conventional signs, methods for representing elevations, and other cartographic technicalities. He urged that geography be studied "in the modern spirit of scientific exactness." In the course as a whole he had much more to say of the physical aspects of the earth than of its historical and human geography. The second lecture considered

> I. The Earth, as upheaved and moulded into the various forms of reliefs . . . II. The Ocean, as determinative of horizontal dimensions or continental contours . . . III. The Atmosphere, as mediator between the sea and the land. IV. The manifestation of varied structure on a small scale in any region, as distinct as in the continental masses. V. Mountain-Structure as dependent for its forms on the character of the constituent rocks . . .

The succeeding lectures bore largely on the physical features of the lands, often with references to their relation to human life. The eleventh talk, for example, on Palestine, dealt chiefly with physical structure, relief, and drainage and concluded with a "Comparison of California and Palestine in their structure,—suggested by Prof. Brewer and others," but apparently without mention of the resemblances between the two regions in climate and vegetation. Indeed, nowhere in the course does Gilman seem to have had much to say about the geography of either climates or plants, subjects in which his contemporaries Guyot, Whitney, and George Perkins Marsh were better versed. On the other hand, he devoted most of the lecture on the Mediterranean region to the results of recent oceanographic (or, more properly, thalassographic) investigations there; oceanography was a "live" subject in the mid-nineteenth century.

In the final lecture, "On the Forces Which Are Now Modifying the Surface of the Earth," he discussed, among other things, "Indications of secular changes produced by astral or cosmical influences," the agency of water (whether "destructive" or "constructive"), of fire, of the atmosphere, of organic life ("Coral islands as studied by Darwin and Dana"), and of man. Under the last heading he commented on the "Opening of Canals and changing of river courses," on "Railroads, as performing the transit work of rivers," on the "Destruction of Forests; *e.g.* results in France," on "Draining of Marshes. Low-lands, etc.; *e.g.* Harlaem lake," on "Restriction of Rivers, (as of the Po, the Mississippi, etc.), and of the Ocean (as in Holland) by dykes," and, finally, on "Culture of the Soil." He must have based some of this on Marsh's *Man and Nature*, to which he gives a reference, thus showing that, unlike a good many American geographers of the early years of the present century, he was familiar with "the most important and original American geographical work of the nineteenth century."[13]

In lecturing on the United States Gilman divided the country into

> seven distinct regions, extending from North to South; 1. Atlantic harbors,—2. Adjacent low-lands,—3. Appalachian mountains,—4. The Great Basin [that is, the central lowland, including the Mississippi Valley and the Great Lakes region],—5. The Cordilleras, from the Rocky Mountains to the Sierras,—6. The Pacific valleys, (California and Willamette),—7. The Pacific seaboard, (including the Coast Ranges).

This foreshadowed the more detailed and systematic essays of Gannett, Bowman, Fenneman, and others in the subdivision of the country into physiographic provinces. Gilman discussed the "natural advantages of each region," and in the talks on different parts of the country he suggested breakdowns of the major areas.

For all but two of the Princeton lectures Gilman gave short lists of references, of interest no less for what they do *not* include than for what they do. The works mentioned were not college textbooks, but books, periodicals, and maps intended for mature readers.

(As bearing on geography as a whole he mentioned Ritter's "Lectures on Comparative Geography" and "Geographical Studies," Guyot's "Earth and Man," Humboldt's "Cosmos" and "Views of Nature," Reclus's "La terre," and Marsh's "Man and Nature." He also called attention to "Petermann's Mittheilungen, 1855–71, with numerous maps," to the *American Journal of Science*, the *American Naturalist*, and the "Journals of the Geographical Societies of London, Berlin, Paris, New York, etc." For geology and physical geography he referred to Dana's "Manual of Geology," Lyell's "Principles of Geology," Agassiz's "Geological Sketches," and "Tyndall on Glaciers.")

For his general lecture on the United States he gave no references whatsoever, but for the three regional lectures on this country he referred to "Guyot on the Appalachian System," the reports and maps of various federal and state geological, topographical, and other surveys, and "Recent Travels." In other words, the student was directed to the original sources, and Gilman's failure to refer to a single comprehensive synthetic work on the geography of the United States or of any subdivision thereof was not due to inadvertence. In his address before the American Geographical Society in the same year he had said: "We look in vain for anything which is satisfactory" in the way of "literary or descriptive generalizations of the knowledge" attained through the surveys and presented on the maps.[14] About twenty years were to pass before Whitney and Shaler filled this gap by getting out books on the regional geography of the United States.[15]

The fourth lecture, "On Some of the Geographical Problems of Today," opened with comments on "Agencies at work. (*a*) the explorers and investigators in the field. (*b*) the journalists, cartographers, and students [but not the 'geographers'] at home." Most of the talk bore on sixteen "Interesting Problems." Six years later, in a letter to the president of the American Geographical Society, Gilman recommended certain themes and speakers for a proposed celebration of the Society's twenty-fifth anniversary (Table 1). As in his "Geographical Notices" and addresses of 1871 and 1872 before the Society, the emphasis here was almost wholly on the empirical aspects of geography—the gathering of facts through explorations and surveys and their representation on maps. Theoretical studies are implied only by the headings "The Physics of the Globe" and "The laws of climate," and history figures only with regard to the history of geographical discovery. Among the individuals named, Guyot alone would be considered a "geographer" at the present time. Nearly all the others were naval officers, geologists, missionaries, archeologists, or other explorers or travelers.

THE RELATION OF GEOGRAPHY TO HISTORY

Dr. Flexner's comment that Gilman was "not an investigator" would be deserved only if "investigation" were confined to field work and laboratory work. Flexner himself pointed out that during Gilman's European journey of 1853–1855 he "accumulated meticulously large stores of knowledge regarding education, the history of learning and science, the achievements of great scholars and scientists, the development of educational institutions at every level." If this is not "investigation," what is? Gilman kept up this inquisitive ardor throughout his life. It prompted Judge Henry D. Harlan, at the exercises held in President Gilman's memory in 1908, to speak of his "enlightened interest in subjects less directly connected with immediate utility—geographical exploration, archaeological research, biographical and historical inquiry"; and Mrs. Gilman wrote of her late husband as a "great reader . . . His books were indeed his tools, and he handled them with accuracy and skill."

Gilman's writings and the record of his career open up many interesting matters with respect to the relations between geography and history, among them (1) his views on the need of studying history as an aid to the study of geography; (2) the bearing of certain of his writings on the history of geographical exploration and geographical ideas; and (3) his service on the Venezuela boundary commission.

Table 1. Geographical Problems and Topics Suggested by Dr. Gilman, 1871 and 1877.[a]

Geographical Problems, 1871	Geographical Topics, Agencies, and Recommended Speakers, 1877
The Physics of the Globe	
The telegraphic determinations of longitude	
The measurement of the figure of the earth	
The laws of climate	Continental meteorology Smithsonian Institution, Joseph Henry, J. H. Coffin, C. A. Schott, U.S. Signal Service
The progress of topographical Surveys	State surveys California: J. D. Whitney, New York: J. T. Gardner, Mass., N. J., Penn., etc.
Hydrographic measurements, and especially "Deep sea Soundings"	Deep-sea soundings A. D. Bache, M. F. Maury, J. M. Brooke, G. E. Belknap, North Pacific Expedition
The history of early European discovery on the American continent	
The existence of an open Polar Sea,[b] and the possibility of approaching it	Arctic E. K. Kane, I. I. Hayes, C. F. Hall
Alaska	Alaska W. H. Dall
The Western Mountains of North America	Cordilleras Pacific Railroad explorations, F. V. Hayden, J. W. Powell, G. M. Wheeler, Clarence King, etc. Measurement of Peaks, etc. A. H. Guyot, Clarence King, W. H. Brewer, etc. Rivers and Lakes Gen. A. A. Humphreys
Routes of transit from Europe to Asia—including the American isthmus investigations	Isthmus of Darien Com. Daniel Ammen
The Basin of the Amazon	
The Lake region of Central Africa, with relation to the overflow of the Nile	Africa Charles Livingstone, H. M. Stanley, Charles Chaillé-Long
The "Recovery" of Palestine	Palestine Edward Robinson, W. F. Lynch, W. M. Thomson, J. T. Barclay, Palestine Exploration Society
The structure of Inner High Asia, and possible routes of transit from the interior to the Indian coasts	
Central Australia	
	Japan M. C. Perry, F. L. Hawks, W. E. Griffis, etc.
	Travels, etc. E. G. Squier, Bayard Taylor

The Study of History in Relation to the Study of Geography.[16] As Acting School Visitor in New Haven in the late fifties young Gilman "suggested that observation, drawing, and history be associated with geography" in the higher classes. Some years later, worried over political difficulties at the University of California, he wrote his friend Andrew D. White:

> At the present, my mind turns more to the direction of editorial life,—either in the newspaper line, or in establishing a monthly to be called "Earth and Man,"—and to be devoted to the discussion of modern social problems,—with reference both to the physical and outward circumstances of human society and to the historical and institutional antecedents. I merely give you a hint of the scope,—but you will quickly expand it. There is no such journal in the world. The graphic methods of illustrating social and historical papers could be most efficiently introduced. It might be made a journal of anthropology,—not of man's body only, but of all his social progress. Such work as Walker is doing for the U.S. Census could be expanded and multiplied indefinitely. History and political economy might be treated on a scientific basis.

(In 1857 at the University of Michigan, White "gave a short course of lectures on physical geography, showing some of its more striking effects on history; then another course on political geography, with a similar purpose." As president of Cornell, he gave a general course on the history of European civilization, in which he emphasized the importance of geography and chronology and which he illustrated with the aid of German historical and physical wall maps.[17])

Later geographical studies in this country might have been very different had Gilman carried out this plan. The title of the proposed journal was that of Guyot's much-read book, and though Gilman wrote of it as a "journal of anthropology," his idea was quite as

---

[a] The "problems," with their corresponding "topics" and so on, have been rearranged in a more or less logical order. "Problems" are from Gilman, *On the Structure of the Earth* (see below, p. 316, note 12). "Topics" are from an unpublished MS. memorandum in the library of the American Geographical Society, in which no initials are given; I have supplied these or first names. Articles will be found in the *Dictionary of American Biography* on all the individuals listed except the Reverend Charles Livingstone (1821–1873), who is probably meant rather than his more famous brother, David (whom he accompanied in Africa, 1858–1863, and of whose death, also in 1873, Gilman could hardly have been unaware, although he might well not have known of Charles's), and Dr. James T. Barclay, an American medical missionary, whose book on Jerusalem, *The City of the Great King*, was published in Philadelphia in 1858.

[b] See Chapter 6.

much for a journal of historical and human geography. (Once before, Gilman had been tempted to give up his academic career for editorial work of a geographical nature. In 1860 he was invited to take charge of a department on a New York daily to deal with "what I most prefer, the relation of foreign countries to our own, not only European but all others, in which I should hope to make available all my geographical studies.")

In each of the seven regional lectures in his course at Princeton, Gilman had something to say of history. In the one on the Rhine and vicinity, after briefly comparing Rhine, Rhone, Po, and Danube, and giving particulars on the Rhine "as a natural Water-course," he enlarged on its functions "as a route of transit, for trade, armies, and tourists, from Sea to Summit," "as the seat of social influence and culture," and "as a Political Boundary." In the lectures on the United States such topics as "The industry of the region [the Appalachians and the Atlantic Seaboard] as determined by physical causes," "The metropolitan position of New York and reasons for its permanent ascendancy," and the Mississippi Valley "as promoting the Union of the States" indicate an environmental-deterministic slant. We have seen, however, that Gilman was also alive to the concept that geography should reckon with "man's role in changing the face of the earth." In the controversies over man's relation to the earth Gilman would seem to have favored holding to a middle ground. At Princeton he branded two tendencies as "both faulty: (*a*) To overlook and disregard all physical limitations to national progress, largely exemplified in the writings of the early historians and political geographers. (*b*) To undervalue the power of man in subduing the earth, and in overcoming obstacles to his progress." With these he contrasted "(*c*) The intermediate position of the modern German writers" (presumably Humboldt and Ritter, to whom he referred elsewhere in his synopsis).

In 1898, at a time when the Spanish-American War had turned the attention of Americans toward the Pacific, Gilman returned to Princeton and delivered an address on "Books and Politics" at the dedication of a new library building. Declaring that "the question of to-day, the question of the decade, it may be the question of the twentieth century, is the attitude of the United States toward the islands of the sea, *de insulis nuper repertis*," he urged that it was a question for "universities and university men to illuminate by the experience of mankind." (In 1886, Gilman had called attention to the fact that Professor Francis Lieber of Columbia University, the great political philosopher, "in a letter to Secretary Seward, at the close of the Civil War, presented a strong plea for the reference of international disputes to universities.") "Go to your books, young men," Gilman advised, "and study geography and history. Resort to the Library by whose reorganisation you are now enriched. Begin the study of Oceana [and] with your geography do not fail to read political history."

The History of Geography. The vast field of the history of geography has been cultivated unevenly, and seemingly fertile parts have been abandoned to weeds. For example, the history of scientific field work and of documentary research as contributing to geographical knowledge would seem to have received less attention than it might deserve, and, indeed, an interesting survey might be made of the scope of different studies in the history of geography and of its neglected areas. The parts in which the public and the historians alike have tended to take the most interest are those concerning the discovery, exploration, and mapping of unknown and little-known areas—something perhaps more characteristic of the United States than of Europe, because we have been nearer in time and in place to *terrae incognitae* than the modern European nations have been. Certainly, these were the parts of the history of geography that Gilman found most appealing.

A few of his "Geographical Notices" bore on current historical studies relating to the discovery and early cartography of America. The richest and most luminous material, however, is in his charming book, *The Life of James Dwight Dana: Scientific Explorer, Mineralogist, Geologist, Zoologist, Professor in Yale University.* Dana had taken a prominent part in the United States Exploring Expedition (1838–1842) under Lieutenant Charles Wilkes, and nearly a third of the biography is devoted to the expedition—its origin, organization, personalities, routes, adventures, and scientific achievements, especially as relating to Dana. The *Life* is an eloquent and authentic tribute to the memory of a friend whom Gilman greatly admired. "If any one in our day can be called a cosmographer, Dana may have that title." "To a great extent [the volume] is based upon Dana's own writings." By including in it letters Dana had exchanged with other scientists—Darwin, Agassiz, Guyot, Geikie, Gray—and comments by others on Dana's work and ideas, Gilman made the book into a useful collection of "original source materials" relating to the development of geology, biology, and the doctrine of evolution during the late nineteenth century. Especially notable are some of the letters and comments regarding the positions that Dana, Gilman, and others took in the discussions then raging over the relation of science and theology. These matters, of course, are not extraneous to the history of geography, unless geography must be deemed immune to the impact of nongeographical ideas.

(In the fifties Dana had engaged in a controversial discussion with Professor Tayler Lewis, who had "defended the literal interpretation of the word 'days' in the first chapter of Genesis, and cast aspersions on the teachings of science and scientific men." Gilman did not feel it "worth while, forty years later, to review the merits of this controversy." In 1897, speaking of the fifty years of scientific work just completed at the Sheffield Scientific School, Gilman said: "Not a word was spoken in disparagement of classical culture, nor a word of religious controversy." And seven years later, at the University of Wisconsin, he rejoiced that the "pulpit, no longer speaking of science in derogatory terms, is almost ready to say that science is the handmaid of religion.")

In his pregnant studies of the dynamics of volcanism and of the growth of coral reefs, begun during the Exploring Expedition and continued long after, in his investigations of the structural and regional geology of New England, and in his larger surmises about the origins of the continents and the history of rock formations, Dana's thought was creative, independent, and replete with geographical implications; and when he sought to defend geological science against those who would interpret the first chapter of Genesis literally, his views were in accord with the strongest intellectual currents of his time. On the other hand, in seeking to reconcile the scriptural account of the Creation with the geological interpretation of the earth's history, he was more old-fashioned. Gilman's comment discloses something of Gilman's own position as well:

It was largely under the influence of Guyot that Dana continued to discuss the Mosaic cosmogony. These two friends, impressed by the Bible lessons of their youth, endeavored to see in the poetical expressions of the first chapter of Genesis exact statements of those natural phenomena which the eye of science recognizes in the development of the universe. It is easy for us to see that they were fettered by a mode of interpreting the Hebrew Scriptures that is not now tenable, and that they were supported in this method not only by the traditions of early life, but also by the dominant theology of the communities in which they dwelt.[18]

The Venezuela-British Guiana Boundary Case.[19] At several times in the world's history geographers or historians or historians and geographers working together have influenced the course of events in a large way. The writings of Richard Eden and Richard Hakluyt, it has been said, were a major cause of the "expansion of England" in the reign of Elizabeth I, and those of Captain A. T. Mahan "stimulated a naval race among the great powers."[20] Mackinder's "Heartland" hypothesis inspired Haushofer, who in turn inspired Hitler. The counsel of Haskins, Seymour, Lord, Bowman, and other historians and geographers at the Paris Peace Conference of 1919–1920 was reflected in the redrawing of the boundaries of Central European states.[21] The work of the boundary commission on which Gilman served in 1896–1897, if less momentous in its consequences, was of the same general order.

In 1895 a long-smoldering dispute between Great Britain and Venezuela over the latter's boundary with British Guiana burst into flame. In the United States the British claims were feared as menacing the Monroe Doctrine, and when Congress asked President Cleveland to name a commission to study the issue there was a "violent reaction," war was threatened, and the stock market declined. Cleveland appointed three lawyers, Andrew D. White, and Daniel Coit Gilman to serve on the commission. In his *Autobiography* White recorded that G. L. Burr, historian (of Cornell University), Justin Winsor, librarian of Harvard, and J. F. Jameson, historian (of Brown University), rendered expert services in the matter of history, geography, and maps. Marcus Baker, of the United States Geological Survey, "also aided us from day to day in mapping out any territory that we especially wished to study." Professor Burr was sent to Europe, where he "drew treasures from the archives of the Hague" and in London "soon showed his qualities . . . and these were acknowledged even by some of the leading British geographers, who had at first seemed inclined to indulge in what the Germans might call 'tendency' geography."[22]

(The notes in the great cooperative *Narrative and Critical History of America* of Justin Winsor [1831–1897] are an invaluable source of bibliographical and cartographical information on the discovery and mapping of this continent.[23] Baker [1849–1903], topographer, geographer, and authority on Alaska, served as secretary of the U.S. Board on Geographic Names and was one of the founders of the National Geographic Society. Burr [1857–1938] gave a course in historical geography at Cornell.[24] Studies might well be made of Winsor and Burr as geographically minded historians.)

With, at last, the cooling of passions, a treaty was concluded early in 1897, referring the issue to arbitration, and the boundary commission was disbanded. The Tribunal of Arbitration met in 1899 and on October 3 of that year rendered a decision on the whole favorable to the British claims. Meanwhile, in 1897, the United States government had published the commission's *Report and Accompanying Papers*, in four volumes, a monumental product of collaborative, if hasty, scholarship, in which a multitude of books of history, geography, and travel and of maps are listed and discussed. Although most of this is of specialized scope, the "Report upon the Cartographical Testimony of Geographers," by Severo Mallet-Prevost, secretary of the commission, is of slightly greater general interest (Mallet-Prevost [1860–1948; B.S., University of Pennsylvania, 1881] was a lawyer and at one time president of the Pan-American Society). Since Winsor, in his broader survey of the maps of the area, did not try "to classify the various boundary lines; nor to trace their genealogy; nor to ascertain their meaning," Mallet-Prevost undertook to do so:

> It is a task which involves the interpretation of maps, and which seeks to read therein the thoughts and intentions of their authors . . . The human mind often works in unconscious obedience to motives which, though but feebly apprehended at the time, are yet potent to determine a particular line of action. *Geographers are not free from the operation of this rule* [italics mine]; and if a subsequent study of their work shall at times disclose their intentions with a clearness of which they were themselves possibly not fully conscious, the fact remains that,

consciously or unconsciously, their work was shaped by those intentions, and that we are warranted in basing our conclusions upon that assumption.[25]

According to A. D. White, "President Gilman . . . was given charge of the whole matter of map-seeking and -making . . . His experience as a geographical student made his work especially valuable, and his influence is to be seen throughout the whole fourteen volumes of historical and geographical work which the Commission furnished to the Arbitration Tribunal at Paris."[26] At the memorial service for President Gilman in 1908, Charles J. Bonaparte, Attorney General of the United States, said: "By the wise choice of President Cleveland he aided in enlightening the foreign policy of our country, [and] in safeguarding the peace of the world."[27]

GILMAN'S INFLUENCE ON AMERICAN GEOGRAPHY

Although Gilman did not contribute many original ideas to geography or establish any school or department or professorship of geography at the Johns Hopkins University, his career there was not without influence on the development of American geographical scholarship.

For one thing, geography was by no means untaught at Johns Hopkins during his administration, and its teaching received his support and encouragement. Of the departments then in operation, those of geology and of history did the most geographical work.

In his inaugural address in 1876 President Gilman called attention to the imminent need of "an accurate survey of the area of the United States, corresponding with the ordnance and geographical surveys of Great Britain, France, Switzerland, and Germany." He spoke of the difficulty that the heads of our governmental surveys were then encountering in "finding men qualified enough to carry forward efficiently such work in all its manifold departments, astronomical, geodetical, topographical, meteorological, geological, zoölogical, botanical, economical," and he suggested that "if our University can provide instruction in these departments of physical research, looking forward to the future development not only of Maryland and the Atlantic seaboard, but also of the entire land, it will do a good service." In pursuit of these ideas a Department of Geology was gradually organized, with a professor of geology, Dr. William B. Clark, in charge (1894–1917). Collaborating with other organizations, it engaged in a number of quasi-geographical undertakings, including the Maryland Geological Survey, established in 1896. The Geological Survey, as summarized by Gilman, undertook topographic mapping, the training of young men as surveyors, and "the diffusion of knowledge concerning the characteristics and resources" of Maryland. Members of the Johns Hopkins Geology Department took part in the resurveying of the state boundaries, and one wrote a history of the Mason-Dixon Line.[28]

In the Department of History and Political Science, "physical and historical geography" was made "the basis of instruction in historical and political science," and a "Geographical and Statistical Bureau" was established, which maintained a collection of maps, atlases, and other geographical and statistical works.[29] Three distinguished historians who took their doctorates at Johns Hopkins in President Gilman's time were later to influence the development of geographical studies in this country in a variety of ways—John Franklin Jameson, Frederick Jackson Turner, and Charles Homer Haskins.

(As Director of the Department of Historical Research at the Carnegie Institution of Washington from 1905 to 1928 Jameson [1859–1937] was the leading spirit in the planning and supervision of the work that led to the publication in 1932 of Charles O. Paullin's *Atlas of the Historical Geography of the United States*.[30] Turner's famous frontier hypothesis has, of course, given rise to one of the strongest controversial currents in the stream of American historico-geographical thought. In the words of Samuel Eliot Morison, "Johns Hopkins University, long the friendly rival and pacemaker of the Harvard Graduate School, gave us one of her greatest sons in 1902 in the person of Charles H. Haskins"[31] [1870–1937; see above pp. 3–4]).

More potent, however, in its influence on the later development of American geography than this teaching of geography at Johns Hopkins was a revolutionary change which the example of that university helped bring about. Of this, Professor Jameson wrote in 1932:

The Johns Hopkins University, dedicated primarily to research and the training of young men for research—an entirely novel dedication in that day—began instruction in 1876; and some new ingredient in the *Zeitgeist*, or in the academic atmosphere, brought it about that other universities speedily caught the enthusiasm for graduate instruction, for the professional training of college and university teachers, and thereby for that introduction to the processes of investigation without which the teacher is ill-qualified for independent thinking.[32]

Specifically, the change has shown itself chiefly in the borrowing from Germany of the seminar, or seminary, method of training students to do research, in the setting up of the Ph.D. as almost a *sine qua non* for appointments to good teaching positions in respectable institutions, in the recognition of an unwritten law in many institutions that promotion is contingent on the continued conduct of research and publication of its results, and in the organization of professional societies and the founding of scholarly journals.[33] This state of affairs, which spread to the field of geography in the early decades of the present century, is now almost as familiar to American geographers as the air they breathe, and by some of the younger ones as much taken for granted as though it were a part of the eternal order of being. Actually, it stems largely from the influence of the Johns Hopkins University of President Gilman's day, and therefore from *his* influence; for this was by far the most powerful factor in determining the policies and standards of that institution.

The effects have been felt especially in certain of the "larger and better-established state universities"[34]—for example, Michigan, Wisconsin, and California—and also in two privately endowed universities founded after the founding of Johns Hopkins—Clark in 1888 and the University of Chicago in 1891. These are the institutions that have trained more geographers in recent years than any other universities in the country.[35] Hence it is not too far-fetched to say that most American geographers who may read this paper would probably have been quite different in their intellectual outlook and scholarly habits had it not been for Daniel Coit Gilman.

CHAPTER 12

Miss Semple's "Influences of Geographic Environment"

NOTES TOWARD A BIBLIOBIOGRAPHY

BOOKS are not unlike people, and some books, like some people, deserve biographies. A "Life and Times of *Influences of Geographic Environment*," if well written, appreciative but not uncritical, and racy but not unfair, would interest American and British readers who enjoy thinking about the nature of geography. Professor Tatham's dictum that *Influences* "will always stand out as one of the great formative works in the geography of the English-speaking world"[1] is not exaggerated (except possibly in the "always"). For this reason I have attempted here to touch on some of the matters with which a full-length bibliobiography of *Influences* might deal, and, in the last section, on a broader question.

FAMILY BACKGROUND AND PERSONALITY OF "INFLUENCES"

A *bibliobiography* is a biography of a book—something quite different from a *biobibliography*[2]—and as such should deal with its subject's ancestry, character, personality, and career.

(If *biobibliography* means "a kind of bibliography combining biographical matter" [Webster], why not *bibliobiography* for a biography of a book, *cartobiography* for the biography of a map, and so on? I know of no comprehensive, full-length bibliobiography of a geographical work, though many short or incomplete studies could be cited. Biographies of books are usually presented incidentally to other subjects, as in biographies of the authors, histories of ideas, and introductions to new editions. In our field more bibliobiographical attention has perhaps been devoted to Ptolemy's *Geography* and Marco Polo's *Travels* than to any other single work.)

Influences belongs to an old family, the Environmentalists, whose lineage has been traced down from the writings of Homer, Hippocrates, Herodotus, and other ancients.[3] The family's history has often been sketched,[4] but its intricate genealogy has never been fully explored. Despite some eccentric and perhaps even disreputable members, I am inclined to rank the family among the F.F.G.'s (First Families of Geography). Neither *Influences* nor its siblings and cousins were the first of the family to settle in this country (the siblings being Miss Semple's other books[5] and the cousins notably the writings of Ellsworth Huntington, R. DeC. Ward, and Griffith Taylor of her generation, and of the somewhat younger Roderick Peattie). There were earnest American Environmentalists in the eighteenth century and even earlier.[6] However, I have not run across any evidence that *Influences* was directly descended from them. Most of its immediate forebears can be identified through the notes. There are some 1450 notes, almost entirely brief references to the sources, with many repeats. Without a count, it seems safe to estimate that Miss Semple consulted about a thousand different works for *Influences*. The notes give the impression that use was made predominantly of serious works of travel or exploration, and of secondary works of comprehensive or even encyclopedic scope in the fields of history, geography, and anthropology. Many of the items of information presented as facts are based on interpretations already somewhat removed from the original observations they reflect. There is an almost complete lack of critical discussion of the relative reliability of the sources consulted.

The title page explains that *Influences* is based on "Ratzel's System of Anthropo-geography," and the preface pays generous tribute to Ratzel's inspiration. Despite her otherwise meticulous documentation, nowhere does Miss Semple indicate specifically the parts of the *Anthropogeographie* that she followed. Thus it is not possible to distinguish between the facts and ideas for which she, and those for which Ratzel, deserves the credit.

(Chisholm, reviewing *Influences* in 1912, wrote of it as "enriched" by a wealth of observation not in Ratzel.[7] Sauer, on the other hand, in 1940 criticized it obliquely: "There is far more in the unknown Ratzel than in the well-publicized one,"[8] and in an unpublished paper Jan O. M. Broek has developed the same idea, even to the extent of suggesting that *Influences* "entombed" Ratzel in the minds of American geographers by causing them to associate him with deterministic extremes not characteristic of his thought as a whole when all his voluminous writings are fairly considered. Harriet Wanklyn, a British biographer of Ratzel, has maintained that Miss Semple's "rendering after the passage of time, and in another language, gave an impression which hardly did justice to Ratzel's total and developing approach to environmental influences . . . Yet if Ellen Semple's rendering of a problematical piece of writing seems (forty [*sic*] years later) to mislead, who has done any better? What would many Anglo-Saxon readers know at all of Ratzel's *Anthropogeography* if it were not for her interpretation of it? And what had she at the time in the way of contemporary and relevant American scholarship to guide her?"[9])

The character and mentality of *Influences* are so well known to American and British geographers that I need not linger over them. This opinion is based on a questionnaire sent in the spring of 1961 to fifty-eight geographers (forty-four American, thirteen British, and one French), ranging in age from twenty-five to eighty-five. Fifty-three replied, and of these only three admitted to not having read the book, though the others, it is clear, had done so with differing degrees of frequency, recency, intensity, enjoyment, and totality. Among other things, the questionnaire asked what the respondents knew of uses that have been made of *Influences* in the teaching of geography in American and British universities and colleges, and the replies, though scrappy, implied a surprisingly widespread use not only before but since the Second World War. This leads me to think that most readers of this article will probably know enough about *Influences* at firsthand to agree that the book is written in an opulent style and makes absorbing reading for those who are not too predisposed against it either by their own innate critical spirit or by what teachers or reading may have told them is the Way to Think. "Categorillas" abound in Semple (see above, p. 160), but they won't hurt you as long as you don't allow yourself to be irritated or taken in by them. Here are two examples. "Genuine nomadic peoples show *no alteration* in their manners, customs, or mode of life from millennium to millennium" (p. 509; italics mine). Since the Arabian Bedouins' manners, customs, and mode of life have been considerably altered within my memory, this would seem categorillic. So also would the assertion that the shepherd folk of the desertic and steppe regions of the Old World "receive from the immense monotony of their environment the impression of unity. Therefore *all of them*, upon outgrowing their primitive fetish and nature worship, gravitate *inevitably* into monotheism" (p. 512; italics mine). It is a moot point whether the beginnings of monotheism were due to the influences of a desert environment.[10]

In *Influences* such adverbs as "inevitably," "always," and "everywhere" are favored, but rare indeed are such expressions as "perhaps," "possibly," "on the other hand," or "it might seem." Notwithstanding the avowals to the contrary in the preface, the book has an "Olympian air"—to borrow a term used by Andrew H. Clark in answer to my questionnaire.

("Small, naturally defined areas . . . *always* harbor small but markedly individual peoples" (p. 172); "territorial expansion is *always* preceded by an extension of the circle of influence which a people exerts" (p. 187); "islanders are *always* coast dwellers with a limited hinterland" (p. 421); "*wherever* an energetic seafaring people with marked commercial or colonizing bent make a highway of the deep, they give rise to this distinction of coast and inland people on *whatever* shores they touch" (p. 273; all italics mine). Restored to their context some of these assertions might seem less categorillic.)

THE CAREER OF "INFLUENCES"

The "career" of a book begins, ordinarily, immediately upon its birth, and the most vigorous and vital years are the years of infancy, as was true of *Influences*. A book responds to its environment by multiplying in number of copies more or less proportionately to its ability to make friends and interest people. Qualitatively, however, it remains the same (unless, of course, there are sudden mutations when new editions are published) with a constancy that may be embarrassing to the author and refreshing or disappointing to the reader. When you go back to a book that charmed you when you were young, only to find that some of the old charm is gone, it may be like revisiting at the age of seventy the enchanted hilltop pastures where you played at the age of ten, to find them overgrown with brush and dead trees. But the *book*, you may be sure, is unchanged; there the trees have died only in your own mind, the brush has accumulated in your own imagination. Moreover, a book seldom "dies." Although untouched for years, as long as a copy exists anywhere a book, like a bear in winter, continues to "live" dormantly. There is always a chance that some alert scholar will "discover" it and enhance his reputation by mentioning it in a footnote or even arranging for its republication in paperback. Such republication could happen to *Influences*, now that the gentle breezes of "Neoenvironmentalism" are beginning to stir.

In dealing with a man's career, a biography must take account of two sets of variables: a changing man and the changing reactions of his environment *to him*. Since the content of an unrevised book does not change, its bibliobiography needs to consider only the environment. Thus a full-length bibliobiography of Miss Semple's book would presumably consist in the main of a thorough record, analysis, and interpretation of changes in the uses and effects of the book and in the attitudes of readers toward it.

During the last fifty years the leading idea developed in *Influences* has aroused more discussion and controversy among English-speaking geographers than has any other one leading idea in our field (a categorilla?). An account of this discussion in all its ramifications would make an interesting case study in that burgeoning new subject, the History of Ideas. To do full justice to the subject would call for an even longer book than *Influences* itself, of which the main theme would be "Who Killed Cock Robin (or Tried to), and Did He Succeed?" The metaphor is not inept; for some of the would-be assassins do seem like twittering sparrows as compared with the robin himself in all his magnificent self-assurance. The book, of course, would have to consider the sparrows' actions severally and collectively; whether the arrows were clean or poisoned, and how the attempted assassinations have been proclaimed—whether with indecisive chirps or pontifical croakings.

In its early years did *Influences* actually exert as much influence as has been alleged? If so, why? Why did a tide of adverse criticism set in? To what extent have published and unpublished attitudes toward it been based on objective, fair-minded, independent weighing of the evidence as provided by the text and by accurate first-hand knowledge of the nature and needs of scholarship and education? Has acceptance or rejection been to any extent a manifestation of a sheeplike following the leader or of delight in contemplating the profundity of one's own scholarship? Have middle-aged and elder geographers been prone to criticize *Influences* unfavorably because their revered but now palpably naïve professors were so enthusiastic about it? Have younger geographers ever been tempted to look on it with approval simply out of contrariness, because they were taught to think it was not all it should be? These are not wholly unanswerable questions, and, if they could be answered with some measure of factual accuracy, the answers might be worth reading.

The following comments by Peter Laslett of Cambridge University on a discussion of "forgotten sources in the history of science" seem pertinent in this connection: "The rediscovery of what unfashionable predecessors knew is not a habit confined to historians . . . But . . . there is a particular temptation . . . to retell the story of the past as our generation ought to hear it told (or wants to hear it told). This is, I suggest, one reason why it is that what is already known gets buried, why fashions of interpretations seem to proceed as if no one knew anything until this latest and most interesting of all points of view made itself manifest. We overlook what our predecessors discovered because we are so busy preaching our doctrine in opposition to the doctrine which they were preaching."[11]

When *Influences* first came out, it was acclaimed by reviewers in this country, one of whom wrote that it was "unquestionably the most scholarly contribution to the literature of geography that has yet been produced in America"; it was also commended for "the enormous breadth of reading" it disclosed, and for its methodical character and "completeness."[12] And in Britain Professor Chisholm was hardly less enthusiastic.[13] "Scholarly" and "complete," like "meaningful," "significant," and "geographical," are ambiguous terms. One is seldom altogether sure whether "scholarly" should be interpreted quantitatively or qualitatively or both. Can one be scholarly with little knowledge? Unscholarly with gigantic erudition? Scholarship, moreover, is sometimes unconsciously equated with "what I know about."

From the outset, politely adverse comments on *Influences* were not lacking, but they became less polite as the years passed. (Of late, however, there has been a return toward politeness.) Chisholm objected mildly to the omission of Ratzel's materials on the vegetal and animal worlds and on the structural works of man, and also to the inclusion of Ratzel's discussion of geographic location (*Lage*) as leading "only to obscurity"[14] (the latter is the part that two of my younger questionnees, Fred Lukermann and Philip W. Porter, found the most original and worth while).

The *Journal of Geography* reviewer in 1911 felt that "the author's tendency toward the use of figure personification" was "somewhat unfortunate in a scientific treatise."[15] For example: "She [the earth] has entered into his [man's] bone and tissue, into his mind and soul. On the mountains she has given him leg muscles of iron to climb the slope; along the coast she has left these weak and flabby" (p. 1); "deserts and steppes lay an arresting hand on progress" (p. 509). But such usage seems to be harmless rhetoric rather than "notorious" evidence of a "concept of an animistic purposeful Nature," as more recently suggested.[16] Need the writer of scientific works be under constraint to use only such figures of speech as even the most literal-minded reader will not take literally? Perhaps.[17]

In 1912 the historian O. G. Libby, reviewing *Influences* in the *American Historical Review*, welcomed it as bridging the gap between geography and history and approved of its "constant emphasis . . . on the complexity of the subject," but he provided an array of specific examples of historical conditions and events that he attributed to other causes than the direct environmental factors to which Miss Semple ascribed them.[18] And historians, for the most part, have probably been prone to look askance at the book. Four British respondents to my questionnaire, whose competence in history as well as in geography is unquestioned, were particularly critical: J. N. L. Baker of Oxford University, G. R. Crone of the Royal Geographical Society, H. C. Darby of University College, London, and H. J. Fleure, dean of British geographers.

Entranced with *Influences* in 1912, when just beginning graduate study in European history, I spoke of my enthusiasm to a much older friend, the late Professor Ephraim Emerton, who taught medieval history at Harvard. He wrote me a letter, which I have kept in my copy of the book ever since and from which the following is copied:

> The fundamental idea that the movements of man on this planet have been—more or less—determined by geographic conditions is undoubtedly sound. The multitude of illustrations are suggestive and—so far as they go—are worth while. I should say it is a good book for a beginner in history . . . only he ought to be very clear in his mind that what he learns here must be modified at every turn by other than geographic considerations. The trouble with the book is that, when all is said and done, it gives you only a peek at the real problem of History—i.e. to understand the actions and motives of organized human life.

Whether or not this is *the* real problem of history, Professor Emerton's remarks express what has been the essence of much of the subsequent dissatisfaction with *Influences*—its rather monistic attitude toward a highly complex subject.

About ten years later the heavy artillery opened fire. Lucien Febvre attacked Ratzel and Miss Semple in his *La Terre et l'évolution humaine*, of which Mark Jefferson wrote in the *Geographical Review* for January 1923: "Good service is done in demanding critical thought . . . It is futile to hope for enlightenment from vague generalizations about Man, Nature, and Influence. That is the fault here found with Ratzel and his disciples, especially Miss Semple." Of her alleged misinterpretations of certain facts, Jefferson quoted Febvre as saying, "Do not object, for it is useless making objections to a believer armed with a dogma that may not be discussed."[19] Yet Febvre himself could have been overdogmatic about Miss Semple's dogmatism. In an unsigned review of Vidal de la Blache's *Principes de géographie humaine* in the same number of the *Geographical Review*, I perpetrated a monster categorilla. Ratzel and his disciples, I wrote, "in their enthusiasm to explain the influences of environment . . . have been blinded to the recognition of *all other influences*" (italics as of 1962). This was untrue and unfair both to Ratzel and to Miss Semple. It was, however, preceded by the following less obnoxious remarks:

> No longer do we try to explain the infinitely complex story of mankind in simple terms. Leadership or social conflict, religion or the quest

for food, intellectual force or environment—no one of these is sufficient. If some of the details of historical events lend themselves to proof, the interpretation of these details, the very soul of history, assuredly does not. Here all that gives us confidence in a man's work are openmindedness, the power of wise and fair selection, and that sublimated common sense sometimes known as the critical spirit. Vidal de la Blache's great contribution to human geography was to bring to it these qualities of historical-mindedness and to show that they are the essentials of geographical-mindedness as well. Their lack among its devotees has been one of the obstacles to the development of human geography.[20]

Although this does not seem altogether unreasonable, it would call for some definition of terms to make much sense. It reflects a dominant philosophy of history among Historians (with a capital *H*) of that period—namely that all Philosophies of History are suspect and that History should be studied empirically and inductively, the facts being allowed to speak for themselves. This is (more or less) what was meant by "historical-mindedness." Few historians, unless I am mistaken, now take such a stand. They have been awakened to the circumstance that the facts cannot possibly speak for themselves. There has to be some sort of chaperonage in the form of preconceived notions, frames of reference, and the like. But otherwise, if "historical-mindedness" and "geographical-mindedness" are construed to mean, among other things, awareness of the extreme complexity of the circumstances with which history and geography deal and reluctance to subscribe to any one master explanation for anything within their provinces, my declaration of 1923 expresses what I still think. At all events, geographers and historians face methodological problems that have much in common, and it would be beneficial to both parties if they knew a little more about what has been going on in each other's minds.[21]

(Commenting "On the History of Psychology in General" in 1961, C. A. Mace made the following rather startling remarks: "There are two sorts of 'philosophy of history.' One is the philosophy that there *is* a philosophy of history . . . The other . . . is that there is no philosophy of history. The historian is content merely to ascertain what actually happened in history, when what happened was just one thing after another. There is no philosophy of geography—the geographer just describes *what* is *where;* so the historian just describes *what* happened *when* . . . Both these theories as regards history are correct—if stated with care."[22])

Geographers seem particularly enamored of categorillas when

they write about the history of geographical thought. At least, in dealing with historical problems outside the substantive fields in which they have special competence, they are not always quite as cautious as when cultivating those fields themselves. Richard Hartshorne has made this clear with reference to the history of methodological ideas in geography, and it seems to me that many geographers' discussions of environmentalism, whether pro or con, have carried overtones of assurance in harmony with *Influences*. The brand of environmentalism expounded there has been classed as "crude"[23] to differentiate it from the "refined" or "mild" or "sophisticated" kinds now supposedly coming to the fore. In view of the dedicated, if not always discriminating, research that went into the compilation of Miss Semple's books, it is a little sad to find that, when mentioned in published discussions of the past quarter century, they have so often been mentioned as exhibits of how geographers ought *not* to think. The study expended on them has been acknowledged but has been damned with faint praise as encyclopedic.[24] However, the most marked characteristic of recent and current discussions of environmentalism[25] has been a growing awareness of the need for correlation between the thought of geographers and that of workers in the realms of history, the social sciences, and even physics, philosophy, and logic.[26] This is where the sophistication is coming in. "Neoenvironmentalism" is a product of this neosophistication. "In not a few ways and places, and to a significant degree, human activity is affected by external environment—that is all that neoenvironmentalism claims, and it does not rule out causative effects of human effort. But the attempt to rule out environmentalism leads to difficulties,"[27] wrote Professor Spate in 1960; and neoenvironmentalism may yet help restore *Influences* to esteem as a "classic"—the sort of esteem that sophisticates bestow on the works of art of primitive peoples. In reply to my questionnaire, Professor H. C. Darby wrote: "On balance, I would be inclined to give it the sort of affection one gives to some out-dated and out-moded classic."

In my questionnaire I asked whether, in the opinion of the questionee, the over-all effect of *Influences* on the development of modern geography was *negligible, detrimental, neutral* (implying a more or less even balance between beneficial and detrimental), *beneficial*, or "*other*." The question, of course, was not intended as a means of arriving at a value judgment concerning the intrinsic merits of the book; one doesn't judge such things in such a manner. It was a sampling of opinion only (and a very small sampling, at that) regarding the impact of the book on our branch of learning. Of the forty-three respondents who ventured to express an opinion, about half favored *beneficial;* a fifth, *neutral;* less than a tenth, *negligible;* and the other "votes" were distributed between *detrimental, stimulating,* and *other*. There were four detrimentalists.

"INFLUENCES" ON A METAPHORICAL MAP

Robert S. Platt, Harold and Margaret Sprout, and others[28] have sought to classify and arrange in logical order hypotheses concerning the relation of man to his environment, which last I shall personify as "Earth." Where should the master hypothesis that Miss Semple developed in *Influences*, and some of its kin, be located on a metaphorical map? Miss Semple was criticized for using "figure personification," a species of metaphor. Metaphorical maps form another species.

Whereas an imaginative map shows a nonexistent region (for example, a Utopia) as it might be mapped if it existed, a metaphorical map is a map, either existent or imagined, of an unmappable conception. Like Shelley's lines addressed to the moon,

> Art thou pale for weariness
> Of climbing heaven, and gazing on the earth . . . ?

metaphorical maps are not intended to be taken literally. Poets have never hesitated to be as metaphorical as they please, since they assume a lower level of literal-mindedness on the part of their readers than scientists do. Yet much may be said for more and better metaphorical maps, metaphorical statistics, and metaphors in general, not only in poetry but in scholarship.

Respectable metaphorical maps have grids of metaphorical coordinates. The area shown by the grid of my "map," somewhat like the known world of the early Greeks, is confined within a half hemisphere between the equator and the North Pole, bounded by the meridians of 90° W and 90° E and consequently bisected by a central prime meridian of 0° (metaphorical Greenwich; Fig. 7). Were any man-milieu hypothesis to exhibit total or *Absolute Substantive Rationality* (this will be explained below), it would be located on the prime meridian, but since no product of the human

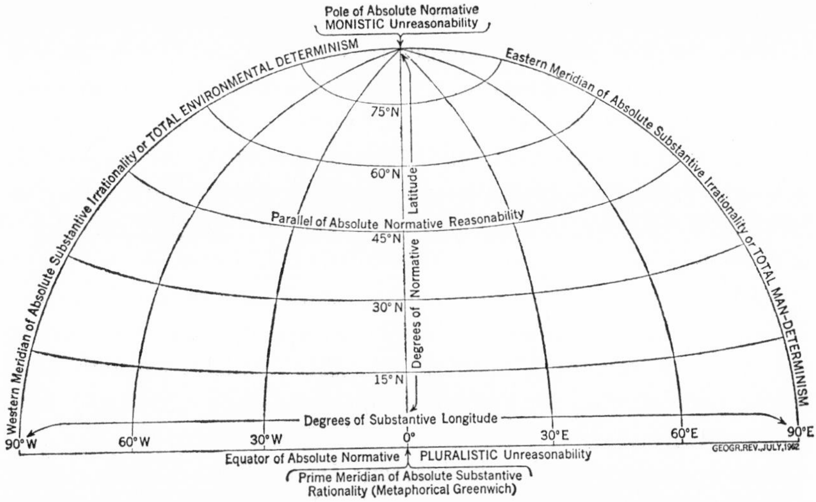

FIG. 7. Grid for a "map" on which the positions of man-milieu hypotheses could be shown if it were possible to determine them (it isn't). The substantive rationality of a hypothesis would fix its longitude, the normative reasonability its latitude. Absolutely irrational and unreasonable hypotheses would be located outside the area of the "map." Note that the coordinates stand for the inherent rationality of the hypotheses and for the reasonability with which they are advocated, and *not* for the relative degrees of determinism and monism as such. It is assumed, however, that with hypotheses of this sort irrationality tends to increase as the extremes of determinism, and unreasonability to increase as the extremes of either monistic advocacy or pluralistic indifference are approached. If, however, it were demonstrated once and for all that a hypothesis of environmental determinism was completely rational, it would therefore be completely reasonable to advocate that hypothesis normatively to the exclusion of all other hypotheses (absolute monism). Under these circumstances the hypothesis would be located at longitude 0°, latitude 45° N.

mind has yet achieved such a state or is likely to achieve it, the vicinity of the prime meridian is uninhabited—that is, there are no hypotheses there. As you go east and west, however, you encounter hypotheses of increasing substantive irrationality until you get to the vicinity of the two 90° meridians, where the "terrain" is again uninhabited because nobody was ever absolutely irrational. These meridians mark the edge of the shuddery unknown where *Absolute Irrational Absurdity* would characterize hypotheses if there were any.

Now as to latitude. Any man-milieu hypothesis propounded with, or accepted as having, *Absolute Normative Reasonability* would lie along the 45th parallel, halfway between the pole and the equator, but here also, for obvious reasons, there is an uninhabited strip. And, as with "longitude," *Absolute* (though actually non-existent) *Normative Unreasonability* obtains both at the pole and at the equator. Socrates and other ancient Greeks must have been thinking along these lines when they developed the philosophy of Nothingtoomuchness. They favored keeping thought and action as close as possible to longitude 0°, latitude 45°.

Such being the grid of our map, we must now try to locate the man-milieu hypotheses. This could be done only with metaphorical chronometer, sextant, loran, and the like—namely, through tests for factual accuracy supplemented by value judgments of soundness and reasonability. These are so variable that every geographer's chart would probably look different from all the others. What observations would have to be made? What, in fact, *are* "substantive rationality" and "normative reasonability?"

By the former I mean the inherent factual accuracy and logical soundness of a hypothesis; by the latter, the reasonability of its proponent's advocacy of it in relation to other possible hypotheses concerning the same problem.

"Moosilauke is a mountain." "Murder is a crime." "Cows can fly." "Geography is human ecology." These are simple substantive assertions, though of varying rationality and factual accuracy. By prefixing "I think that" we convert them into substantive expressions of opinion—which man-milieu hypotheses are.

"You should climb Moosilauke." "Thou shalt not kill." "Cows should fly." "Geography should be human ecology." These, too, although they are expressions of opinion, are also *normative* in that each not only implies a surmise but also suggests an obligation or a desire. However, the strength of the obligation or desire—the *degree of normativity*—differs. The obligation not to kill is considerably more reasonable and compelling than the desires regarding geography and cows.

"Murder is a crime" and "Geography is human ecology," though substantive statements in form, are also normative by implication: the former is a way of saying "Thou shalt not kill," the latter (standing alone) a misleading way of saying "I think that geography ought to be human ecology."

Confusion has often arisen through failure to distinguish clearly between substantive assertions of fact ("Moosilauke is a mountain") and those of opinion ("Islanders are always coast dwellers with limited hinterlands"). This principle is well understood and has often been recognized in geographical writings. Another source of confusion, however, which would seem to be less well understood, springs from failure to distinguish clearly between the substantive and the normative attributes of geographical ideas of all kinds. The hidden normative nature of much that appears on the surface to be innocently substantive may be extremely misleading (*Normative*, which implies "laying down the law," should not, of course, be confused with *nomothetic*, or trying to find out what the law [if any] is.[29])

Since man-milieu hypotheses are concerned with Man in relation to Earth, differences among them that are significant for metaphorical cartographical purposes may be sought in the relative amounts of power and influence they assign to Man and Earth with respect to each other. (God, being the over-all arranger, is equally responsible regardless of whether Earth or Man is deemed the primary operating agent; this relieves us of theological worries.) Now, it would seem that any line of hypothetical reasoning, if pursued to its uttermost conclusion with relentless, inexorable logic and disregard for facts, will end up as complete nonsense—the Absolute Irrational Absurdity characteristic of the nonexistent hypotheses along the bounding meridians of our chart. Let us, therefore, consider the eastern edge of the chart as the locale of hypotheses of *Total Environmental Nondeterminism*, or, to put it positively, of *Total Man-Determinism*, and the western edge as the locale of the diametrically opposite hypotheses of *Total Environmental Determinism* (or *Total Man-Nondeterminism*). Thus along longitude 90° E Man would be capable of doing anything whatsoever he pleased both to himself and to long-suffering Earth, provided only that he exercised his freedom of will to the limit—which might have its good points, despite the excessive responsibility. Opposite, along 90° W, Man would be envisaged, not simply as Miss Semple conceived of him, "the child of the earth," who has "entered into his bone and tissue, into his mind and soul" (p. 1), but as a pawn and plaything, or even a part, of an Earth who remorselessly dictates everything that he thinks and does down to the minutest detail and thus relieves him of *all* responsibility; and this, too, might have its advantages. The various man-milieu hypotheses that geographers and others have expounded all lie between these limits, with, however, considerable gaps owing to human inability to achieve either absolute rationality or absolute irrationality.

We might, accordingly, arrange some of the hypotheses more or less as follows in order from east to west (though it should be understood that there is nothing elegantly mathematical, impeccably cartographical, rigorously logical, methodologically sophisticated, or conspicuously accurate about this arrangement; also that no commitments are made with regard to relative proximity to either the prime meridian or the bounding meridians).

Farthest to the east would lie the hypotheses of old-fashioned geographic-theologic teleology, according to which God designed and created Earth expressly for Man's habitation, exploitation, and delight. No geographer, to my knowledge, subscribes to this in public today, though in the past such revered figures as Thomas Jefferson, Ritter, Morse, Maury, and Guyot were among its adherents (see below, pp. 257–285). Farther west would come the doctrine of "Marshism," as it might be called after George Perkins Marsh, the doctrine that Man is actively engaged in changing the face of Earth, often in an appallingly destructive manner. American geographers gave little serious attention to "Marshism" until about the time of World War II, but since then it has risen high in their esteem.[30]

(Miss Semple refers to Marsh's book, *The Earth As Modified by Human Action*, at least twice in *Influences:* "Civilized man spreading everywhere and turning all parts of the earth's surface to his uses, has succeeded to some extent in reducing its physical differences. The earth as modified by human action is a conspicuous fact of historical development" (p. 120). Reclamation of marshlands and submerged areas in the Netherlands is cited as "the greatest geographical transformation that man has brought about on the earth's surface" (p. 324). I have found no references to Marsh in Miss Semple's other two books.)

Still farther west would be the hypothesis of possibilism,[31] according to which Earth offers Man an immense range of possibilities, among which he is free to choose. Beyond this would lie the domain of the various hypotheses of environmentalism proper, which differ according to their interpretations of the manner and intensity with which the mind and will of Man intervene between an undeniably influencing environment and his responses to its influences. The Sprouts class man-milieu hypotheses more or less in order from "east" to "west" as Cognitive Behaviorism, Probabilism, Possibilism, Mild Environmentalism, and Environmental Determinism. Miss Semple's professions in her preface they class as reflecting Mild Environmentalism, but her practice in the text as more deterministic.[32] They hold that "neither Huntington nor any other alleged determinist known to us has denied man's capacity to choose among alternative courses of action. We have never discovered any interpretation of history that even closely approaches rigorous environmental determinism."[33] R. S. Platt thus grouped "terms expressing environmental relations in a scale from strong to weak: environmental determinism implying absolute cause and effect; environmental control implying less than absolute determinism; environmental influence implying active if not determinative natural force; environmental response implying that nature speaks and man answers; possibilism implying certain inherent possibilities from which to choose; and environmental adjustment implying that man may choose from what he understands to be available. Miss Semple's approach . . . is near the middle of the scale."[34]

In the old days it was easier to find geodetic latitude than geodetic longitude. On our chart, however, the normative "latitude" of a man-milieu hypothesis is harder to fix than its substantive "longitude." Indeed, such position finding is largely a matter of guesswork, where one geographer's guess is almost as good as another's. Without any universally recognized measure of Normative Reasonability, *relative monism* (nonpluralism) if measurable might provide a scale. (At first thought, "relative monism" might seem a contradiction in terms, like "relative monogamy" or "relative monotheism." Monism, however—at least as here conceived—is like monotony, which *can* be relative.) *Monism* here means "having a single preferred hypothesis or notion pertaining to a particular matter"; its opposite, *pluralism,* is the fruit of willingness to admit the possible applicability of more than one hypothesis or notion. Thus only absolute monism and absolute pluralism are mutually exclusive in the sense that monogamy and polygamy are mutually exclusive. Relative monism varies inversely with relative pluralism: 85 percent of the one is 15 percent of the other.

A high degree of monistic confidence is not necessarily normative; one may believe strongly in something without obligating or desiring others to share in one's belief. Nor is a low degree of monism necessarily nonnormative; there have been propagandists for tolerance and broad-mindedness who have themselves been narrow-minded. In general, however, in matters geographical (and moral), the more monistically one regards one's chosen belief, the more normatively one is tempted to urge others to accept it, and vice versa. But in each case *Absolute Uncompromising Normative Monism* and its opposite, *Absolute Compromising Normative Pluralism*, are, like *Absolute Substantive Irrationality*, equally and absolutely absurd in dealing with such complex realities as those of the man-milieu relation.

Although they have touched on the subject many times, our authorities, so far, have refrained from trying to classify and arrange man-milieu hypotheses according to their normative attributes. On this basis, any man-milieu hypothesis, whether a definition or a whole system of methodology, for which anything approaching all-embracing, all-else-excluding sufficiency (plus obligatory or greatly desired acceptance) was claimed with regard to the subject to which it pertained would be located high in the "Arctic" regions near the Pole of Absolute Normative Monistic Unreasonability. As one moved southward, one would find hypotheses being advocated with increasing indifference to the opposing or conflicting claims of other pertinent hypotheses, until, in the steamy doldrums, all man-milieu hypotheses would be esteemed of approximately equal validity in an atmosphere of easygoing tolerance. The Equator of Absolute Normative Pluralistic Unreasonability would be the locale of hypotheses whose proponents had no convictions whatsoever. The reader may speculate on what the Southern Hemisphere would be like.

Thus with metaphorical instruments of precision every man-milieu hypothesis, every definition, and every book and article on the subject could be pinpointed on our chart. But the instruments are nonexistent, and fortunately, because otherwise geographers would have little to argue about. Yet even without them it is possible and desirable to distinguish "longitude" from "latitude" and to perceive, if not to measure, differences in relative position. Failure to do so has been the cause of a great deal of seasickness among those who have sailed the misty seas of geographical thought.

Today it is probable that most geographers would agree in locating Miss Semple's great book in high latitudes far to the west.

CHAPTER 13

Notes on Measuring and Counting in Early American Geography

Kearsarge, Mount, solitary peak (2,937 ft.) in . . . S central N. H., 7 mi. SE of New London.[1]

THE ACTUAL height of Mount Kearsarge and its distance from New London are *geographic* attributes of that mountain, regardless of what anybody may know or think about them. Stated as "2,937 ft." and as "7 mi." they are items of *geographical* information (on the distinction between geographic and geographical see above, p. x). In both cases they are *geoquantities*, in that they pertain to terrestrial (geo-) circumstances.

GEOQUANTITIES, GEOMAGNITUDES, AND "QUANTITATIVITY"

Any quantity expressed graphically or in numbers in such a way as to permit us to compare it with other similarly expressed quantities is a *magnitude*, or *geomagnitude*[2] in the case of a geoquantity. Magnitudes are based on measurement, counting, weighing, or guessing. Since there is no accepted suitable adjective to distinguish items of information that *are* from those that *are not* derived from accurately determined magnitudes, *quantitative* has come to be used for the former. Accordingly we should be inclined to regard the figure "2,937 ft." for the height of Mount Kearsarge as *more* quantitative (and therefore as more "scientific," perhaps) than President Dwight's "12,729 ft." for the height of Mount Washington (see above, p. 150), though both are equally precise.

By the same token, although medieval maps show geomagnitudes graphically and medieval geographical writings present them numerically, such information strikes us as unscientific because it is unquantitative, in the sense of being inaccurately quantitative (rather than nonquantitative, in the sense of being devoid of all quantitativity). In contrast, many expositions of present-day geography have been rendered so exceedingly quantitative in appearance as to take on a "new look" that is baffling to old-fashioned geographers who still venture to believe that, just as all that glitters is not gold, so all that is quantitative is not science. In a delightful footnote which may be commended to mathematically minded geographers, George Perkins Marsh wrote: "The pretended exactness of statistical tables is generally little better than an imposture."[3]

In his *Encyclopaedia*, which was used at Harvard in the late seventeenth and early eighteenth centuries,[4] the erudite Protestant German scholar Alstedius (Johann Heinrich Alsted, 1588–1638) devoted a fairly long section to proving, with the help of John of Holywood (thirteenth century) and Christoph Clavius (sixteenth century), that the grains of sea sand on the earth are not infinite in number, as thought before Archimedes' time, but that it would require *far less* than 1000000000000000000000000000000000000000-000000000000000 (it would now be written 10^{54}) of them to fill the whole world "right up to the concave firmament."[5] This is one example of geomagnitudes that have been known to Americans, and it has a medieval flavor. (Cotton Mather, who was a bit medieval himself, was so much impressed by the "prodigious learning" of the "incomparable Alstedius" that he excoriated those "silly and flashy men" [Harvard students, no doubt] who had called Alstedius "All's Tedious."[6] A recent map and an article in American geographical journals bearing the titles, respectively, "Adjusted Rate-of-Rate-of-Change ('E') in the Sheep/Swine Ratio, Trios of Twenty-year Periods, by County, Nova Scotia" and "Order Neighbor Statistics for a Class of Random Patterns in Multidimensional Space"[7] are more modern. The present paper illustrates geoquantitativity of a kind intermediate in time and nature between that of Alstedius and these two up-to-date examples.

The movement of geographical understanding since the Middle Ages has been carried forward largely by man's ability to extend and supplement his native faculties by devising and employing aids to his physical mobility and mental agility, his accuracy of observation, and his retentiveness of memory. This has been particularly true of the quantification of geography as shown by the increasing use of and reliance upon accurately ascertained geomagnitudes. Spurts ahead, from time to time, have followed the discovery of new ways in which different types of quantitative aid may be used in combinations—for example, the compass (and later the chronometer) with the ship, the telescope with the angle measurer, the telegraph with the barometer and thermometer, the camera with the airplane, the transistor with the satellite. Without such aids in such combinations, man's geographical understanding would be inferior to that of many a noble animal who possesses excellent native endowments for field observation.[8] Eagles, with their marvelous eyesight and mighty wings, would make superb geographers if they could only count higher, remember more, and compute better.

In *The American Geography* (1789) Jedidiah Morse wrote: "Geography is a science describing the surface of the earth,"[9] but in later editions he called it "a branch of mixed mathematics,"[10] a view that probably stems from the *Geographia Generalis* of Varenius (Bernhard Varen, 1622–1650) where geography is similarly defined.[11] Bartholomaeus Keckermann (1571–1609), whose conceptions of geography influenced Varenius[12] and whose works were also studied at Harvard in the seventeenth century,[13] classed *Geographia* as a subclass of a branch of Primary or Theoretical Mathematics, and *Geodesia*, concerned with *Enthymetrica* (longitudes), *Embadrometrica* (latitudes), and *Stereometrica* (*profunditates*: heights and depths),[14] as one of the two main branches of Secondary or Practical Mathematics. Varenius and Morse, however, classed geography as "mixed" mathematics, as distinguished, presumably, from the "pure" or unmixed variety. In any case, in Morse's geographies, as in other Early American works of the kind, the quantitative ingredients are mixed in among the more abundant nonquantitative elements like raisins in a bread pudding. The miscellaneous examples that follow of some of these "raisins" will be grouped according to whether they are *geometric, nongeometric*, or *semigeometric*.[15]

("Measurement has long been considered a hallmark of science properly practiced, and once a new discipline has developed a mathematical discourse, it has almost immediately laid claim, at least in the language of its most enthusiastic disciples, to the significant status—science!"[16] With these words the editor of *Isis: An International Review Devoted to the History of Science and its Cultural Influences* opened the June 1961 number of that admirable journal, a number devoted entirely to nine papers emanating from a Conference on the History of Quantification in the Sciences held in November 1959 under the auspices of the Social Science Research Council with the support of the National Science Foundation. The meeting "was attended by some thirty scholars representing eleven academic disciplines: the history of science, physics, chemistry, biology, botany, mathematics, psychology, sociology, economics, political science and anthropology," and the published papers, besides dealing with generalities, bear upon quantification in medieval and modern physics, in chemistry, in medical science, in psychology, in economics, in sociology, and in biology. For reasons unknown to me, geography was left out in the cold. Maybe it is not a science or not scientific *enough;* maybe those who got up the conference forgot all about it; maybe it was invited but was too modest to attend.)

GEOMETRIC GEOMAGNITUDES

Geometric geomagnitudes pertain to distances and directions as initially measured (or estimated with reference to measurements) in relation to the (theoretically) smooth, rounded surface of the terrestrial spheroid (earth), and also to areas and volumes derived from such initial measurements or estimates. The art or practice of making such measurements and estimates could be called *geomensuration.* (This bastard combination of Greek and Latin roots means literally "earth-measurement." *Geometry* also means "earth-measurement" and according to Proclus (c. A.D. 450) "is said by many[17] to have been invented among the Egyptians, its origin being due to the measurement of plots of land. This was necessary there because of the rising of the Nile, which obliterated the boundaries appertaining to separate owners."[18] At an early date, however, the term had been preempted to designate a branch of mathematics having no necessary geographical implications. Hence, notwithstanding its questionable parentage, "geomensuration" would seem an appropriate and useful designation for what "geometry" means etymologically.)

One could perhaps claim without exaggeration that geomensuration has furnished geography with the skeletal frameworks without which the latter would have assumed the consistency of the jellyfish; also, that in so doing geomensuration has contributed much to the orderly conduct of human affairs, including that of man's most disorderly behavior, war—imagine fighting the Battle of Normandy without maps! (The analogy of the jellyfish was suggested by the appearance of certain medieval maps whose bony structure seems jellylike, if not wholly nonexistent.)

Four interlocking quantitative geographical arts are based principally upon geomensuration[19]—geodesy, surveying, navigation, and cartography—and since their beginnings they have all been concerned predominantly with the measuring and estimating of horizontal distances and directions (*horizontal geomensuration*). In modern times the requirements of navigation (whether on the water, in the air, or in outer space), meteorology, oceanography, geology, and engineering have led to an almost equal concern with *vertical geomensuration* (Keckermann's *stereometrica*). Also, since about 1750 geographical, demographic, social, economic, climatological, and other studies that make large use of statistics have created new demands for the measurement and estimation of areas (*areal geomensuration;* strictly this calls for qualification as "horizontal-areal," but, since vertical areas are of negligible concern to all but window washers, house painters, and possibly a geomorphologist or two who might grow excited over the area of a cliff face, it seems simpler to omit the qualification).

In this paper, apart from the remarks in the immediately following parenthetical paragraphs, I shall bypass the history of vertical geomensuration as well as the alluring topic of *volumetric geomensuration.*[20] My geometric "raisins" relate solely to horizontal geomensuration and will be grouped according to whether they have had to do with distances and directions or with areas.

(Although it will be by passed here, the development of vertical geomensuration has been neglected in the study of the history of geography and would almost seem to cry out for historical treatment—if not necessarily as a unit and in its entirety, at least as it has evolved in particular regions during particular periods. It could be broken down topically—and more or less logically—according to whether the measurements and estimates have been *upward* (*altimetry*, *hypsometry*) or *downward* (*bathymetry*), and also according to whether they have pertained to:

(1) the height of the face of the land above sea level or other datum planes (*hypsometry*);

(2) the height of the atmosphere[21] or of any part thereof (*altimetry*);[22]

(3) the depth of the floors of oceans, seas, lakes, rivers, swamps, and other bodies of water (*subaqueous bathymetry;* usually called simply *bathymetry*);

(4) depth as determined with reference to points, lines, or surfaces, real or assumed, within any body of water (*intra-aqueous*, or simply *aqueous*, *bathymetry*);

(5) depth as determined with reference to points, lines, or surfaces, real or assumed, within the solid earth or lithosphere (*subterranean*, or *lithospheric*, *bathymetry*).

(In writings on the history of hypsometry—such few as there are—mountain heights have figured somewhat as heroes and kings in the old-time chronicles or as playwrights and star actors in the history of the stage. They have received an undue share of attention as compared with that bestowed upon lowly and inconspicuous levels of far greater extent, scientific worth, and human use.

(Hypsometry, as witnessed by ancient Greek measurements of mountain heights, and subaqueous bathymetry, as witnessed by Saint Paul's experience on the way to Rome ["they . . . sounded, and found it twenty fathoms: and when they had gone a little further, they sounded again, and found it fifteen fathoms"[23]] had both been "born" in ancient times, but like the elderly baby in the *Bab Ballads* they were still very much in their infancy in this country until after the close of the Early American period. As for the three other kinds of vertical geomensuration, they could hardly be said even to have been born, if we disregard a few sporadic figures and guesses. Likewise, very little had been achieved toward the accurate and effective mapping of hypsometric or of subaqueous bathymetric relief. Until about 1830 the former was shown on American maps exclusively in a sketchy manner by hachuring and the like,[24] and the latter by the sprinkling over of certain shallow coastal waters on marine charts with numerals indicating depths in feet or fathoms.[25] Unless I am dreadfully mistaken, the mapping of atmospheric altimetry and of aqueous and subterranean bathymetry was still unheard of.

(From among the immense complex of causes that have helped revolutionize vertical geomensuration since the Early American period—say since about 1830—three might be singled out as of outstanding importance: the coming of the railroads in the mid-nineteenth century, the development of aviation a hundred years later, and the prodigious growth of the petroleum industry during the past three or four decades. By creating altogether unprecedented demands for measured geoquantities, the railroads were largely responsible for the revolutionizing of hypsometry, the airplane for that of atmospheric altimetry, and the petroleum industry for that of subterranean bathymetry. Although beyond the scope of this paper, the following observation in Marsh's *Man and Nature* (1864) deserves notice: "The inclination and elevation of [railroad] lines constitute known hypsometrical sections, which give numerous points of departure for the measurement of higher and lower stations, and of course for determining the relief and depression of surface, the slope of the beds of watercourses, and many other not less important questions."[26])

HORIZONTAL POSITIONS, DISTANCES, AND DIRECTIONS

Evans's Map of 1755. Not many published maps have been accompanied by such an admirable printed text explaining how they were compiled and what they show as is found in the "Analysis" (or first "Geographical Essay")[27] accompanying Lewis Evans's *General Map of the Middle British Colonies in America* (1755). Here Evans devoted the equivalent in length of some four pages of the kind you are reading to a critical discussion of the determinations of latitude and longitude and of the various surveys and maps upon which his map was based. This was seldom done in his time and is still rarely done even for the most scholarly of compiled maps, and it alone qualifies Evans's work for honor as a notable landmark in pre-Revolutionary American geography. Evans pointed out that, because

> different Parts of this Map are done with very different Proportion of Exactness, Justice to the Public, requires my distinguishing the Degree of Credit every Part deserves; and to make some Recompense for the Defects of those Places, where no actual Surveys have been yet made, by giving such a Description as the Nature of the Subject will admit; which may, at this Time, be of as much Consequence as the nicest

Surveys destitute of this Advantage . . . I have particularly endeavoured to give these Parts, which are done from Computations, another Appearance than those among the Settlements, where I had actual Surveys to assist me; lest the Reader be deceived by an Appearance of Accuracy, where it was impossible to attain it.[28]

This suggests the "relative-reliability" diagrams and special symbols for conjectural features that appear on some of the better compiled maps of today. Although Evans did not employ these specific devices, the parts of his map that were based on surveys may be distinguished from the less reliable parts by their more intricate drainage pattern and greater density of place names. (In a pamphlet [1612] descriptive of the so-called "Captain John Smith" map of Virginia we read: "In which Mappe observe this, that as farre as you see the little Crosses either Rivers, Mountains, or other places haue bene discovered, the rest was had by Informacion of the Salvadges, and are sett down according to their Instrucions."[29])

During the period before 1880 most of the larger surveying operations in this country were sporadic undertakings conducted for the purpose of running county, provincial, or state boundary lines,[30] or in connection with the administration and sale of public land.[31] Far-flung governmental surveys, designed primarily as a basis for mapping on large scales, were fairly well advanced in Europe during the last half of the eighteenth century,[32] but lagged in the United States[33] until well on in the nineteenth. They were foreshadowed in the stress that Jefferson laid upon the necessity of accurate surveying and mapping in his instructions to Meriwether Lewis for the Lewis and Clark Expedition and in his proposals that led to the founding of the Coast Survey in 1807.[34]

Difficulties of Early American Surveying. Evans had explained that surveys in America were beset with special difficulties little appreciated by the European. The country east of the Mississippi was almost "every where covered with Woods . . . Here are no Churches, Towers, Houses or peaked Mountains to be seen from afar, no Means of obtaining the Bearings or Distances of Places, but by the Compass, and actual Mensuration with the Chain."[35] A half century later Dr. Jeremy Belknap and Dr. Samuel Williams also commented on the troubles of surveying with chain and compass. In his *History of New Hampshire* Belknap thus described the rough-and-ready surveys conducted in the laying out of roads and township boundaries in the backwoods during the late eighteenth century:

> Some allow one in thirty, for the swagging of the chain. The length of a man's arm to every half chain, has been allowed for inequality of surface. The half chain is most convenient in thick woods; but some have very absurdly used a line; and if any allowance is made for its contraction by moisture, it must be arbitrary. Surveyors are often sworn to go according to their best skill and judgment; this they may do with great sincerity, and yet, for want of better skill, may commit egregious mistakes. The variation of the needle, has not in general been attended to with that caution which it demands, and from this negligence, many errors have arisen. It was once proposed, in the General Assembly, that durable monuments should be erected in convenient places, on a true meridian; by which all surveyors should be obliged to regulate their compasses; few of them, at that time, being skilled in the method of finding the variation by the sun's amplitude; but the proposal was rejected.[36]

Similarly Dr. Williams tells us in his *Natural and Civil History of Vermont* that, when the northern boundary of Massachusetts was run in 1742, the magnetic declination was less than the surveyor had been instructed to assume, and as a consequence the line was deviated to the north of the parallel of latitude it was supposed to follow, occasioning a "loss of 59,873 acres to Newhampshire; and of 133,897 acres to Vermont."[37] Had the boundary been surveyed correctly, the northwest corner of Massachusetts would have been some 3½ miles south of its present position and both Williams College and North Adams, Massachusetts, would now be in Vermont.

The mischance that deprived his adopted state of some 134,000 acres may help explain an appendix entitled "An Account of the Variation of the Magnetic Needle, in the Eastern States" in Dr. Williams's book. Here, after a brief historical introduction, the author points out that

> the Magnetic Needle can never give to the surveyor who follows its directions, a straight or an accurate line. And it ought not to be used at all, where the business requires great accuracy and precision. It is however scarcely practicable in America, to substitute any thing better, in the room of it: Most of the lines which have been already run by surveyors, were run by the Needle; this is much the most convenient instrument that can be carried, or used in the woods.[38]

As is often done today, Williams and his contemporaries used the term "variation" loosely both for the declination of the compass and for changes in the declination. He ascribed the discovery of the diurnal variation (change) to "G. Graham in 1723."[39] By "annual variation" he appears to have had in mind the progressive secular change from year to year rather than the small rhythmic annual fluctuations with the seasons, although the latter had already been discovered.[40]

To his "Account" Williams subjoined a table of "Magnetic Observations made in Canada, and the Eastern States of America," which shows in the 1809 edition of his book the results of a total of 33 observations (10 by Williams himself) at 24 stations (3 in Canada, 2 in New York, 19 in New England) at different dates between 1605 and 1806—a worthy effort to gather together such scattered data as could be found.[41] Williams did not try to plot these on a map.

(Evans on his map of 1775 included a "Magnetic Meridian 1750," but without explaining the information on which it was based [it would seem to have been consistent with that of Williams for New York and Montreal: Montreal, 10°38′W in 1749, New York, 6°22′W in 1750]. In the Preface to the *Analysis* Evans wrote: "To render this Map useful in Commerce, and the Ascertaining the Boundaries of Lands, the Time of High-Water, at the Full and Change of the Moon, and the Variation of the Magnetical Needle are laid down. But as these deserve particular Explanations, I have for want of Room, concluded to treat of them at large in a separate Essay."[42] I have not located any such essay.)

The precise lines followed by many of our town, county, and state boundaries, and even by the international boundary between Vermont and the Province of Quebec, stand today as monuments to early surveyors' mistakes, some due to carelessness and some to hard liquor (the latter understandable in view of the discomforts that had to be endured), but most of them unavoidable because of the nature of the terrain and the inadequacy of the instruments used. Dr. Williams, for example, was commissioned by the Governor of Vermont in 1806 to examine the positions of the monuments marking the eastern portion of the boundary between Vermont and Quebec. He found that the line as run in 1772 was in "great error," being so far to the south that Vermont was deprived of 401,973½ acres, equal to 17 44/100 townships.[43] Williams himself, however, was not a very accurate surveyor; "After taking a number of celestial observations" he located the 45th parallel (which the international boundary between Vermont and Quebec was supposed to follow) approximately 14 and 7 miles too far north, respectively, at the Connecticut River and at Lake Memphremagog.[44]

(A study might well be made of the influence of alcohol on boundaries and thereby, in turn, upon subsequent human affairs. Preparatory to the joint survey of the Virginia–North Carolina boundary in 1728, the Virginian commissioners wrote to the North Carolinians: "We shall be provided with as much Wine and Rum as will enable us and our men to drink every night to the Success of the following Day, and because we understand there are many Gentiles on your frontier who never have had an opportunity to being Baptised we shall have a Chaplain to make them Christians."[45] The bill of "sundries" in connection with the running of the provincial boundary between New York (now Vermont) and Quebec in 1771 totaled £146 *6s. 6½d.*, of which £29 *16s.* 11*d.* was for Madeira, rum, brandy, and wine. *Precisely* how much this affected the boundary might be difficult to determine. The late Dr. Mayo, however, was surprised that "some have marvelled that the line they surveyed was so far from straight!"[46])

If the survey of the northern boundary of Massachusetts was one of the less accurate—and hence more typical—of the larger surveying operations in this country during the Colonial period, the surveys connected with the establishment of the northern and eastern boundaries of Maryland were probably the most accurate. Charles Mason and Jeremiah Dixon, skilled English geodetic surveyors, were invited to come to America to run the lines when the Americans at work on them had bogged down. During the years 1763–1767 Mason and Dixon established both the north-south Maryland-Delaware line and a part of the east-west Maryland-Pennsylvania boundary (the "Mason-Dixon Line").[47] A "survey made 130 years later with modern instruments and methods" showed that "the position of [the latter] at the northeast corner of Maryland differed only 2.3″ (180 feet) from that determined by them."[48] About 90 miles west of this corner their line is some 710 feet too far to the south,[49] representing an error of about one-fortieth that registered by the position of the straight portion of the northern boundary of Massachusetts at the same distance from its eastern end near Lowell.

As a side issue to the survey of the north-south boundary, Mason and Dixon, under the sponsorship of the Royal Society, measured an arc of meridian, an operation designed to contribute toward a better conception of the exact size and shape of the earth. Their work yielded an overestimate of about 2800 feet in the length of a degree of latitude,[50] or roughly 0.9 percent. This error, which exceeded those made in the famous measurements of arcs of meridian by Norwood in England (1635), Picard in France (1669–70), Maupertuis in Lapland (1736), and La Condamine in Ecuador (1735–44) (see below, p. 246, Table I) may have been due partly to the use of wooden rods (rather than of triangulation from an accurately measured line) in measuring the length of the line.[51]

AREAL ESTIMATES AND MEASUREMENTS

I am aware of the damp souls of housemaids
Sprouting despondently at area gates.[52]
—T. S. Eliot

"Area," referring to the basement of a house or "areaway" leading thereto, is here used in a sense close to its original Latin meaning, "building lot" or "court." The word has come to be applied to surfaces and their sizes; thus we may find ourselves writing "the area of this area is 5 acres" (see above, p. 43). Although "area" was given all these meanings in England in the eighteenth century, some American writers of that time seem to have preferred such terms as "superficial content," "extent," or even "dimensions" for areal magnitudes large enough to be of geographical interest. Williams mentions 10,237½ square miles as the "superficial area contained within the boundaries of Vermont" but states that "the dimensions of America, compared with the dimensions of Asia, Africa, and Europe" are as 141 to 249.[53] Such indefiniteness suggests that these magnitudes were still regarded more generally as curiosities than as having any particular scientific or practical use—no doubt owing to the difficulty of measuring them accurately and the scarcity of satisfactory available nongeometric statistical data with which the areas could be combined to yield semigeometric information of significance.[54]

The Earth's Total Area. Cotton Mather informs us that "according to the accurate observations of the English Norwood and the French Picart the Ambit of our Globe will be 24,930 [English or statute] miles"; and hence, that the surface must be 197,331,392 square miles, the "solid content" 261,631,995,920 cubic miles, or "30,000,000,000,000,000,000,000 cubic feet."[55] In a table included near the beginning of his *American Universal Geography* (1793), Morse gave much the same figure for the area of the earth's surface, rounded however to 199,000,000 square miles,[56] and it, too, stemmed ultimately from Norwood's measurement of an arc of meridian between London and York (1635). If the U.S. Hydrographic Office's estimate of 196,950,284 square statute miles is taken as "correct," Mather's and Morse's figures were overestimates of about 0.2 percent and 1 percent, respectively. These data contrast favorably with the overestimate of 44 percent made by the medieval Arab geographer Abū-l-fīda (*c.* A.D. 1300) and the underestimate of 76 percent by the Englishman Gerard Malynes (1636). When fairly correct data like Norwood's and Picard's for the length of a degree became available in the seventeenth century, corresponding approximations of the earth's total surficial area could be reached by assuming the earth to be a perfect sphere and solving a simple problem in solid geometry.

Various estimates of the earth's area may be summarized thus:

	Square English Miles	Error (percent)	Degree (English statute miles)
Abū-l-fīda, *c.* 1300	283,015,000	+43.7	82.
Snellius, 1615	184,700,000	− 6.23	66.75
Alstedius, 1630	197,280,000	+ 0.14	69.6
Malynes, 1636	46,560,000	−76.36	
Norwood, 1635 / Morse, 1793	[199,550,000]	[+ 1.32]	[69.55]
Picard, 1669–1670 / Mather, 1721	197,331,392	+ 0.19	69.13
Templeman, 1729 / Guthrie, 1794	199,260,000	+ 1.17	69.50
Guthrie, 1771	199,512,595	+ 1.29	69.54
Morse, 1789	199,859,866	+ 1.48	69.60
Morse, 1793–1819	199,000,000	+ 1.03	
Morse, 1796	197,502,336	+ 0.28	69.2
U.S. Navy H.O.	196,950,284		68.71 at equator 69.51 at pole

Abū-l-fīda (1273–1331), who calculated the area of the earth at 20,360,000 square parasangs by assuming the circumference to be 8000 parasangs and applying the formulas $A = C^2/\pi$ and $\pi = 3\frac{1}{7}$ states that his figure for the circumference was derived from Ptolemy's calculation, according to which $1° = 66\frac{2}{3}$ miles $=$ $22\frac{2}{9}$ parasangs.[57] If the Arab mile in this case was 2000 meters,[58] Abū-l-fīda's area would be about 283,015,000 English statute miles.

After completing his famous triangulation in the Netherlands in 1615 to measure an arc of meridian, Willibrord Snellius estimated the area of the earth at 33,507,717,000,000 square Rhenish rods (*Ruten*).[59] Snellius converted his own figure in *Ruten* into 18,811,353 square Dutch miles, but in doing so made a mistake in arithmetic, which apparently went undetected until 1904, when Schmiedeberg discovered it. The correct figure should have been 14,892,321.61. Assuming 28,500 rds = 55,100 *toises*[60] and 1 *toise* = 6.397 feet (Table I, note *b*, p. 247, below), Snellius' estimate amounts to some 184,700,000 square statute miles.

In a work on commercial law published in London (1636), Gerard Malynes had included a chapter in which the areas of the earth, continents, and various countries and provinces were given, inaccurately, in acres. He estimated the earth's area as 29,803,575,000 acres,[61] or (at 640 acres to the square mile) less than a quarter the correct amount.

The area of a perfect sphere having a circumference of 360×69.09 miles, calculated from the equation $A = C^2/\pi$, would be very close to the actual area of the earth. The length of a degree is 69.09 statute miles at about north latitude 48°. Hence, almost no part of the error in a calculation of the earth's area based on a degree length determined near this latitude and on the assumption that the earth is a perfect sphere could be due to the latter assumption. Had Norwood's degree been correct, instead of 0.6 percent in excess (Table I), the error of +1.32 percent in the area derived from his degree would be reduced to only +0.12 percent, since an error of *a* percent in the length of the degree produces an error of $(200a + a^2)/100$ percent in the area; if, for example, the degree length is 2 percent in excess, the area will be $(200 \times 2 + 4)/100 = 4.04$ percent in excess. In other words, less than one-tenth of Norwood's error of +1.32 percent is due to failure to take account of the actual shape of the earth and more than nine-tenths of the inaccuracy springs from the length of the degree used.

Whereas fairly accurate figures for the area of the whole earth were available in Early American times, the measurement of smaller, irregularly shaped geographic areas had to be made on maps, could not be more accurate than the maps themselves, and involved other difficulties.[62]

Evans on the Settled Area in North America. Though quantitatively minded, Lewis Evans would seem to have presented in his geographical publications no specifically measured data for either altitudes or areas. His "Analysis" (1755), however, contains the following comment:

> Any Person, who knows the Nature of the Soil, and the Extent of our Settlements, will confess, that all the Land, worth the Culture, from New Hampshire to Carolina, and extended as far back as there are Planters settled within three or four Miles of one another, though including nine Colonies, is not equal in Quantity to Half the arable Land in England. All the Whites in the Remainder of the British Colonies on the Continent, scarce amount to 120 000 Souls. How different this from the Conceits of those who would represent some single Colonies as equal to all England. The *Massachusetts*, though made such a Bugbear, as if its Inhabitants were so rich and numerous, as that they might one Day be able to dispute Dominion with England, is not as large as Yorkshire, or has Half so much arable Land.[63]

Actually, the area of Massachusetts (8257 sq. mi.) is somewhat greater than that of Yorkshire (6079). The passage, however, is of less interest in terms of the accuracy of the information that it conveys than as a forerunner of countless American studies to come in the years ahead in which areal magnitudes have been interpreted as bearing upon problems of demography, politics, and economics.

Jefferson on Virginian Areas. At different places in the *Notes on the State of Virginia,*[64] Jefferson gave the areas of certain physiographic and other divisions of Virginia, which may be summarized thus:

	Area (sq. mi.)
(1) Coastal Plain (including 542 sq. mi. in Eastern Shore)	11,205
(2) Piedmont	18,759
(3) Great Valley	11,911
(4) West of the "Alleghaney" to the meridian of the "Great Kanhaway"	22,616

(5) West of that meridian to the Ohio	57,034	121,525
(6) Whole state west of the "Alleghaney"	79,650	
(7) Whole state east of meridian of the "Great Kanhaway"	64,491	
	144,141	
Less (4), which is part of (6)	—22,616	
		121,525

(I have substituted the following physiographic terms for Jefferson's designations: "Coastal Plain" for "Between the sea-coast and the falls of the rivers," "Piedmont" for "Between the falls of the rivers and the Blue ridge of mountains," and "Great Valley" for "Between the Blue ridge and the Alleghaney.")

Jefferson does not explain whether he himself or someone else calculated these areas, but in either case it must have been by measurement on maps. His total (121,525) included the combined areas of what are now Virginia, West Virginia, and Kentucky, of which the total land area as measured by the U.S. Bureau of the Census of 1940 (106,902 sq. mi.) was considerably less than Jefferson's total. This might be due to misunderstanding on Jefferson's part with regard to the southern boundary of Virginia. In describing the boundaries he wrote, "On the West by the Ohio and Missisipi, to latitude 36°30′ North: and on the South *by the line of latitude last mentioned*"[65] (italics mine). He would seem to have overlooked the fact that, owing to errors, the part of the southern boundary as surveyed in 1728 lies several miles north of lat. 36° 30′.[66] Indeed, throughout their entire length from the Atlantic to the Mississippi, the southern boundaries of Virginia and Kentucky lie considerably to the north of that parallel. However, by subtracting his figure for the area west of the Alleghany Mountains (79,650) from his total (121,525) we obtain 41,875 sq. mi. for the area east of the Alleghany, which should and does approximate the 1940 Census figure for the land and water area of Virginia as of today (42,326 sq. mi.)

Areal Data in Morse's Geographies. In Morse's *Geographies* we find (1) a general world table of areas and also a variety of other tables and scattered figures for areas both (2) outside and (3) within the United States.

Although Dr. Morse did not acknowledge its source, his world table (which first appeared in his edition of 1793)[67] was taken almost

A TABLE,

Exhibiting the Superficial Content of the whole Globe, in Square Miles, sixty to a degree, and also of the Seas and Unknown Parts, the Habitable Earth, the Continents; likewise the great Empires, and principal Islands, arranged according to their magnitude.

	Square Miles		Square Miles
The Globe, as some suppose, about	199,000,000	Jamaica	6,000
Seas and unknown parts	160,000,000	Flores	6,000
The habitable World	39,000,000	Ceram	5,400
America	14,000,000	Breton	4,000
Asia	10,500,000	Socrata	3,600
Africa	9,500,000	Candia	3,220
Europe	2,600,000	Porto Rico	3,200
Continent of New Holland	4,000,000	Corsica	2,520
Persian Empire under Darius	1,600,000	Zealand	1,900
Roman Empire in its meridian	1,600,000	St. Jago	1,400
Russian——	4,000,000	Long Island or Manhattan	1,400
Chinese——	1,700,000	Majorca	1,400
Great Mogul's——	1,100,000	Negropont	1,300
United States of America	1,000,000	Teneriffe	1,270
Turkish Empire	950,000	Gothland	1,000
Present Persia	800,000	Madeira	950
Borneo	228,000	St. Michael	920
Madagascar	168,000	Skye	900
Sumatra	129,000	Lewis	886
Japan	118,000	Funen	768
Great Britain	72,900	Yvica	625
Celebes	68,400	Minorca	520
Manilla	58,500	Rhodes	480
Iceland	46,000	Cephalonia	420
Terra del Fuego	42,000	Amboyna	400
Mindinao	39,100	Orkney Pomona	324
Cuba	38,400	Scio	300
Java	38,200	Martinico	260
Hispaniola	36,000	Lemnos	220
Newfoundland	35,500	Corfu	194
Ceylon	27,700	Providence	168
Ireland	27,500	Man	160
Formosa	17,000	Bornholm	160
Anian	12,000	Wight	150
Gilolo	10,400	Malta	150
Sicily	9,400	Barbadoes	140
Timor	7,800	Antigua	100
Sardinia	6,600	St. Christopher's	80
Cyprus	6,300	St. Helena	80
		Guernsey	50
		Rhode Island	50
		Jersey	43
		Bermudas	40

New Guinea, New Zealand, New Caledonia, New Hebrides, Otaheite, Friendly Islands, Marquesas, Easter or David's, Pelew Islands } Islands lately discovered, but not fully explored, and whose dimensions are not exactly known.

unchanged from William Guthrie's *New System of Modern Geography* (1770–) and ultimately from a table in Thomas Templeman's *New Survey of the Globe* (1729). Templeman's "plates" (tables), setting forth areas (in square nautical miles), lengths, breadths, distances, and populations, though not the first of the kind,[68] "surpassed all previous publications in their detailed coverage of the countries of the world"[69] and for nearly a century were extensively copied by the compilers of geographical and statistical works. Morse's general table gives the areas for the globe, the "seas and unknown parts," the "habitable world," the continents, the "Persian Empire under Darius," "The Roman Empire at its meridian," and other empires; likewise the United States and some six dozen islands, from Borneo with 228,000 down to Bermuda with 40 sq. mi.

Although the caption states that the table exhibits "the Superficial Content of the whole Globe, in Square Miles, sixty to a degree," Morse, misled by Guthrie and oblivious of the inconsistency, actually presents the area of the whole earth's surface (199,000,000 sq. mi.) and five other of the larger areas in square *English* miles instead of nautical miles, as implied, as Templeman had presented them, and as is actually the case for other data in Morse's table (see below, page 248, Table II, note d). Except for these six figures and for the fact that Morse increased the area of Russia and added "the United States of America" and "Long Island or Manhattan" (Long Island is obviously meant: Morse's figure is 1400 sq. mi.; the *Columbia-Lippincott Gazetteer* gives 1401 sq. mi.), all of the data in Morse's version of the table stem from Templeman. The discrepancy as regards statute and nautical miles persisted undetected through all five subsequent editions of *The American Universal Geography*. Although the table is given in abridged form in Morse's later editions, Templeman's long-lived figures (in nautical miles) for Borneo, Madagascar, and Sumatra persisted through to the edition of 1819, where they are found along with the 199,000,000 for the whole earth. Whoever edited the table for Mathew Carey's American edition of Guthrie's *Geography* (1794)[70]—and it may have been the famous Dr. Rittenhouse—made the table at least consistent within itself by restoring Templeman's original figures in nautical miles.

For his miscellaneous tables and the scattered areal magnitudes for other parts of the world than the United States Morse borrowed the data with little change from European sources, notably Guthrie, Zimmermann's *Political Survey of the Present State of Europe* (1787), and, in his editions of 1812 and 1819, from Pinkerton.[71] Zimmermann, a German naturalist and geographer, had written the *Political Survey* during a visit to England, in order to acquaint British readers with the nature and some of the findings of the new study of politico-geographical statistics that had been rapidly developing in Germany during the preceding half century or so.

If for the rest of the world Morse merely copied or adapted the areal data of others, for the United States he seems to have computed some of the figures himself. This is implied by the manner in which he discussed certain of the areas and also by the following remarks of the mathematically minded Rev. James Freeman in a critical review of the 1793 edition of Morse's *Geography:*

> In describing countries, it appears to be Mr. M.'s plan, in his first volume, to give as the length and breadth of any state, two numbers which, multiplied together, produce their contents. These numbers are either obtained by calculation, or arbitrarily assumed; and for the most part are very erroneous. Thus he says, that the length of Connecticut is eighty-two miles, which, according to his own account, cannot be true for its length must be either its north or south line, neither of which is eighty-two miles.[72]

Freeman also pointed out that "Mr. M's plan, exceptionable as it is, would be less confused, if he uniformly adhered to it. But he frequently deviates from it without giving any notice to the reader."

Morse actually took as the "length" of Connecticut the approximate mean (82 mi.) of its northern and southern boundaries (72 and 90 mi.), and as the "breadth" the mean (57 mi.) of its eastern and western boundaries (45 and 72 mi.). By multiplying these averages, he obtained the area, 4674 sq. mi.[73] The procedure, if crude, was preferable to that of Pliny, who had "advanced the scheme of using the sum of the length and breadth of continents as a basis for size comparisons."[74]

For the area of the whole United States and certain major subdivisions as they were prior to the Louisiana Purchase, Morse cited figures that Thomas Hutchins, Geographer to the United States, had calculated "from actual measurement of the best maps" in terms of acres.[75] Converted into square miles, some of these data are as follows:

	Per Hutchins (sq. mi.)	"Actual" (sq. mi.)	Hutchins' error (percent)
Water	79,683	93,284	−16
Land	920,327	867,980	+ 6
Gross area	1,000,000	961,264	+ 4

Hutchins's overestimate of the total land area was offset by a much greater proportional underestimate of the water areas. The measurement of water areas has been a thorny problem right down to the present time, partly for geometrical reasons, partly owing to differences in the definition of what should be regarded as "water," and partly because of the risk of overlooking small lakes in little-known regions and especially in glaciated country. Hutchins's total for water (51 million acres = 79,683 sq. mi.) is the sum of 15 items for specific lakes, parts of lakes, bays, and so forth, plus one for "Sundry small lakes and rivers" in the Northwest Territory (301,000 acres) and one for "All the rivers within the thirteen states including the Ohio" (2,000,000 acres). Since he overestimated the size of the part of Lake Superior in the United States by 65 percent, his underestimate for the remaining water areas was considerably in excess of 16 percent.

In the first edition of Morse's *American Geography* (1789) there is a table showing for each state and for other major divisions of North America the length, breadth, population, name of the chief town with its distance and bearing from Philadelphia, the latitude and longitude of the capital, and other facts, *but not the area.*[76] In the text, however, areas are given for eight of the thirteen states; New Hampshire, Rhode Island, the two Carolinas, and Georgia were omitted. While the gaps were filled and some of the data were revised in later editions, the errors remained considerable. Not until the edition of 1812 did Morse see fit to present the areas of all the states, territories, and larger divisions of the country together in one table where they might be conveniently compared. In the edition of 1819 the areas are given for nineteen states and territories and the District of Columbia, those for Maine, Pennsylvania, Virginia, and the seven states and territories to the southward and southwestward of Virginia all being rounded to the nearest thousand. Twelve of these figures are overestimates, varying from 1.5 percent for Pennsylvania to 32 percent for Maryland, and seven are underestimates, from 5 percent for Connecticut to 21 percent for South Carolina.

The German geographer-historian Christoph D. Ebeling, in his monumental, uncompleted work on the United States, published in seven volumes, 1793–1816, not only indicated the area of each State with which he dealt but named the map or other source from which the area was derived, and for several states cited alternative figures, with their sources; but, despite his care, Ebeling's data do not seem to be conspicuously more accurate than those of Morse. Ebeling's work formed a part, or continuation, of the *Neue Erdbeschreibung* of Anton Friedrich Büsching (1724–1793), in which, probably for the first time, areal figures were introduced into "a strictly geographical work as a regular feature in the description of each country."[77] The editor of the English translation (1762) of Büsching must have found areal figures boresome, for he omitted them.[78]

Within states, areal magnitudes of significance pertain to political divisions (counties, townships, parishes, congressional and legislative districts, and so forth), holdings of real estate, bodies of water, or areas of "other" types. Except with regard to real estate, very little would seem to have been done during the Early American period in the way of measuring or estimating such areas. Mention has been made of Hutchins's estimates of water areas and Jefferson's of the major physiographic regions of Virginia. If Jefferson gave no county areas in his *Notes on the State of Virginia*, in a letter dating from 1812 he expressed the opinion that his own Albemarle County contained "about 750 square miles, about 20,000 inhabitants, or 27 to the square mile."[79] Morse in *The American Gazetteer* tells us that Essex County, Massachusetts, is "in length about 38 miles, in breadth 25, and is shaped triangularly, Chelsea being the acute point . . . It is subdivided into 22 townships, which contain 7644 houses and 57, 913 inhabitants; being the most populous of its size of any in the state, having about 135 souls to a square mile."[80] This implies that Dr. Morse had in mind an area of approximately 57,913/135 = 430 sq. mi. (the area of a perfect triangle of the dimensions given would be 475 sq. mi.). Morse gives no other figure in the *Gazetteer* for any one of the 17 counties of Massachusetts (which then included Maine) that would imply that he had tried to ascertain its total area. In the 1789 edition of his *American Geography*, however, there are two statistical tables, one for Massachusetts (including Maine), the other for New Jersey.[81] For each county these tables present a variety of figures, including the number of acres of "improved" and of "unimproved" land, respectively. The term "unimproved" land is not explained; it may apply to land in farms, but it clearly does *not* include all the land in each county not accounted for in the "improved" category. For each state the figures are given to the nearest acre and for five Massachusetts counties to the nearest quarter acre, another example of precision as distinguished from accuracy. The figures may well have been derived from assessment lists drawn up for taxation purposes.

Later Areal Measurements for the United States. Facts of the kinds we have been considering, though hardly of transcendent interest in themselves, at least suggest the rather meager quantity and inferior quality (as judged by present-day standards) of the areal magnitudes concerning their own country that were known to Americans during the early years of independence. The task of measuring such areas accurately and comprehensively for the country as a whole and of publishing the results was one that only the Federal Government has ever been in a position to undertake, and the Government did not reach that position until many decades later, after it had developed the organizations for making the surveys, preparing the maps, and developing the mensurative and computational skills necessary for the purpose. Nor was there any particular incentive for the Government to interest itself in more than crude and incomplete areal estimates and measurements until businessmen, geographers, demographers, and others could demonstrate definite needs for them and practical uses to which they might be put. According to the late M. L. Proudfoot's historical summary[82] of these developments, "the first area table covering the entire United States by States and Territories was released in 1860 by the General Land Office." In 1870 the Land Office's tables for the first time gave "the areas of the counties of one State, namely, Nebraska." The geographer Henry Gannett "laid the foundation for accurate and detailed area measurement in the United States."

This was in 1880 in a publication by the Census Bureau, which for the first time supplied area figures on a county basis for the whole country. In connection with the Sixteenth Census (1940) a complete remeasurement of the previously measured areas was carried out, also under the direction of geographers, and for the first time the areas of all the minor civil divisions (townships, cities, parishes, and so forth) were mapped consistently and their areas measured and published in tables for the whole country.[83]

NONGEOMETRIC GEOMAGNITUDES

CLASSIFICATION

Nongeometric geoquantities and their associated geomagnitudes pertain to the nonspatial attributes of terrestrial phenomena (that is, their attributes other than their locations, sizes, and shapes). They are of two kinds—*temporal*, which bear upon locations and dimensions in time (A.D. 1963; 350 years) and call for no particular comment, and "others." For want of a better term, I shall designate the latter as *topical*.[84]

It seems historically illuminating to group topical geoquantities in two major overlapping categories, depending upon whether they relate to conditions that are (or are conceived to be) continuous or discontinuous in space. The former (*continuous* geoquantities) apply principally to the variations in intensity from place to place of physical phenomena such as heat, moisture, pressure, gravity, magnetism, or electricity, studies of which fall generally within the realm of the physical sciences; the associated magnitudes in this case are normally ascertained by instrumental measurement and consequently may be called *mensurative*. *Discontinuous* or *discrete* topical geoquantities, on the other hand, apply to spatially discontinuous phenomena—in other words, to aggregates ("populations") of individual units, which may be either spatially noncontiguous (trees, houses, islands, congressmen) or contiguous (grains of sand in a pile, congressional districts), and either tangible (congressmen) or conceptual (congressional districts). Study of discrete topical geoquantities falls within the realm of the statistical sciences, among which may be included not only statistics, as such, but all of the natural and social sciences that make large use of statistical data and techniques and, in the opinion of many, derive their "scientific" quality from this circumstance. Here statistical geomagnitudes (the number of cows in Iowa) are determined basically by counting and hence they may be called *enumerative*.[85]

If one of the most conspicuous *negative* changes in the nature of American geography since 1607 has been its decreasingly theological orientation (see below, p. 253), an equally noteworthy *positive* change might well be attributed to the immense accrual of topical geomagnitudes, both mensurative and enumerative, that have become available for geographical uses as a result of developments in the physical sciences that began in the seventeenth century and those in the statistical sciences that began toward the middle of the eighteenth. Except for the mariners' compass, I can think of no instrument developed before 1607 that was capable of being used for the measurement of a continuous topical geoquantity. As regards discrete geoquantities, people, of course, had been taking censuses and making other enumerations for military, administrative, and commercial purposes since the dawn of civilization, but such enterprises were sporadic in their geographic distribution and disparate in the data they yielded, and the latter were generally inaccessible to students. Hence, until modern times (say since 1600) geographical understanding was almost totally nonquantitative except in its horizontally measured geometric aspects. Today, as already intimated, geographical scholarship has become largely concerned with the exposition, explanation, and interpretation of the superabundance of nongeometric topical magnitudes that the advancement of the physical and statistical sciences has placed at its disposal.

A few illustrations will now be given of halting initial Early American attempts during the period between about 1750 and 1820 to deal with matters of climate, zoology, and population in terms of topical geomagnitudes.

EARLY AMERICAN CLIMATIC GEOMAGNITUDES

Galileo invented the thermometer about 1597 and one of his pupils kept temperature records in Florence for sixteen years.[86] Nevertheless, no two thermometers were sufficiently alike to yield comparable readings until the early eighteenth century, and not until the last quarter of that century were the physical principles affecting the construction and use of thermometers, barometers, and other meteorological instruments well-enough understood to arouse serious interest in the possibility of geographical studies of climate based on comprehensive, comparable records.[87] Furthermore, since there is nearly always a time lag between the invention of an instrument or technique of potential geographical utility and its actual use for geographical purposes, geography benefited little from the early instrumental observations. While it has profited enormously from meteorological data secured with standardized instruments and procedures at many stations distributed over large areas, these benefits did not come until after the mid-nineteenth century.[88]

Although a few enthusiasts along our eastern seaboard had been keeping notes on the weather since the earliest days of settlement,[89] the "instrumental" period in American meteorology did not open until the 1730's.[90] Most of the observers were clergymen, farmers, professors, and physicians (especially physicians interested in the relation of weather and climate to health). What may have been the first systematic instrumental meteorological records kept in this country were those of Dr. John Lining at Charleston, South Carolina, begun in 1738 and continued for about eight years by him and from 1750 by Dr. Lionel Chalmers.[91] Lining's observations covered pressure, temperature, humidity, depth of rainfall, and cloudiness. By 1800, records covering periods of varying duration were available for about a dozen places.[92] Jefferson in 1797 and Dr. B. S. Barton in 1807 both advocated plans for simultaneous recordings in different parts of the country,[93] but no such plan was put into effect for another quarter of a century.

Jefferson on Climate. Attempts to discover regional differences in climate and set them forth in quantitative terms are recorded in Jefferson's *Notes on the State of Virginia* and in Williams's book on Vermont. Jefferson made it a practice to keep weather records during much of his life,[94] even on days when he was occupied with more weighty affairs. (In his *Account Book 1776–1778*, the following appears for July 4, 1776: "pd Sparhawk for a thermometer £3-15 and recorded temp. four times. at 6 A M: 68°, 9 A M 72¼°, 1 P M 76°, and at 9 P M 73½°"; on July 8, he "pd Sparhawk 4-10 for a barometer."[95]) His wind vane and other meteorological instruments may still be seen at Monticello. In the *Notes* he appears to have gone about as far as it was possible to go at the time in portraying the climate of Virginia quantitatively—and this was not very far.[96] Most of the quantitative climatic information presented in the *Notes* consists of:

(1) A one-page table setting forth certain results of temperature, rainfall, and wind-direction observations made daily at Williamsburg during the five years 1772–1777.

(2) An explanation that "cotemporary observations" made by Jefferson at Monticello and by his friend, the Rev. James Madison, at Williamsburg during a period of between five and six weeks had indicated that at Monticello the temperature averaged 6⅛° lower and the barometer 0.748 inch lower than at Williamsburg.[97]

(3) The following table of wind directions, "formed by reducing nine months' observations at Monticello" to four principal points and an equal number, "to wit, 421 . . . at Williamsburgh, taking them proportionably from every point":

	N.E.	S.E.	S.W.	N.E.	Total
Williamsburgh	127	61	132	101	421
Monticello	32	91	126	172	421

(In a letter to Jacob Bigelow dated Monticello, April 11, 1818, Jefferson included a table of "the monthly average Thermometer, Rain, and Prevalence of the Several winds stated in days of the year": N 61, NE 29, E 15, SE 16, S 60, SW 66, W 47, NW 71.[98])

Samuel Williams on Climate. On the title page of the Rev. Dr. Samuel Williams's book on Vermont we read under the author's name:

Member of the Meteorological Society in Germany,
of the Philosophical Society in Philadelphia, and
of the Academy of Arts and Sciences in Massachusetts.

Since he listed it first, the doctor was presumably proudest of his membership in the German society, the Societas Meteorologica Palatina at Mannheim, the founding of which in 1780 has been said to have marked "the birthday of modern meteorology." This institution counted members in many different countries, encouraged them to make systematic weather records, established rules for the purpose, and published a journal.[99]

As with the variation of the compass (p. 213, above), Williams sought to present the climate of Vermont against a broader areal background. He furnished four tables setting forth data concerning temperatures for Charleston, S.C., Maryland (lat. 37°), Williamsburg, Va., Philadelphia, Pa., Cambridge, Mass., Rutland and Burlington, Vt., Quebec, and Prince of Wales Fort (Hudson Bay); concerning wind directions, for all but Charleston; rainfall for Charleston, Williamsburg, Cambridge, and Rutland; and "weather" for all but Charleston and Williamsburg.[100]

For each station Williams's table for temperatures (with minor gaps) shows the "mean heat" for each month and for the year, and also the "least" and the "greatest heat." The data for Charleston, Cambridge, Burlington, and Williamsburg were based on five-year records, but those for the other stations on records for one year each; and the dates of the observations range from as early as 1738–1742 for Charleston to 1803–1808 for Burlington. The rainfall table, based on eight years of observation at Charleston (1738–1745), five years at Williamsburg and Cambridge, and one at Rutland (1789), shows the "mean altitude in inches" by months and for the year.

(Governor Drayton in his excellent geographical book on South Carolina [1802] gave no quantitative meteorological data except in three tables, all on one page, showing for Charleston: (1) rainfall in inches in each year, 1795–1801, and to March 27, 1802; (2) "the greatest and least height of Fahrenheit's Thermometer" for each year, 1750–1759, by seasons; and (3) the same by years for the period 1791–1798.[101] Unlike Williams, Drayton gave no mean temperatures—which prompts the observation that popular interest always tends to prefer extremes to means, whether meteorological or otherwise[102] [see above, p. 140]. Although Réaumur had pointed out in the 1730's that records of maximum and minimum observations standing alone were of little scientific value, the French meteorologist Louis Cotte, in 1775, by averaging recorded temperature extremes reached the conclusion that the summer temperature was everywhere the same throughout the earth.[103] Cotte's procedure was comparable in statistical irrationality to Pliny's geometrical irrationality in estimating areas by adding length and breadth [see above, p. 223].)

Williams's wind table shows for each station the number of

observations for each of the eight principal points of the compass (N, NE, E . . .), together with the total; the table for "the State of the Weather" similarly gives the number of observations under the headings "Fair," "Cloudy," "Rain," "Snow," "Hail," "Fog," "Thunder," "Aur. Bor.," and "Hazy," and the total. Since the data for winds and weather are not expressed comparably for the different stations, it is not easy to make much out of them. Jefferson, as may be seen, partly eliminated this difficulty in his little table reproduced above, but did not do so by the method almost universally employed today, that of expressing the individual magnitudes as percentages of totals.

(The idea of *percentage* may have been known in India as early as the seventh century B.C.[104] Percentages, together with the symbol %, were in use in Italy in the thirteenth century after Christ,[105] and the allied concept of decimal fractions was developed in the sixteenth century and probably first systematized by the Dutch mathematician, Simon Stevin, in 1585.[106] The American Isaac Greenwood [1702–1745] devoted twenty pages to decimals in a textbook published in 1729,[107] decimal coinage was adopted by the Continental Congress in 1785,[108] and the Philadelphian, William Barton, in a demographic study published in 1793, presented a table showing "the proportionate number of Annual Deaths, to 100 Annual Births" for some forty cities in Europe and America.[109] Even so, the examples of the use of the term "per cent" before the nineteenth century given in the *Oxford English Dictionary* [under "Cent"] would seem to refer solely to interest or exchange rates. In the most widely studied American textbooks of arithmetic of the eighteenth century, percentage was not treated as a separate topic and "per cent" appears to have been applied solely to rates of interest.[110] Zimmermann [1787] set forth the "proportion of the numbers of inhabitants in Great Britain and Ireland to the population of other States" by means of such confusing ratios as 10 to 19, 37 to 10, 5 to 1, 100 to 237.[111] The long lag in the adoption of the use of percentages illustrates the "principle of the resistance to novelty," which, like its counterpart, the "principle of the persistence of error," has played an important role in the history of geography—as also in that of most human affairs.)

If Jefferson and Williams were pioneers in the United States in the broad statistical study of climatic differences from place to place, Williams was also an American pioneer in another type of climatic research that has come greatly to the fore during the last twenty or thirty years—"microclimatology," or the detailed investigation, by observation and experiment, of atmospheric differences and changes within small areas as affecting and affected by plants, animals, and men. Williams supplemented his more conventional weather records with thermometer observations taken in open fields and pastures and the nearby woods, at different times of year in holes bored in trees, and in wells and springs, and also with ingenious experiments designed to shed light on the quantity of water vapor, air, and heat absorbed and thrown off by trees.[112] Comparison of the temperature readings in the woods and an adjacent pasture gave him a clue to a theory of climatic change, and, like Lewis Evans and Thomas Jefferson, he thought that the temperature of well and spring water gave a good indication of the mean annual atmospheric temperature.

Evans had written in this regard:

> The mean of our heat & Cold in the plains [of Pennsylvania] is 50° or 51°, in the hills 48° or 49°, as we are able to ascertain with great exactness, by examining the State of Water in deep Wells, & in Springs issuing out of the Hills; as we go further southward this mean rises: & I have found in the middle of South Carolina at 66° & 67°.[113]

In a summary of his meteorological journal for the years 1810–1816 Jefferson pointed out that "another index of climate" (besides phenological and other data previously discussed) might be sought in the "temperature of the waters issuing from fountains." An examination of 15 springs in the body of the hill at Monticello had revealed that the water of the coolest was 54½°, the outer air than being at 75°. According to a friend in Maine, in August the temperature of water 4 feet deep in an open well 28 feet deep was 52°, whereas that of the Kennebec River at the same time was 72½°.[114]

From such studies Williams reached a conclusion that the conservationists of today would endorse, namely, that "it is not only from the earth, but from the air and water, that trees and plants derive their nourishment, and increase: And where no waste has been occasioned by man or other animals, it is not impossible that the vegetables may return more to the earth, than they have taken from it; and instead of serving to impoverish, operate to render it more rich and fertile."[115]

Dr. Williams was firmly convinced, along with many others in the country, both at the time and for many years before and after, that the clearing of the forests and the progress of cultivation were bringing about a warmer and a drier climate in eastern North America.[116] This, he believed, was amply shown by comparing present conditions with those described in older works. But, he added,

> although the general effect has been every where apparent, it is not an easy thing to ascertain the *degree*, to which the temperature has changed, in any particular place. When our ancestors first came in to America, thermometers were not invented: And they have not left us any accurate meteorological remarks or observations, from which we can determine the exact degree of cold, which prevailed in their times.

A passage in William Wood's *New England's Prospect* (1634) gave him the clue he wanted. From a comparison of Wood's account of the dates of freezing and melting in Boston Harbor in 1633 with temperature observations that he himself had taken at Harvard from 1780 to 1788, Williams surmised that the increase "of temperature in the winter at Boston, from the year 1630 to the year 1788, must have been from ten to twelve degrees." Between May 23 and November 16, 1789, Williams took simultaneous temperature readings, presumably near Rutland, Vermont, 10 inches below the surface in woods and in a neighboring pasture. He found that in summer the temperature in the pasture was 10 or 11 degrees higher than in the woods, but that in October and November there was no difference. From this he concluded that

> the earth and the air, in the cultivated parts of the country, are heated in consequence of their cultivation, ten or eleven degrees more, than they were in their uncultivated state: It should seem from these observations that the effect, or the degree of heat produced by cultivation, is the same with the change of climate, that has taken place in the eastern part of Massachusetts.[117]

In the second edition of his book on Vermont Williams included an appendix, entitled "Observations on the Change of Climate in Europe and other Places," in which he expressed the view that throughout most of the region extending from Palestine to Germany the climate had become "sixteen or seventeen degrees warmer than it was eighteen centuries ago."[118] Josiah Dwight Whitney, controverting the theory that removal of the forests had caused a change in climate, cited Williams's speculations among other examples of "a tendency to generalize without any sufficient basis of fact"[119] (1894). It is, of course, a canon of statistical inquiry that sound generalizations cannot be based on inadequate sampling.

A ZOOGEOGRAPHICAL CONTROVERSY: QUANTITATIVE CONSIDERATIONS

European disparagement of this continent as unfit for human habitation—the doctrine of a "Miserable New World," as Professor E. T. Martin has called it[120]—aroused the ire of Jefferson and other Americans of his time. They sought to give added force and authority to their defense of America by recourse to quantitative evidence of a sort and to arguments based thereon, although the geography of plants and animals, for obvious reasons, does not lend itself as readily to quantitative treatment as do the physical, demographic, and economic aspects of geography.

Jefferson devoted about forty pages of the *Notes on the State of Virginia*[121] to a refutation of Buffon's and Raynal's theories that the climate of America is hostile to the optimum development of animal and human life. Buffon had maintained that the wild animals of America were in general smaller than those of the Old World, that domestic animals had degenerated in America, and also that there were fewer species here. As evidence, Jefferson presented a table entitled "A comparative View of the Quadrupeds of Europe and of America."[122] It is divided into three parts. The first shows the names of 26 animals that are "aboriginals" of both continents; for 6 of these, the weights of both European and American representatives are shown; for 12, the European weights only; for 2, the American weights only; and for 6, no weights. The second part lists 18 "aboriginals" of Europe only, for 14 of which weights are given, and 74 "aboriginals" of America only, with 34 weights; and the third part lists 8 animals that had been domesticated in both continents, for one of which (the cow) both European and American weights are shown, for 2, the European weights only, and for 4, the American weights only. Thus, for the following 7 animals only are directly comparable weights supplied (although a good deal more evidence, which Jefferson had gone to great pains to gather,[123] is discussed in the text):

		Europe	America
		lb.	lb.
Bear.	Ours	153.7	410
Red deer.	Cerf	288.8	273
Beaver.	Castor	18.5	45
Otter.	Loutre	8.9	12
Martin.	Marte	1.9	6
		oz.	oz.
Flying squirrel	Polatouche	2.2	4
		lb.	lb.
Cow		763	2,500

With admirable caution, Jefferson explained that his comparisons were not intended "to produce a conclusion in favour of the American species, but to justify a suspension of opinion until we are better informed, and a suspicion in the mean time that there is no uniform difference in favour of either [continent]; which is all I pretend."[124] In discussing the same subject, however, Dr. Williams introduced a similar table comparing the weights of fourteen animals found both in Europe and in Vermont; in every instance the weight in Vermont exceeded that in Europe as ascertained by Buffon. Unlike Jefferson, Williams did not hesitate to draw a sweeping conclusion: "The inference is clear, and decisive: It is in America, and not in Europe, that the quadrupeds of a cold climate, attain their greatest magnitude, and highest perfection."[125]

As to the number of species, Jefferson held that his tables, "taken all together," showed that Buffon's claim was "erroneous."

By these [tables] it appears that there are an hundred species [26 + 74] aboriginal of America. Mons. de Buffon supposes about double that number existing on the whole earth. Of these Europe, Asia, and Africa, furnish suppose 126; that is, the 26 common to Europe and America, and about 100 which are not in America at all. The American species then are to those of the rest of the earth, as 100 to 126, or 4 to 5. But the residue of the earth being double the extent of America, the exact proportion would have been but as 4 to 8.[126]

Had the "exact proportion" been maintained, there would have been only 63 species in America or 37 less than the estimated 100. Using essentially the same data, Williams concluded that "in respect then to the different species of quadrupeds, if we are to judge by any enumeration which has yet been made, the greatest force and vigour of nature is found in America."[127] (Williams's estimate was as follows: species peculiar to America, 75; peculiar to the Old World, 100; ratio of area of America to area of the Old World, 141 to 249 [from Guthrie: see below, p. 248] or 75 to 132. Hence, "to preserve an equality," the Old World should have produced 32 more species than was the case.) He did, however, append a footnote of caution: "the enumeration of quadrupeds seems to be too imperfect to afford any accurate calculations of this kind."

THE GEOGRAPHIC DISTRIBUTION OF POPULATION

The quantitative study of the geographic distribution of human population was very difficult until the age of nation-wide population censuses. The first modern censuses of whole populations date from the late seventeenth and early eighteenth centuries.[128] The first nation-wide census of the United States was taken in 1790; and the first censuses of England and France were as late as 1801.

A large amount of miscellaneous statistical information concerning the population of the United States, however, had been gathered before the first national census, on the basis of which scholars in the present century have estimated the total population at various dates[129] and have drawn maps of its distribution.[130] Similar estimates (but no maps) were made in the seventeenth and eighteenth centuries, usually by deduction from the recorded number of militiamen, polls, taxables, families, houses, vital statistics, and the like.

Jefferson on the Population of Virginia. Jefferson's treatment of the population of Virginia in his *Notes* is a good illustration of the complex expedients to which it was necessary to resort in order to obtain a view of the population of a state before the national census figures became available. By a calculation in which he applied some half dozen rather arbitrary assumptions to four precise figures for different parts of the state, Jefferson came up with 567,614 as the total population of Virginia in 1782[131] (actually about 520,000 according to a modern estimate[132]). The only other quantitative item in the *Notes* on the distribution of population *within* the state is a tabulation showing the number of militiamen by counties.[133]

Jefferson's calculation was based on the following *data* (A) and *assumptions* (B):

(A) *Data:*

(1) For the whole of Virginia with the exception of eight counties in the western part:

(*a*) 53,289 free males above 21 years of age;

(*b*) 211,698 slaves of all ages and both sexes;

(*c*) 23,766 persons "not distinguished in the returns but said to be titheable slaves" (that is, slaves over 16).

(2) For the eight western counties: 3,161 militiamen.

(B) *Assumptions* (stated more briefly than Jefferson stated them):

Throughout the entire state:

(1) Half the total population was male;

(2) Half the male population, both free and slave, was under 16.

For the whole of Virginia with the exception of the eight western counties:

(3) Item A.1.*a* constituted three-quarters of the free male population over 16;

(4) The total number of slaves was item A.1.*b* plus double A.1.*c*.

In the eight western counties:

(5) All free males over 16 were militiamen (hence, per items B.1 and B.2, one-quarter of the free population);

(6) The ratio of the total free population to the total slave population was the same as in the rest of the state.

Jefferson explained B.3 thus: "As the number 53,289 omits the males between 16 and 21, we must supply them from conjecture. On a former experiment it had appeared that about one third of our militia, that is of males between 16 and 50, were unmarried. Knowing how early marriage takes place here, we shall not be far wrong in supposing that the unmarried part of our militia are those between 16 and 21. If there be young men who do not marry til after 21, there are as many who marry before that age. But as the men above 50 were not included in the militia, we will suppose the unmarried or those between 16 and 21, to be one fourth of the whole number above 16 [17,763]."

Jefferson does not make clear why he doubled the number of tithable slaves (B.4) and for the western counties reckoned all free males over 16 (instead of excluding those over 50) as militiamen (B.5). Rearranged, the way in which Jefferson arrived at his final figure may be summarized thus (the basic data being shown in italics):

The whole of Virginia with the exception of the eight western counties:		
Free males over 21 (A.1.*a*): *53,289*		(*a*)
Free males over 16 (B.3): 4/3 of 53,289	71,052	(*b*)
Free females over 16 (B.1):	71,052	(*c*)
Free population over 16: (*b*) + (*c*)	142,104	(*d*)
Total free population (B.2): 2 × (d)	284,208	(*e*)
Slave population (A.1.*b*): *211,698*		
Plus 2 × *23,766* (B.6) 47,532	259,230	(*f*)
Total population (*e*) + (*f*)	543,438	(*g*)
The eight western counties:		
Militia (A.2): *3,161*		
Free population (B.5): 4 × militia	12,644	(*h*)
Slave population (B.6): (*f*)/(*e*) times (*g*)	11,532	(*i*)
Total population: (*g*) + (*h*) + (*i*)	567,614	

THE UNITED STATES CENSUS AS A SOURCE OF GEOGRAPHICAL DATA

The Constitution of the United States called for an "enumeration" of the population every ten years as a basis for the apportionment among the states of members of the House of Representatives and, also, of direct taxes. This launched the United States Census. The published returns of the first count (1790) were tabulated by states, counties, and "minor civil divisions," and presented for each of these areal units six figures bearing exclusively upon population: (1) total free whites; (2) free white males, (3) free white females; (4) other free persons; (5) slaves; (6) grand total. Subsequent censuses were enlarged in scope to embrace an immense range of topics, relating not only to the population but to many other aspects of our national life and economy, and, by producing comparable statistics for the whole country, became a primary force in promoting the quantification of American geography.[134] For example, soon after the first census Morse and others began systematically to include in their geographies, gazetteers, and similar works population data for states, counties, towns, and smaller places, although it does not seem to have occurred to anyone in this country to map such data until after the mid-nineteenth century.[135] Morse also introduced in his *American Universal Geography* (1793, 1796) a table for the whole country showing for each state the number of males and females (only in Massachusetts, Rhode Island, and Connecticut were the latter in the majority), but he had to explain that this was restricted to the whites because the census did not provide the necessary breakdown by sexes for the slaves and free colored persons. "A conclusive argument has been derived against polygamy,"[136] he explained, from what was believed to be a worldwide tendency for human males to be born in greater numbers than females.

POPULATION TRENDS GEOGRAPHICALLY CONSIDERED

The growth of population and differences from place to place in the rates of increase were also matters of especial interest to Americans during the early years of independence (as they are today with "the population explosion") and they were discussed at length by Jefferson, Franklin, and writers of lesser fame. Figuring that the population of Virginia had doubled every 27¼ years[137] from 1674 to 1772 and projecting this trend into the future,[138] Jefferson, on the basis of his estimate of the total for 1781, predicted a population of 4,540,912 in the year 1862¾, provided there would be no immigration.[139] The actual census figure for Virginia (including what is now West Virginia) in 1860, however, was only 1,596,318. Jefferson was familiar with the principle that Sir Matthew Hale had expounded as early as 1667 that human populations tend to increase by geometric progression.[140]

Jefferson must have calculated his figure thus:

Population: in 1781	567,614
after 27¼ years, $2 \times 567{,}614$	1,135,228
after $2 \times 27\frac{1}{4}$ years, $2^2 \times 567{,}614$	2,270,456
after $3 \times 27\frac{1}{4}$ years, $2^3 \times 567{,}614$	4,540,912

In general terms, this procedure could be expressed thus: if $P =$ population at the beginning of an era, $n =$ number of times the population doubles before the end of the era, and $P_n =$ the population at the end of the era, then

$$P_n = P \cdot 2^n, \tag{1}$$

$$n = \frac{\log P_n - \log P}{\log 2}. \tag{2}$$

If the population doubles at a constant rate every d years, the length of the era will be nd years. (3)

In contrast to Jefferson, Dr. Belknap in calculating the growth of New Hampshire's population assumed an increase by arithmetic progression.[141] This bit of statistical naiveté—characteristic, perhaps, of its time and place—was caught by the Rev. James Freeman, to whom Belknap had submitted the manuscript, and in a critique published as an appendix in the Belknap *History* Freeman made it clear that "the inhabitants of a country augment, as far at least as depends on natural increase, in the same manner as a sum of money put out at compound interest."[142] Freeman subjected Belknap's figures to a careful analysis, from which he concluded that doubling in New Hampshire had taken place in less than 18 years and he predicted a population of 207,865 in 1800 (it was 183,855, according to the Census). In dealing with the growth of population, Belknap and Williams made use of statistics for baptisms, deaths, casualties, and the like, for certain New England towns,[143] but in their time these data, as well as such figures as there were concerning age, sex, place of birth, and other demographic characteristics were so scattered and divergent as to shed little light on regional differences in the conditions to which they pertained; hence they were not of much geographical use.

(A reader of the three preceding paragraphs in galley proof wondered why, if Jefferson's understanding of the mathematics of population growth were less naive than Belknap's, Jefferson went so far astray in his prediction. Jefferson's error was to assume that the population would continue to double every 27¼ years. What actually happened was that the rate of increase slowed down, so that the population of Virginia barely tripled during the 98 years from 1772 to 1860, which would imply an average rate of doubling during that interval of about once every 65 years. Belknap seems to have ventured no prediction.)

Williams also ventured to estimate the time that had elapsed since the Indians first populated America.[144] He placed their arrival in this continent "with some degree of probability" at a date at least 1800 years before Columbus. This figure he derived from the following assumptions and line of reasoning: (1) the Indian population of America was descended from a single couple and had doubled every 60 years until the time of Columbus; (2) it had not increased since then; (3) "observations that were made in Virginia and Massachusetts," presumably during the period of exploration and early settlement, indicated that the Indian population throughout America at the time of Columbus must have been about one per square mile (a rather broad assumption); (4) hence, the total Indian population must have been about 14,110,874, since geographers (actually Guthrie, 1771, see p. 248, below, Table II) had computed the area of America (both North and South) at that number of square miles; (5) consequently, it would have required at least thirteen and a half centuries for the Indian population to have increased to 14,110,874 (Williams does not explain how he made this calculation, but it conforms with his data and assumption that doubling had occurred every 60 years, as may be shown by thus applying formulas (2) and (3) above: if $P = 2$, $P_n = 14{,}110{,}874$, and $d = 60$, n will be approximately $6.85/0.30 = 22.8$ and $nd = 22.8 \times 60 = 1368$ years); (6) since this period was not long enough to account for the establishment of the densely populated Indian empires in Mexico and Peru, "the duration of which could be traced back four or five hundred years," this amount should be added, making a total of about 1800 years. This, however, would be only the minimum. "The number and variety" of the Indian "*languages* implies and requires a much longer duration, and an higher antiquity." Thus, despite all the figures, we are left up in the air.

(Jefferson had observed that "the territories of the *Powhatan* confederacy, south of the Patowmac, comprehended about 8000 square miles, 30 tribes, and 2400 warriors. Capt. Smith tells us, that within 60 miles of James town were 5000 people, of whom 1500 were warriors. From this we find the proportion of their warriors to their whole inhabitants, was as 3 to 10. The *Powhatan* confederacy then would consist of about 8000 inhabitants, which was one for every square mile."[145])

SEMIGEOMETRIC GEOMAGNITUDES

DENSITIES OF POPULATION

A semigeometric geomagnitude is a hybrid produced by combining two or more geomagnitudes, of which at least one is geometric and one nongeometric. Probably the most generally familiar magnitudes of this sort are those determined by dividing, say, the number of persons or cows or farms or whatever else in a region by the area of the region and thus obtaining figures indicating the number of persons, cows, farms, or whatever else per unit of area. In the case of people, such geomagnitudes are usually known as "densities of population."

Dr. Williams's argument concerning the peopling of America combined two conceptions that were in the air in his day: the semigeometric one of population density and the semichronologic one of the rate of doubling of a population. (A semichronologic geomagnitude, the temporal counterpart of a semigeometric one, is a hybrid derived from the combining of one or more measured or estimated durations of time [chronologic magnitudes], with one or more nonchronologic geomagnitudes.) I have not discovered what inspired Williams to make the combination. He may have developed the idea independently. Others in Europe, however, had been thinking along similar lines. As early as 1679, Anton van Leeuwenhoek, the pioneer microscopist, had estimated the potential population of the earth at 13,385,000,000 (in what may have been the first estimate of its kind) on the basis of a calculation of the population per square German mile in Holland.[146]

The demographer and statistician Johann Peter Süssmilch (1707–1767)[147] had also arrived at much the same figure and had estimated that the maximum limit would be reached, theoretically and barring the effects of disease and war, in about 400 years. Süssmilch made much use of Templeman's area tables,[148] data from which, as we have seen, found their way into Dr. Morse's *American Geography*. Süssmilch estimated that the earth's maximum potential population was 13,932,000,000, on the assumption that the habitable land areas should support population at an average density of 6000 to the square mile (about 280 to the square English mile). He figured, further, that the earth's actual population was 1,080,000,000 and that the period of doubling was 84 years. (Applying formula (2) on page 241, however, I obtain only about 277 and not 400 years until the limit would be reached.) In his first edition Süssmilch had estimated the potential world population at 4,000,000,000, the actual population at 1,000,000,000, and the period of doubling at 100 years, and therefore, that the limit, barring wars and epidemics, would be reached in 200 years.[149]

In the first edition of *The American Geography* (1789) Morse gave no figures for densities of population in the United States. In the second edition (1793) he presented one figure, 60 per square mile for Massachusetts,[150] and in the 1796 edition he added a figure for Rhode Island (53 per square mile).[151] Not until his sixth edition, in 1812, did he furnish a density figure for every state.[152]

A generation ago, Professor Hermann Wagner of Göttingen pointed out that the "fundamental anthropogeographic concept" of population density was not introduced into geographical science until considerably later than was the case in the politico-statistical literature of the eighteenth century. In the latter, "densities were believed to provide the neatest possible means for comparing the cultures of nations." In the geographical manuals, on the other hand, they remained for many decades little more than a dead letter (*totes Beiwerk*) seldom used for comparisons and even more rarely investigated with a view to determining the causes of the regional differences that they disclose.[153]

(The German geographer and statistician Crome wrote in 1785: "One cannot arrive at a just and true judgment about the condition of a country if one fails to take account of its population, as well as of its areal extent, and to think of them both in close combination with one another."[154] Crome's book contains much of interest on the measurement of areas, the enumeration of population, and densities of population, together with a historical sketch of the development of studies along these lines and bibliographical data.)

The first edition (1789) of Morse's *American Geography* contained a table derived from Zimmermann, which set forth the "number of inhabitants in each square mile" for the several countries of Europe.[155] Morse, however, omitted three columns that appear in the original table showing, respectively, the "proportion of the numbers of inhabitants in Great Britain and Ireland to the population of other states," "the density of population in Great

Britain and Ireland compared to [that] of other countries," and the same for the density of population of England and Wales.[156] While we can hardly blame Morse for not including these in a book for Americans, had he wished to compare our country with the European nations in a similar fashion, he might easily have substituted figures for the United States in place of Zimmermann's for Great Britain and Ireland and for England and Wales. But the subject does not seem to have interested him, for he omitted the table altogether from his subsequent editions. Nevertheless, the special tables that he took from Zimmermann for his edition of 1793 showed the number of "persons in each square mile" for eleven European countries and their principal subdivisions, and thus helped introduce to American readers the conception of density of population, if not the term itself.

In Zimmermann's *Political Survey* a note on the table reads: "By density of the population of a country, I understand the sum of inhabitants upon the same number of square miles. It is plain the density is greater, if upon a very small number of square miles a greater number of people can live. Hence the density of population of two different countries is in direct ratio of their population, and inversed ratio of their areas." This suggests that the term was just coming into use in England. I have not encountered it in American geographical writings of the Early National Period. Wagner[157] points out that the concept did not receive its established designation in German (*Volksdichtigkeit* or *Volksdichte*) until the 1820's.[158] Ebeling refers to it as *Dichtigkeit der Bevölkerung*.[159]

Morse's neglect in dealing with population densities in this country contrasts with Ebeling's care. Writing contemporaneously with Morse, Ebeling included the density of population along with the area and total population of each of the ten American states covered in his manual, and in two instances he likened the densities of population of American states to those of European countries: Connecticut's to Portugal's, Pennsylvania's to Sweden's.[160]

In recent years mathematically minded geographers have become increasingly interested in what I have classed as semigeometric geographical quantities, and not only in such simple ratios as population densities, but also in a great variety of averages, indices, coefficients, and other often complex mathematical expressions that describe divers characteristics of the spatial distributions, associations, and other relations of terrestrial phenomena. Thus the density figures that Dr. Williams manipulated none too critically and that Dr. Morse cited somewhat inadvertently are, like the pussy willows of early spring, notable chiefly as the harbingers of larger growths to come.

If any one conclusion may be drawn from the examples that we have been considering, it is that the actual quantitative geographical information available in our early national period was woefully meager and inaccurate, but that the geographical writers of the time enthusiastically made the most of it for what it was worth and often showed a robust confidence in the validity of large generalizations drawn from a few figures with the aid of a few bold assumptions.

APPENDIX. TABLES

Table I. Length of a degree of latitude as calculated at various times, 1528–1910.[a]

Observer	Place	Date	Length of degree[b]		Approximate error[c] (percent)
			Meters	Statute miles	
Fernel	France	ca.1528	110,642	68.75	−0.6
Snellius	Netherlands	1615	107,430	66.75	−3.4
Norwood	England	1635	111,925[d]	69.55[d]	+0.6
Riccioli and Grimaldi	Northern Italy	1654	121,310	75.38	+9.2
Picard	France	1669–1670	111,258	69.13	<0.1
Jean Dominique Cassini	Dunkirk-Paris	1718	111,060	69.03	−0.2
Jacques Cassini	Paris-Mediterranean	1700–01	111,330	69.60	+0.9
Maupertuis	Lapland	1736	111,990	69.59	+0.4
La Condamine	Quito	1735–1744	110,650	68.76	+0.1
Mason and Dixon	Pennsylvania-Delaware	1764	110,875[e]	68.89[e]	+0.9
Hayford	At pole	1910[c]	111,700[b]	69.41[f]	
	At equator		110,576[b]	68.71[f]	

[a] This table is intended (i) to enable the reader to compare Mason and Dixon's degree-length with earlier measurements, and (ii) to furnish a background for a better understanding of the areal measurements of the earth's surface discussed on pp. 216–219, above. Incidentally, the table sheds a little light—if not new light—on the history of a great discovery: that the earth is not a perfect sphere. According to the two Cassinis' measurements, a degree of latitude in the northern part of France was nearly 0.6 mile shorter than in the southern part. From this it was inferred that the length of a degree decreases progressively from the equator

to the poles and hence that, contrary to Newton's inferences from gravity measurements, the earth is not an oblate spheroid flattened at the poles, but a prolate spheroid elongated toward the poles. The measurements of arcs by La Condamine and Bouguer at the equator and by Maupertuis in Lapland demonstrated the inaccuracy of the Cassinis' southern arc and confirmed Newton's theory. See A. R. Clarke, *Geodesy* (Oxford, 1888), pp. 2–14; C. L. Crandall, *Textbook on Geodesy and Least Squares* (New York: Wiley, 1907), pp. 2–8.

[b] Except as indicated below, the lengths of the degrees are those given by O. Peschel, *Geschichte der Erdkunde* (see below, Chapter 1, n. 5), pp. 394–397, 657–661, in toises, transformed (1) into meters according to the ratio 57,057 toises to 111,250 meters (*ibid.*, pp. 397, 658) or 1 toise = 1.9497 meters, and (2) into statute miles, reckoning 5280 feet per mile, 3.2808 feet per meter, and therefore 6.397 feet per toise. (Ebeling's data on the Mason and Dixon survey indicate 6.395 feet per toise; see note e.)

[c] The error has been calculated in each case by comparing the early measurement as shown in meters with the length of the degree at the latitude in question as indicated in the Royal Geographical Society's *New Geodetic Tables* (R.G.S., Technical Series, No. 4, 1927). These tables are based on a modification of J. F. Hayford's determination of the figure of the earth of 1910 established as "standard" by the International Geodetic Congress at Madrid, 1924. A curve derived from these tables and showing the variation in the length of a degree from equator to pole accompanies an unsigned note, "Maupertuis and the Flattening of the Earth," in the *Geographical Journal 98* (1941), 291–293. On this curve are also plotted early degree-length estimates in meters approximately as follows: Norwood, 111,920; Picard, 111,250; Cassini (Dunkirk-Paris), 111,020; Cassini (Paris-Mediterranean), 111,280; Maupertuis, 111,950; La Condamine, 111,580. While these figures differ somewhat from those presented above, the greatest difference (La Condamine, 70 meters) would be reflected by a difference of only 0.06 percent in the approximate error.

[d] Peschel, *Geschichte der Erdkunde*, p. 395n, citing Maupertuis, *Figure de la terre* (Amsterdam, 1738), p. VIII, states that Norwood calculated the length of the degree at "367,196 Füss (feet), d. h. 57,300 Toisen oder um 250 Toisen zu viel." Now 57,300 toises transformed as explained in note b, would equal 111,720 meters, or 69.42 miles; 367,196 feet, however, equal 111,925 meters (69.55 miles), a figure which conforms closely with the 111,920 meters shown for Norwood in note c.

[e] From Mason and Dixon's data, the Royal Astronomer, Nevil Maskelyne, found the degree length to be 363,763 feet; see T. D. Cope, "Charles Mason and Jeremiah Dixon," *Scientific Monthly 62* (1946), 549. C. D. Ebeling, however, gave for Mason and Dixon's degree length 363,771 English feet = 68.896 English miles = 14.9491 geographical miles = 56,888 Paris toises (*Erdbeschreibung* [see below, Chapter 13, n. 72], V (1799), 6).

[f] Meters from R.G.S. *New Geodetic Tables,* miles from meters as explained in note b.

Table II. Certain areas as shown in a table appearing in the works of Templeman, Guthrie, and Morse[a]

Region	Templeman, 1729 Square nautical miles[b]	Templeman, 1729 Error[c] (per-cent)	Guthrie, 1771 Square statute miles[d]	Morse, 1793 Square statute miles[d]	Morse, 1793 Error[c] (per-cent)
Globe	148,510,627	+ 1.2	199,512,595	199,000,000	+ 1.03
Seas and unknown	117,843,821		160,522,026	160,000,000	
Habitable world	30,666,806		38,990,569	39,000,000	
Europe	2,749,349	− 1.63	4,456,065	2,600,000[b]	−30.67
Asia	10,257,487	−20.42	10,768,823	10,500,000	−37.87
Africa	8,506,208	− 0.76	9,654,807	9,500,000	−17.39
North America	3,699,087	−46.58			
South America	5,454,675	+ 7.62			
America	9,153,762	−23.72	14,110,874	14,000,000	−11.62
New Holland				4,000,000	+47.45
Roman Empire	1,610,000		1,610,000[b]	1,600,000[b]	
U.S.A.				1,000,000[e]	+12.09[f]
Borneo	228,000	+ 5.49	228,000[b]	228,000[b]	+ 6.45
Great Britain	72,926	+10.26	72,926[b]	72,900[b]	+10.26

[a] Square miles from Thomas Templeman, *A New Survey of the Globe* (London, 1729; see p. 222, above), pl. 29; William Guthrie, *A New System of Geography . . .* (London, 3rd edn., 1771), p. 33; Jedidiah Morse, *The American Universal Geography*, I (Boston, 1793), p. 61 (see also I [1796], pp. [61]–[62]; I (1819), p. 67). Errors, except as regards that of the United States, are calculated from the areas given in square statute miles in *World Almanac* (1963). See above, p. 221, for the complete table as given in Morse (1793).

[b] Square nautical miles at 60^2 to the square degree.

[c] In reckoning the errors for the areas as given in square nautical miles, these areas were first converted into square statute miles by multiplying them by $(69.5/60)^2$ in the light of Templeman's statement that there are "in reality" 69½ miles in a degree (Templeman, *New Survey*, p. ii; see also M. J. Proudfoot, *Measurement of Geographic Area* [see below, Chapter 13, n. 54], p. 59). The difference between each area in square statute miles as thus ascertained and the area as given in the *World Almanac* for 1963 was then taken as the amount of error and this amount is shown in the table as a percentage of the *World Almanac* figure.

[d] The table shows that (1) Morse's areas for the Globe, Seas and Unknown, Habitable World, Asia, Africa, and America are merely those of Guthrie in round numbers, (2) Templeman's and Guthrie's figures for the Globe are the sums of their figures for the Seas and Unknown plus the Habitable World, and (3) the last are the sums of their figures for the four continents. Since the area of the earth's surface according to the *World Almanac* (1964, p. 454) is 196,951,072 square statute (or English) miles, the foregoing figures in Guthrie's and Morse's tables must stand for statute miles and not nautical miles. These figures in Guthrie are probably due to a well-meaning attempt to bring the first part of Templeman's table up to date, in which areas expressed in square statute miles were introduced and the inconsistency with the title and the rest of the table was overlooked. Had Morse been more critical, he would have compared his figure

199,000,000 with data that he himself presented elsewhere and seen that it *must* stand for square statute miles. Thus, in his *American Universal Geography* (I [1793], p. 45), Morse explains that "Mr. Norwood found, by accurately measuring from London to York in 1635, that one degree contained 69½ statute miles, nearly; consequently, if the whole 360 degrees be multiplied by 69½ we shall find the circuit of the whole earth, in measured miles to be 25,020. The accurate measure is 25,038," calculated from a more accurate statement of Norwood's estimate of the length of a degree at 69.55 miles; see Table I. From the last figure, Morse could have derived an area for the earth's surface of about 199,550,000 square statute miles. In his 1796 edition (I, p. 53) Morse explains that "if a degree be 69.2 miles, the circumference is 24,912" and "the superficial content is 197,502,336 square miles."

[e] Morse, *American Universal Geography* (1819), I, pp. 209–210, gives the area as of 1810 as 1,968,990 square miles.

[f] Error calculated from area as of 1790 given in U.S. Bureau of the Census, *Historical Statistics of the United States* (Washington, 1960), p. 25. The corresponding error in Morse's figure for 1810 (see preceding note) is +14.45 percent.

CHAPTER 14

Notes on Early American Geopiety

The mountains skipped like rams, and the little hills like lambs.

Fire and hail; snow and vapour; stormy wind fulfilling his word:
Mountains, and all hills; fruitful trees, and all cedars:
Beasts, and all cattle; creeping things, and flying fowl:
Kings of the earth, and all people; princes, and all judges of the earth . . .
Let them praise the name of the LORD.

The high hills are a refuge for the wild goats; and the rocks for the conies.
The LORD . . . looketh on the earth, and it trembleth: he toucheth the hills and they smoke . . .
Let the sinners be consumed out of the earth, and let the wicked be no more. Bless thou the LORD, O my soul. Praise ye the LORD.[1]

THESE GLORIOUS verses express pious emotion evoked by the wonder and the terror of the earth in all its diversity.[2]

SOME EXAMPLES OF EARLY AMERICAN GEOPIETY

In somewhat the same spirit as that of the Psalmist, though less rapturously, Jonathan Edwards wrote:

He that is travelling up a very high mountain, if he goes on climbing, will at length get to that height and eminence as at last not only to have his prospect vastly large, but he will get above the clouds and winds, and where he will enjoy a perpetual serenity and calm. This may encourage Christians constantly and steadfastly to climb the Christian hill.

That high towers or other high places are commonly smitten with thunder, and mountainous places more subject to terrible thunder and lightening, shews how that pride and self-exaltation does particularly excite God's wrath.[3]

In his *Essay on the Invention of the Art of Making Very Good if Not the Best Iron from Black Sea Sand* (1762) Jared Eliot of Connecticut compared the circulation of iron from ore through smelting into manufactured articles, and then through rusting back into ore, with the circulation of water from the seas via clouds, springs, and rivers back to the seas; he also appended a comment on how this shows that "the Doctrine of the Resurrection is neither inconsistent with, or contrary to true Principles of Philosophy [that is, science], as founded on Observation and Experience."[4]

GEOPIETY, GEOTHEOLOGY, AND GEOTELEOLOGY

The foregoing quotations express emotional, or thoughtful, or thoughtfully emotional piety aroused by awareness of terrestrial diversity of the kind of which geography is also a form of awareness—circumstances, that is, of *geodiversity* (see above, p. ix). There being no established term for geographical piety of this sort, I have coined one, *geopiety* (adj., *geopious*), in which Greek and Latin roots are unconventionally but perhaps not too unhappily married (Greek with Greek might be *geohosiety*). Geopiety could be regarded as a province in a larger kingdom of *georeligion* (or *geoeusebia?*), and the latter, in turn, as part of the still greater unnamed empire where religion and geography meet. Similarly, geopiety could be viewed as including various counties and other districts, among them *geotheology*[5] and *geoteleology*, and it is mainly with the last-named that this paper will deal. Unlike the boundaries of political units, those separating the various portions of the empire have not been accurately determined and demarcated and in many areas there is an overlapping of jurisdictions.

(Georeligion is religious awareness [that is, perception, emotion, cognition, belief, ratiocination, and so on] concerning any manifestation of geodiversity. It is *religion* that has to do with geographic actualities. *Geography* having to do with religious actualities [that is, with the geodiversity of religion and associated phenomena] has usually been called "the geography of religion" or, more ambiguously, "religious geography" ["religious" here referring to the subject matter rather than to the inherent nature of the geography in question]. Shorter terms for geography of this sort might be either *religiogeography* (*eusebiogeography?*) or *theogeography*. [A "theogeographer" need not be religious.]

(Georeligion and theogeography are not the only kingdoms in the empire where religion and geography meet.[5] Geography and religion borrow from, influence, or use one another, or meet in other ways than those implied by my definitions of these two kingdoms. Since there is no established terminology for the different parts of the empire, one must either resort to coined or specifically defined terms to identify them, or else use a great many words and end up in complete bafflement. Indeed, the logical relations of the different parts of the empire cannot be perceived at all clearly without employing the language of symbolic logic. This, however, is not necessary here.

(The awareness constituting geopiety [and, more broadly, georeligiousness in general] could be called geopious awareness, and by extension the adjective geopious could be applied also to those who experience such awareness and to the actualities that inspire them to experience it. Thus we might conceive of such geopious places as Mecca, Lhasa, Jerusalem, sacred groves, and so forth as arousing geopious emotions on the part of geopious persons. Such sentiments are analogous to those with which the celestiopious [for example, sun worshipers] contemplate heavenly objects, the iconopious contemplate icons and idols, the anemopious contemplate spirits, the anthropopious saints, heroes, or anthropomorphic gods, the scientopious science. Anemopiety or anthropopiety may lend an aura of geopiety to localities with which spirits, gods, saints, or heroes have been or are believed to be associated.[6])

Like Mussorgsky, I shall present some "pictures at an exhibition" and not a sonata or symphony. The pictures were assembled during a rambling historic tour through a small part of the Province of Geopiety. Collected out of American geographical and semigeographical writings of the period before the Civil War, they fall far short of representing all of the many kinds of geopiety there are and have been among different religions, or even those to which expression was given in this country during the Early American period. Bearing, as they do, almost exclusively upon the teleologically explanatory side of Early American geopiety, they do scant justice to its no less important and interesting spiritual, adorational, worshipful, prayerful, aesthetic, and other sides. They do, however, impinge inescapably upon certain of its moral implications, although from a geographical rather than an ethical point of view.

THE EVAPORATION OF AMERICAN GEOPIETY

For present purposes I am venturing to define theology as intellectual piety rather than as study of the nature of piety (the latter, which might not be at all pious, would bear the same relation to piety that *geosophy* bears to geography; see above, p. 83). From this it would follow that geotheology is intellectual geopiety.

Before the Civil War geopious expressions, both naïve (or nonintellectualized) and in more sophisticated geotheological forms, were frequent in American geographical writings, if not in the purely descriptive passages, certainly in those that sought to explain the origins or to interpret the influences of terrestrial circumstances. Indeed, during Puritan times, or, say, into the mid-third of the eighteenth century, American geographical understanding as manifested in print (especially in New England) absorbed piety from the surrounding intellectual atmosphere as a towel does moisture from a down-East fog. Since then there has been a gradual secular change in the intellectual climate, reflected in a progressive if spasmodic decrease in the "humid" components of openly expressed piety in scholarship of other than a specifically theological nature.[7] This advancing desiccation has affected urban and university communities more than others, and there still are, of course, large parts of the country where the humidity has not declined much; indeed, with the religious revival and revivalism of the last two or three decades, there may have been a perceptible rise in the average humidity of the country as a whole. The over-all effect of the theological drought, however, has been to evaporate the patent (if not the latent) piety out of American geographical scholarship and education to such an extent that American geography for many years has been as theologically dry as towels are likely to be aqueously dry at high noon in Death Valley.[8]

The drought has occasioned a certain obliviousness on the part of the majority of living American geographers to the history of geopiety and to the possible bearing of this history upon the nature of some of their own ideas. When they have written about the past development of geographical thought, they have tended to disregard the parts that religion in general and piety in particular have played in the process. Consequently, study of the subject has been left almost wholly to nongeographers, and these, though they have written a good deal about it, have not considered it as a field in itself or in relation to the history of geography, but as incidental to other scientific or humanistic interests.

(In this connection the following remarks by S. Pines seem pertinent: "The history of science, in contradistinction for instance to many approaches to the history of philosophy, has, as it were, a built-in principle of valuation. In assessing scientific achievements of the past, it tends to judge their value by their similarity or opposition to the conceptions of modern science."[9] There are those who regard the history of geography as of interest with reference solely to what the geography of earlier times has contributed to that of the present.)

THE NATURE OF EARLY AMERICAN GEOTELEOLOGY

It should become evident as we go along that the pictures in this "exhibition" are representative of different schools of Early-American thought as regards the nature of God. The God of the Puritans who not infrequently intervened to upset the orderly operation of his own laws—the God envisaged by Morton, the Mathers, and Jonathan Edwards—was very different from the law-abiding God of the Jeffersonians, and yet more different from Thomas Ewbank's God the Mechanician. But, though of different schools, the pictures (like landscapes as distinguished from portraits, still lifes, and so forth) are all of one general type in having a common subject, *geoteleology*. This form of geopiety calls for explanation.

In a textbook on physics and cosmology much studied at Harvard (and, to a lesser degree, at Yale) during the period between 1687 and about 1728, Charles Morton (1626/27–1698) concisely defined the larger concept of which geoteleology is a part. The "end" of the world, he wrote, is primarily God's glory and secondarily the use of man.[10] Rather than mere termination, by "end" he meant the goal (purpose; "end sought") or what is implied by the Greek *telos*. While the concept of teleology is not necessarily restricted to the operation of God's will toward the fulfillment of the ends that God has or has had in view, this is what is usually meant by the term, and on this basis geoteleology could be defined, after Morton, as the concept or doctrine of divine providential dispositions of terrestrial circumstances primarily for God's glory and secondarily for the use of man.

Enlarging upon the view that the world's primary "end" was God's glory, Morton observed that men do not fly like birds, dive like fishes, become invisible like angels, "because the infinitely wise God has ordered for everything what is best: his will and wisdom being the measure of all perfection," and Morton brought his book to an end with the couplet:

> The World of Parts perfection is Possesst
> and Every part of Vertues which are best.[11]

John Bartram concluded a technical discussion of the mechanics of the natural increase in soil fertility on somewhat the same note, with the obiter dictum: "We see ye works of providence adapted so as mutually to assist each other."[12] Jonathan Edwards felt that "God does purposely make and order one thing to be in agreeableness and harmony with another."[13] And Adam Seybert wrote in 1798 that marshes and swamps (more normally regarded as the breeding places of disease) "appear . . . to have been instituted by the Author of Nature" in order to prevent the deleterious overcharging of the atmosphere with oxygen. "I am of opinion that ere long marshes will be looked upon by mankind as gifts from Heaven to prolong the life and happiness of the greatest portion of the Animal kingdom."[14] No doubt many other examples could be given bearing witness to the emphasis laid in Early American writings upon the universal benevolence of God. Man, however, was deemed to be the principal beneficiary of that benevolence. (Insofar as geoteleology relates to man as God's tenant on the earth, it should perhaps be qualified as "human" or "anthropic" or "anthropogeographic." Whatever its proper designation, I shall call it "geoteleology.")

As in geology, the geoteleological dispensations of Providence may be considered *dynamically*, as forces or processes; *historically*, as events; or *regionally*, as arrangements of conditions and processes in space. The principal dynamic forces and processes considered in geology are physical, chemical, and biological; comparable geoteleologic forces might be classed in terms of God's intent as *awardative*, *punitive*, and *corrective*. The God of the Old Testament and the Puritans—and it is with that God rather than with the God of Jeffersonians or the several Gods of latter-day Americans that we are mainly concerned—was looked upon as having created and as operating and modifying the world with reference to man somewhat, if not exactly, as a kindly father strives to "create" and arrange matters with reference to his children—(1) awardatively, by providing in advance for their material and spiritual well-being or by rewarding them (or promising to do so) for virtuous behavior; (2) punitively, by punishing, or threatening or providing for the future punishment of, vicious behavior; (3) and correctively, with a view to effecting an improvement of such behavior. In the eyes of some Puritans, however, especially those who had come under Calvinistic influences, the punishment, unlike that administered by a kindly father who tries not to show partiality toward his different children, was held corrective with regard to the elect only; for the rest of mankind it was to be worse than exterminative, namely, eternal damnation.

Thus the rigors of the New England winter, a geophysical phenomenon, inspired in Cotton Mather these geopious thoughts:

> But if the *Winter* brings much of MERCY to us, it brings much of Hardships too. Pliny calls the *Snow*, and the *Ice*, the *Punishments of the Mountains*. We who dwell in a *plain* region, as well as they who dwell upon the Rigid and Ragged Edges of such *Mountains* would be sore *Punished* by the Hardships of the Winter, if the Mercy of our God should not relieve us.[15]

From a decade or so later we read in Mather's *Diary* for November 14, 1711: "I am now getting on my Winter-Garments. I would do it with agreeable dispositions of Piety." And when the days began to lengthen, the cold—and also the piety—began to strengthen. Under date of January 16, 1711/12 Mather wrote: "It is now very extreme Winter: and we are now in the Extremity of it," which occasioned him to "form certain Supplicationes Hyemales, or Winter-Desires and Prayers," among them these:

My Garments. Lord, I am not afraid of the Winter because of my double Cloathing . . . lett me also be cloathed with thy spirit . . . Oh glorious, Oh durable Cloathing.

My fuel. Lord, enable me to warm all that are about me with holy Dispensations . . . And, oh save me and mine from the eternal burnings.[16]

From the *historical* point of view the operation of these providential forces and processes, together with the events and conditions engendered thereby, may be classed as either long-term or short-term. The remainder of this paper will be devoted mainly to illustrating, first, the former in a section in which ideas concerning the enduring geographic consequences of the Deluge and of the End of the World will be considered, and then the latter in a section in which punitive and awardative providences will be treated separately. The last part of the paper will deal with *regional* geoteleology.

THE MAJOR GEOTELEOLOGIC PROVIDENCES

The major geoteleologic providences were the original Creation of the World during the six days as described in the first chapter of Genesis, the Deluge as described in the seventh and eighth chapters, and the End of the World as foretold principally in the book of Revelation. The Creation was awardative, designed to provide a world for God's glory and man's use and delight; the Flood and the End of the World were essentially punitive, but, for some, corrective or awardative. The accounts of these three events provided a temporal and spatial framework for nearly all theories concerning the history of the universe, both as it has been in the past and will be in the future, that were developed in Christian communities until the nineteenth century. During the seventeenth and eighteenth centuries, the advancement of the sciences—in particular of astronomy, geology, anthropology, archeology and geography—was generally conceived as furnishing data that could be built into the framework, somewhat as walls, floors, doors, and so on are built into the structure of a skyscraper; and it was normally thought that this could be done without enlarging the frame. By the eighteenth century, however, it had already become obvious to some that major temporal and spatial enlargements were necessary, and much of the so-called "warfare of science with theology" in the period since 1700 has been concerned with the questions of whether or not such enlargements were religiously and ethically justifiable, scientifically needed, and practically possible without a complete alteration of the whole structure and nature of the edifice. The history of how these changes gradually came to be made, however, has been told too often and is far too complex either to require or to permit repetition here.[17]

THE DELUGE AND ITS GEOGRAPHIC CONSEQUENCES

Of the Biblical accounts of the three major geoteleologic providences I shall consider only that of Noah's Flood in any detail, because the story of the Flood was by far the most productive of geographical speculations, because so much has already been written concerning the history of interpretations—geological and otherwise—of the story of the Creation, and because I have not attempted to study either this history or that of eschatological beliefs.

Theorizing about the Flood has probably been going on ever since men had sufficiently recovered from its effects to begin again to theorize about anything.[18] None too explicit, the account in Genesis has been variously interpreted. To most Early American interpreters, however, the Deluge was "universal," in that it covered the entire surface of the earth, including the highest mountains, required a prodigious quantity of water, and exterminated the entire human race (except for Noah and his family) and every other "living substance . . . which was upon the face of the ground." The scriptural narrative and the fact that it *was* scriptural gave explorers and travelers a theologically respectable, prefabricated explanation of things they observed in the field but might not have dreamed of so explaining had not the Biblical tradition stamped the story of Noah indelibly upon their minds. After about the year 1600, the story also inspired ingenious men of learning, endowed with varying degrees of familiarity and competence in dealing with the newly unfolding revelations of science, to try their hands at applying them to explanations of precisely how the wrath of God was made manifest in the Deluge and how the latter, in turn, affected the face of the earth and its inhabitants. These ingenious speculations were carried on mainly in Europe, and so far as we are concerned in this paper, mainly in Britain during the last thirty years of the seventeenth century, and, naturally, they gave rise to prolonged and bitter controversies.[19] Indeed, the first formulation of these hypotheses might be likened to the fountainheads of rivers of ideas, which gushed forth from a common source to fructify broad realms of British and American science and literature for generations, somewhat as the four rivers of Paradise of medieval geographical lore fructified broad kingdoms of the world—the flaw in this metaphor being the lack of resemblance between the turbulent atmosphere around the sources of these rivers of thought and the calm airs of the Terrestrial Paradise.

The earliest American poetess of any talent, Mrs. Anne Bradstreet, in allegorical verses on *The Four Elements* (1650) had Water (personified) comment on various floods and then exclaim:

> But these are trifles to the flood of Noe,
> Then wholly perish'd Earth's ignoble race,
> And to this day impairs her beauteous face . . .[20]

She was inspired, no doubt, by the very old notion that God's aquatic punishment of earth's ignoble race had extended even to the innocent earth itself.[21] A generation or so later this idea was given imaginative expression by three renowned English "world-mongers" or "world-makers," Thomas Burnet (1681), John Woodward (1695), and William Whiston (1696).[22]

The English "World Mongers" (1681–1696). Burnet, along with Varenius and many others before him,[23] held that the antediluvian world was smooth; he likened it to an egg, spinning in space on an axis perpendicular to the plane of the ecliptic (hence in those days there was eternal spring in mid-latitudes); while the first cause of the Flood was God's wrath, its efficient or "scientific" cause was the breaking open of the earth's crust by the sun's heat. This allowed the waters to escape from inside the earth, "threw the axis of the globe aslant . . . [and] begot the disfiguring mountains and the deep hollows into which the waters retreated to form the oceans."[24] Woodward's antediluvian earth, unlike that of Burnet, was essentially similar to the earth we know, with mountains, oceans, and so on. It, too, was destroyed and remade by the Flood, which in this case dissolved much of it. Upon the subsidence of the waters, the matter held in suspension was laid down in strata and then subjected to further mountain-making processes through thermal action. In both hypotheses, the Flood left the earth a total wreck, a heap of hideous ruins. Even more ingenious and spectacular was Whiston's theory (1696). It postulated that the chaos out of which God created the earth was the debris of a comet and that another comet had brought on the Flood by coming close to the earth, spraying it with water from its own atmosphere, and opening fissures in the earth's crust which released water from the abyss in the earth's interior.

Physical Geodiluviology: the Colonial Period. In his *Brief Retrospect of the Eighteenth Century* (1803), Samuel Miller, our first American predecessor of those now ambiguously called "intellectual historians," devoted some forty pages to a survey of the theories of these three great "world-mongers" and of some twenty or more others, from Burnet (1681) to Kirwan (1799), mostly European. The eighteenth century "teemed" with them, wrote Miller, adding that "these plans, to say nothing of the impious nature and tendency of some of them, have generally rather resembled philosophical dreams, than the conceptions of sober and waking reason."[25] In his *Philosophical Grammar*, which was used as a textbook in America in the middle and latter half of the eighteenth century, Benjamin Martin gave four "rules for philosophizing" based on Newton's *Principia.*[26] The fourth rule was the Baconian one, to the effect that propositions and conclusions derived from actual experiments "must be insisted upon" in preference to "hypotheses or received suppositions" until "some other phaenomena either render them more accurate or liable to exception." On applying this rule, Martin felt that "Mr. Whiston's and all other World-mongers' Systems and theories dissolve into a philosophical nothing, which want actual experiments to support them," "experiment" here implying tests by observation or "experience." In a footnote Martin quoted Dr. Keill: "the Deluge was the immediate Work of the Divine Power . . . and no Secondary Power, without the Interposition of Omnipotence, could have brought such an effect to pass."

To my knowledge, the history of the dissemination of European "world-mongering" theories in America and of their development by Americans has never been written.[27] Samuel Miller records that copies of Burnet's, Woodward's, and Whiston's books were included in the famous gift of 800 volumes by Jeremiah Dummer to Yale in 1714.[28] Cotton Mather wrote in his *Christian Philosopher* (1721):

> The vain Colts of Asses that fain would be wise have cavilled at the unequal surface of the earth, have open'd against the Mountains, as if they were superfluous excrescenses; but warts deforming the face of the Earth, and proofs that the earth is but a heap of rubbish and ruins. Pliny had more of religion in him.[29]

This would seem to reflect Burnet's thought, although Burnet is not mentioned by name. Cotton Mather, however, was familiar with Woodward's book.[30] Harvard College had a copy of Burnet in 1723 and of Whiston in 1725, Whiston was used by Isaac Greenwood and John Winthrop when they were professors at Harvard in the period 1738–1779,[31] and Jonathan Edwards read both Burnet and Whiston.[32] Yet it is probably safe to say that no Early American ever worked out an original Flood theory of his own in the grand manner.

From the early eighteenth century to the mid-nineteenth, however, a great many Americans have followed with interest and participated in the current discussions of the causes and effects of the Flood, and many field observations in this country were interpreted as proof that it had actually occurred on a larger or smaller scale.

In Colonial times Lawson thought that "in Carolina (the part I now treat of) are the fairest marks of a Deluge (that at some time has made strange Alterations as to the Station that country was then in) that I ever saw, or, I think, read of in any History."[33] Catesby was similarly impressed, noting that there is no part of the Globe where the signs of a deluge more evidently appear than in the Northern Continent of America,[34] and especially in Carolina. Jones, in Virginia, thought that the universality of the Flood was proved "when we remark that in most places, at a great depth and far distant from the sea, are many beds of strange shells, and bones, and teeth of fish and beasts vastly different from any land or water animals now found in these or any other parts of the world," and he considered them a "token and relict of Noah's flood."[35] Jared Eliot asserted that peat

> is reasonably supposed to be made of the Wood which grew before the Flood [and] affords us an Evidence . . . that the . . . Scripture account of Noah's Flood is true . . .

When the Trees and Shrubs were torn down by the mighty commotion of Wind and Water, these were hurried to and fro, and at length lodged by their own Weight and Intanglement into low Grounds, and in time was converted to what is now called Peat . . . [A man to whom Eliot showed] what appeared to be some fowl's dung [surmised that it might possibly be] the Dung of that very Dove which came out of the Ark.[36]

There is an echo of Burnet in a letter that Franklin wrote to Jared Eliot in 1747; having mentioned shells high up in the Appalachians, Franklin exclaimed: "It is certainly the wreck of a world we live on!"[37]

Lewis Evans, in an inscription on his map of 1749 relating to the "Endless Mountains" (Folded Appalachians) of Pennsylvania and to the neighboring part of New Jersey, presents briefly the nearest approach that I have encountered in Colonial American literature to an original theory concerning the effects of the Flood upon the face of the earth. It would seem to imply two successive "Creations," followed by a Deluge of less than mountain-covering depth, and it testifies to familiarity with Woodward's theories. I shall quote it, in part, for the reader to interpret as he may:

[The Endless Mountains] furnish endless Funds for Systems and Theories of the World, but the most obvious to me was, That this Earth was made of the Ruins of another, at the Creation. Bones and Shells which escaped the Fate of softer animal Substances, we find mixt with the old Materials and elegantly preserved in the loose Stones and rocky Bases of the highest of these Hills. These Mountains existed in their present elevated height before the Deluge, but not so bare of Soil as now. The further Ridges which are much the largest and highest, proceeding from the inclination of the whole toward the sea, are of very rich Land, even on the Tops; while the very Vallies, on the hither side seem swept of all the Soil. Their Height no doubt rendered them less exposed to that general Devastation, and preserved them unhurt, while the Soil and the loose Parts of lower Hills and Vallies, agitated by a greater Weight of Water, were borne away suspended in the dashing Waves, and thrown downwards in Stratas[38] of different Kinds . . . Dr. Woodward from infinite Examples, discover'd, that this World had been in a State of Dissolution. But the Power he ascribes to the Water of Deluge is too much a Miracle to obtain Belief. We have here glaring Marks of a Deluge of far more recent Date, which the Compass of Britain might not perhaps have furnished the Dr. with.[39]

Evans's doubts as to the universality and miraculousness of the Flood were in keeping with the ideas of the Enlightenment that began to inundate this country during the mid-eighteenth century. Against these a reaction set in toward the end of the century and continued well on into the nineteenth. Thus Dr. Jedidiah Morse (1789) took Evans to task for questioning

the reality of the flood, of which Moses has given us an account. But Mr. Evans thinks this too great a miracle to obtain belief. But whether is it a greater miracle for the Creator to alter a globe of earth by a deluge when made, or to create one new from the ruins of another? The former certainly is not less credible than the latter. "These mountains," says our author, "existed in their present elevated height before the deluge, but not so bare of soil as now." How Mr. Evans came to be so circumstantially acquainted with these pretended facts, is difficult to determine, unless we suppose him to have been an Antediluvian, and to have surveyed them accurately before the convulsions of the deluge; and until we can be fully assured of this, we must be excused in not assenting to his opinion, and in adhering to the old philosophy of Moses and his advocates.[40]

Physical Geodiluviology: the Early National Period. In the original edition of his *Notes on the State of Virginia* (Paris, 1785), Jefferson suggested a chemical theory to explain the origin of the shells high up in the mountains. This was inconsistent with belief in the universal coverage of Noah's Flood, and since it "raised a storm of criticism" he saw that it was omitted from later editions.[41] To account for the shells, in the London edition of 1787 Jefferson cited the hypothesis of the Noachian Deluge along with two other hypotheses (one of them Voltaire's), but concluded by stating that "the three hypotheses are equally unsatisfactory; and we must be contented to acknowledge, that this great phaenomenon is as yet unsolved. Ignorance is preferable to error; and he is less remote from the truth who believes nothing, than he who believes what is wrong."[42] As to the Deluge, Jefferson held that there was not enough water in the atmosphere to raise the sea level more than 52½ feet. "In Virginia this would be a very small proportion even of the champaign country, the banks of our tide-waters being frequently, if not generally, of a greater height. Deluges beyond this extent, then, as for instance to the North mountain or to Kentucky, seem out of the laws of nature."[43]

Jefferson, however, had many worthy contemporaries who regarded literal, or more or less literal, belief in the Bible (for "literalness" is variable[44]), as preferable to belief in nothing, and who were also convinced, as the earlier observers had been, that the actual occurrence of Noah's Deluge was sustained by the results of empirical observations. Samuel Miller thought that "every mountain and every valley lifts up its voice to confirm" "the account of the Deluge in the sacred writings."[45] Hugh Williamson argued that the water for the Flood was miraculously created. It came from the Southern Hemisphere, leaving behind it the shattered remains of a once much greater southern continent. The Ark was swept back to Mount Ararat against the current by northerly winds. The Flood so mingled the waters all over the world as to reduce them to an even temperature "30° or 40° above freezing point." Hence there was no ice over the seas for a long while after the cataclysm. Since then there has been a gradual cooling, with the formation of ice in the polar regions. "No fact in natural history is more certain than that there was more heat, less cold, in high northern latitudes, in the eighth or ninth century than there is at present," a proposition well established today as regards the climate of Iceland and Greenland.[46] The deterioration in the northern climate explains (according to Williamson) "why certain countries, neither desirable nor productive today, were formerly *officina gentium*,[47] the very nursery of nations; and why, in the process of time, it became necessary for those very people to migrate by thousands in quest of better habitations."[48] This foreshadows Ellsworth Huntington's "pulse-of-Asia" theory of a hundred years later (Huntington's theory, however, was postulated upon increasing drought).[49]

Another of Jefferson's younger contemporaries, Benjamin Silliman, Sr., of Yale, who did a great deal for American science in general and for geology in particular, was nevertheless in favor of an essentially literal acceptance of the scriptural account of Noah's Flood; and for opposing this and other similar orthodoxies Dr. Thomas Cooper, a college president, was brought to trial before the South Carolina legislature in 1832.[50] Nor was it until well after the mid-century that Agassiz's glacial hypothesis gained general acceptance over against the older view that the erratic boulders, "drift," and other evidences of an ice age were due to a mighty flood carrying waste-laden icebergs and ice floes far and wide overland. In 1856, J. P. Lesley, a great figure in the history of American geology and geomorphology, maintained that "a cataclysmic deluge wrought our topography" in the northeastern United States and that the "Niagara Chasm" and "all our north and south thorough-cut valleys and ravines" (for example, of the Finger Lakes) were due to it.[51]

Biological Geodiluviology. If the book of Genesis leaves much to the imagination with regard to the physical effects of the Flood, it is hardly more specific regarding the biological effects. The circumstance that "every living substance was destroyed which was upon the face of the ground, both man and cattle, and the creeping things, and the fowl of the heaven," together with the Ark's provision for selective survival, suggests that the repopulating of the earth by men and animals must have emanated from Mount Ararat.[52] Fish and aquatic vegetation appear to have been able to fend for themselves. How land-growing vegetation and flying insects survived is not clear; being neither creeping things nor fowl of the air, the insects may have escaped without having had to seek shelter in the Ark.

(Acosta had "attempted to answer the problems created for Christian believers in the universal deluge by the existence of animals in America which could not be found in Europe" [Smallwood]. He wrote: "Although all beasts came out of the Arke, yet by a naturall instinct and the providence of heaven, diverse kindes dispersed themselves in diverse regions, where they found themselves so well, as they woulde not parte; or if they departed, they did not preserve themselves, but in the processe of time, perished wholly."[53] Smallwood points out that Acosta believed "that the dispersal of the animals and birds to America had been made possible by the actual contiguity of the main lands of the Indies and the New World,"[54] a connection that was shown on many maps until late in the sixteenth century and not finally altogether discarded, despite the long-lived belief in the mythical "Strait of Anian," until Bering's voyages in the early eighteenth century. According to Athanasius Kircher (1602–1680),[55] among the passengers of the Ark there were neither fish nor "reptiles that spawned from putrefaction, because there was enough of that after the Flood."[56] This might account for the insects. In his treatise on the natural history of North Carolina, Lawson included alligators, snakes, terrapins, toads, and lizards among insects.[57])

Lewis Evans had the hardihood to declare that "the brute Creation as well as the Human of America are originally of this Continent, & plainly discover that they never came from the other."[58] As regards the "brute creation," this might have been an echo of the theory of Abraham Milius (seventeenth century) that the Flood did not extend to the mountains of America, which were presumed to be higher than those of the Old World and therefore might have served as a refuge for the descendants of the plants, birds, and animals.[59]

THE END OF THE WORLD: GEOESCHATOLOGY

Although the geographical implications of the End of the World are, indeed, all-inclusive, the book of Revelation is less specific about this forthcoming event than is the book of Genesis about the Creation and the Flood—and this has permitted more scope for the imagination. Burnet thought that the immediately activating agency would be fire erupting out of the interior of the earth; Whiston, that yet another comet would set off the great combustion. "The Conflagration," Burnet wrote, "will begin at the City of Rome, and the Roman territory," "for reasons which were" (as Professor Perry Miller observed) "at least to a Protestant wholly understandable," since "the most sulphurous soil and the most fiery mountains (*viz.* Vesuvius) exist in Italy."[60] According to Jonathan Edwards:

> The torrents and floods of liquid fire that are sometimes vomited out from the lower parts of the earth, the belly of hell, by the mouths of volcanos, indicate or shadow forth what is in hell, *viz.*, as it were, a lake of fire and brimstone, deluges of fire and wrath to overwhelm wicked men, and mighty cataracts of wrath to come pouring down out of heaven on the heads of wicked men, as mighty torrents of liquid fire have sometimes come pouring down from Mount Etna and Vesuvius on cities and villages below. Such things do forebode the general conflagration.[61]

We read in Cotton Mather's *Magnalia*, however:

> The learned Joseph Mede [or Mead (1586–1638), Biblical scholar and commentator on the Millennium] conjectures that the American Hemi-

sphere will escape the conflagration of the earth, which we expect at the descent of our Lord Jesus Christ from Heaven; and that the people here will not have a share in the blessedness which the renovated world shall enjoy, during the thousand years of holy rest promised unto the Church of God.[62]

Since Cotton Mather's day, or certainly since Jonathan Edwards's, interest in the End of the World has gradually waned almost to the vanishing point in this country except among the adherents of a few fundamentalist sects. As Miller put it, eschatology has become "virtually a lost art."[63] Those with a taste for the out-of-the-way might do well to explore the almost unvisited domains of the history of *geoeschatology* and *eschatologicogeography*. The former, a branch of geotheology, might have to do with geographic aspects of the earth's final coming to an end, as a matter of active belief, the latter, a branch of theogeography, might deal with the geographic distribution and regional manifestations of divers eschatological beliefs. The best original sources would be hell-fire sermons.

SOME LESSER GEOTELEOLOGIC PROVIDENCES

If the Creation, the Deluge, and the End of the World have been or will be the supreme manifestations of geoteleologic action, lesser manifestations have frequently been noted in American writings. I shall present some examples mostly from the Colonial period, first from the realm of punitive and then from that of awardative geotheology. An illuminating study in comparative states of mind might be made of the relative amounts of attention devoted to the punitive and to the awardative, respectively, both in general and as between different parts of the country (say New England and Virginia) at different times. I have made no such study, and my examples are drawn, rather lopsidedly, chiefly from New England. Divine portents or warnings of scourges to come, as well as actually inflicted punishments, are classed as punitive.

PUNITIVE GEOTELEOLOGY

Charles Morton, distinguishing between the upper and the inferior regions of the earth, explained that the former, being "more lax and open," is entered by the sun's heat, and showers and moisture, thus giving place for the roots of vegetables. Since it is "more manifestly Useful for the Dayly support of man," it "fell more expressly under the curse of man's Sin, which curse consists of sterility of the good and fertility of the bad," the former implying the nonproduction of that which is good without much labor and industry, and the latter, the free production of what is comparatively evil, such as weeds that hinder the fruition of the former.[64] No gardener is unaware of this.

Since comets neither kill or injure nor perform recognizably helpful services to human beings, they are today regarded as curiosities, on a par, perhaps, with the duck-billed platypus. In former times, as we have seen in mentioning Whiston, they stood high in the esteem of the learned, the superstitious, and the learnedly superstitious. Increase Mather wrote a book, *Kometographia*, published in 1683, in which he expressed distress that "people were misinterpreting [comets], were trying not to regard [them] as voices of heaven foretelling approaching judgment; he took pains to show how they might, even by purely mechanical means, produce droughts, caterpillars, tempests, inundations, and epidemics, but whether or not they had such effects, he was convinced that they were portents of God's displeasure."[65] Surely Whiston's theory, published eight years later, must have interested him, if he knew of it; it may, however, have been too heretical for his approval.

In his *Brief History of Epidemic and Pestilential Diseases* (1799) Noah Webster, of dictionary fame, attributed epidemics to the state of the atmosphere as affected by the moon, comets, earthquakes, volcanic eruptions, electricity, and so forth. He felt, however, that the general purpose of the evils afflicting man, including diseases, was "to create and preserve that sense of obligation and accountability to God which is the germ of piety and moral excellence,"[66] but this was about all he said on this aspect of the subject. In keeping with the spirit of the Enlightenment, he refrained from dilating on divine wrath as a cause of epidemics,[67] as he might well have done had he lived in Cotton Mather's time (it should be said, though, to Cotton Mather's credit that he took a remarkably enlightened attitude toward the scientific study and combating of diseases rather than advocating a supine, if "pious," acceptance of such afflictions[68]).

Cotton Mather wrote: "I take earthquakes to be very moving preachers unto wordly-minded men,"[69] and the play on words may have been intentional. Between them, the New England earthquakes of October 29, 1727, and November 18, 1755, inspired the publication of at least 33 sermons.[70] *Seismotheology*[71] was an even livelier subject than *cometotheology* in Colonial times.

The great floods during the summer of 1683, especially of the Connecticut River but also of "the mighty river Danow (the biggest in Europe)," and in southern France, Virginia, and Jamaica, inspired Increase Mather to write: "There is an awful intimation of Divine displeasure in this matter . . . thus doth the great God, 'who sits King upon the Floods for ever,' make the world see how many wayes he hath to punish them, when it shall seem good unto him."[72] Aridity was also a manifestation of divine displeasure: Cotton Mather asserts that clouds are "carried about by the winds as to be so equally dispersed, that no part of the earth wants convenient showers, unless when it pleases GOD, for the punishment of sinful People to withhold Rain."[73] Professor Kittredge showed that when, in seventeenth- and eighteenth-century England and New England, the theory that storms were raised by devils was combated in the pulpit, the argument was likely to be that they are sent by God as a punishment for sin. Kittredge gives the title of a tract published in London in 1704, the work of John Hussey, pastor of the Congregational Church in Cambridge, Massachusetts. Abridged, the title was: *A Warning from the Winds. A Sermon preach'd upon Wednesday, January xix, 170¾, Being the day of Publick Humiliation, for the late terrible, and awakening storm of wind, sent in great rebuke upon this kingdom . . . as a punishment of that General Contempt, in England cast under Gospel-Light, upon the work of the Holy Ghost . . . to which is subnected a Laborious Exercitation upon Eph. 2. 2 About the Airy Oracles, Sibyl-Prophetesses, Idolatry, and Sacrifices of the elder pagan times.* In the sermon the preacher cried: "Let God have the honor of these blasts: Entertain not a thought that the winds were raised by Satan, Witches, Cunning-Men, or Conjuration"[74]—which brings us to the subject of *geodemonology*.

Geodemonology. Demonology is to devils and the diabolical what herpetology is to reptiles and the reptilian. We have this on the authority of the *Bestiary* (12th century), which maintains that "the Devil . . . is the most enormous of all reptiles."[75] Furthermore, somewhat as herpetology is a branch of vertebrate zoölogy, so is demonology a branch of punitive theology. This seems reasonable because the status of being a devil (which differs from merely acting like one) has usually been regarded as a consequence of punishment and also because the principal occupation of devils is to punish nondevils. Geodemonology, accordingly, is that branch of geotheology which relates to geographic actualities when conceived as either devils, or persons or populations possessed by devils, or else as haunts of devils, or as otherwise exhibiting diabolic or demonic attributes. The *Inferno*, *Paradise Lost*, and *Moby Dick* are (among other things) treatises on geodemonology.

Dr. Increase Mather explained that Satan and his minions (demons, witches, and the like) "are limited by the providence of God, so as that they cannot hurt any man or creature, much less any servant of his [one of the elect ?] without a commission from him whose kingdom is over all."[76] This raises larger issues concerning the relation between God and Satan that lie outside the immediate scope of the history of geography; our sole concern here is with *geographic* manifestations of the diabolical.

Dr. Mather, whose *Essay for the Recording of Illustrious Providences* was published in 1685, the year in which he became President of Harvard College, thought that "it is not heresie to believe that Satan has sometimes a great operation in causing thunder-storms," that both scripture and "histories" have "abundantly" confirmed Satan's powers in the air and ability to cause storms, that "there are devils infesting this lower world"—and that they are not mere symbols of human sin and temptation. We usually think of the lesser demons or witches of the Mathers' day as operating on a rather close and intimate personal basis within narrow topographical limits (for instance, those of Salem), but Dr. Mather cites a case where the bewitched sister in Hartford, Connecticut, of a man who was killed by lightning 40 miles away at Northampton, Massachusetts, was immediately informed of "that terrible accident" by the "demons which disturbed her."[77] Cotton Mather, in a chapter of the *Magnalia* entitled "Ceraunius: Relating Remarkables done by Thunder," makes it clear that, though the natural causes of thunder were known to us [*sic!* B. Franklin and his kite came more than fifty years later], " 'tis likely that evil angels may have a particular energy and employment, oftentimes in the mischiefs done by thunder . . . Satan, let loose by God, can do wonders in the air . . . New England hath been a countrey signalized with mischiefs done by thunders, as much as perhaps most in the world. If things that are smitten by lightning, were to be esteemed sacred, this were a sacred country."[78] In his *Wonders of the Invisible World* he thus explained, in part, at least, the prevalence of witchcraft in New England: "Where will the Devil show most malice but where he is hated and hateth most."[79] Churches and ministers' houses were susceptible because "the daemons have a particular spite at houses that are set a-part for the peculiar service of God."[80] If New England's most learned scholars held such notions, so also did the Indians. Alsop records that the Susquehannocks owned "no other Deity than the Devil. The Priests . . . oft-times raise great Tempests when they have any weighty matter or design in hand, and by blustering storms inquire of their Infernal God (the Devil) *How matters shall go with them either in publick or private.*"[81] This was a mode of divination surpassing the fondest dreams of the Roman soothsayers.

Cotton Mather wrote of the many disasters and afflictions that fell upon certain irreligious English who sought to people and improve the parts of New England to the northward of "New-Plymouth," "until there was a plantation erected upon the nobler designs of christianity" (Boston, of course). Although some fine settlements were established in the northeast regions (Marblehead), the settlers were more interested in catching fish than "to approve themselves a religious people." Hence they were wiped out by catastrophies. "Yea, so many fatalities attended the adventurers in their essays, that they began to suspect that the Indian sorcerers had laid the place under some fascination; and that the English could not prosper upon such enchanted ground, so that they were almost afraid of adventuring any more."[82] This implies a belief (not necessarily shared by Cotton Mather himself) that the Indians' devils could injure white people. The same idea was entertained in New Hampshire in 1784. Dr. Jeremy Belknap records that during his trip to the White Mountains in that year "the good women" of Eaton and Conway, "understanding there were 3 ministers in the company, were in hopes we should lay the spirits [presumably evil] which have been supposed to hover about the White Mountains, an opinion very probably derived from the Indians, who thought these mountains the habitation of some invisible beings, and never attempted to ascend them."[83]

Not only did they believe in the physical existence of devils, a respectable belief in Puritan times, but the Indians were often imputed to be devil-worshipers or devil-possessed and hence diabolical in themselves, which was not respectable. After citing Mede's views regarding the Conflagration that would usher in the Millennium (pp. 266–267, above), Cotton Mather explained the remainder of Mede's theory thus:

> The inhabitants of these regions [the American hemisphere] who were originally Scytheans, and therein a notable fulfilment of the prophecy about the enlargement of Japhet, will be the Gog and Magog whom the devil will seduce to invade the New-Jerusalem, with the envious hope to gain the angelical circumstances of the people there. All this is but conjecture.[84]

And in William Hubbard's *General History of New England* we are informed that Mr. Mede's opinion

> carryes the greatest probability of truth with it . . . that when the devill was putt out of his throne in the other parts of the world, and that the mouth of all his oracles were stopped in Europe, Asia, and Africa hee seduced a company of silly wretches to follow his conduct into this unknown part of the world, where hee might lye hid and not be disturbed in the idolatrous and abdominable, or rather diabolicall service he expected from those his followers; for here are noe foote stepes of any religion before the English came, butt merely diabolicall.[85]

Mede's theory came out of the geographical lore of the Middle Ages, in which, on maps and miniatures and in many a volume, the dread tribes of Gog and Magog were described or depicted as having been enclosed by Alexander the Great within a mighty wall. Here they would remain until, at the Crack of Doom, they would break out, and, led by Antichrist (Satan), would ravage the world.[86]

While both Hubbard and Cotton Mather expressed scientific restraint with regard to specific theories as to whence the Indians came and what routes they followed, they were unshaken in the conviction that the Red Men were "diabolicall," an uncharitable attitude as compared with that of the German Pastorius in Pennsylvania, who contrasted the peace-loving character of the Indians of that region with the ruthlessness of the Europeans.[87]

AWARDATIVE GEOTELEOLOGY

In 1721 Cotton Mather's *The Christian Philosopher: A Collection of the Best Discoveries in Nature, with Religious Improvements* was published in London. It was based on a fairly enlightened and comprehensive knowledge of contemporary scientific movements in Britain. In it many of the observed facts of physical geography are explained teleologically as due to provisions made by the Deity especially for the benefit of mankind. For example, Mather explains that were it not for the tides, the ocean would stagnate and all places toward the shore would become a Mephitis (source of foul exhalations).[88] (Morton, however, had attributed this idea to the "Antients" and had questioned it on the ground that there are no tides in "Greatest part of the ocean" and yet it "dont stink us out of the world."[89]) Cotton Mather cites Dr. Cheyne's view that it is lucky we do not have two moons, "else they would cause tides when in their conjunctions to rise to the tops of the mountains, and in their quadratures we should have no tides at all." "The Spherical Figure of Our Globe has numerous and marvelous Conveniences . . . How incommodious must an Angular Figure have been." It is also fortunate that the axis of the globe is steady and "not carelessly tumbling this way and that, as it might happen." Gravity is "a most noble Contrivance (as Mr. Derham observes) to keep the several Globes of the Universe from shattering to pieces" and our globe in particular from being "spirtled into the circumambient Space." Cotton Mather also rallied to the defense of mountains, having, as we have seen, excoriated the "vain Colts of Asses" who would disparage them. Mather felt that, instead of being evidences of God's diluvial punishment of mankind, they had been created especially to serve men in various ways, among others as "the Bulwarks of Nature, set up at the charge of the Almighty; the Scorns and Curbs of the most victorious armies."[90] (Hugh Jones, another Colonial writer, averred that Providence had erected the Appalachian ranges to protect the British colonists against the French and Indians.[91]) Cotton Mather likewise thought it "very remarkable that our compassionate God has furnished all Regions with plants peculiarly adapted for the relief of the diseases that are most common in those regions."[92] Jared Eliot felt that the widespread distribution of peat "in all parts of the Country, yea all over the world," was not only testimony to the truth of the Biblical account of the Flood, but "evidence of the Care, Wisdom and Goodness of Providence, in preserving as it were in Pickle, the Wood which grew before the Flood for our use, and that the Ruins of the old World should supply the Wants and Wastes of the present."[93]

Somewhat as theological preconceptions haunted geological thought long after most geologists had ceased to look to a literal reading of the book of Genesis as a source of authentic geological facts, so also geoteleological comments added touches of color in American geographical descriptions of scenery and places long after the times when most writers had ceased to use them otherwise than as metaphors. In a memorial addressed to the Secretary of the Navy by the American Geographical and Statistical Society in 1857, a proposal to build a railroad across the Syrian Desert was advocated: "It would almost seem that it is for this that God created the vast plains of El-Hamad, as their formation is most remarkably adapted for this purpose" (a few lines previously, we read: "The desert . . . as has been well said, 'Was made by man, not by God' ", and the self-same desert is meant).[94] Two years later the Rev. Thomas Starr King described the amorphous Mount Hayes (overlooking Gorham, New Hampshire) as "the chair set by the Creator at the proper distance and angle to appreciate [Mount Washington's] kingly prominence."[95] (It should be remembered, though, that it is not always easy to distinguish a metaphor from a nonmetaphor, or statement meant to be taken literally, and that this difficulty has been a fertile soil in the production of rank and rankling religio-scientific controversies.)

Awardative Selection. Those who regard themselves as elect or chosen people are often tempted to attribute to God a disposition to extend to their nation or kin special theoawardative consideration (as implied, for example, by the words "Gott mit Uns," or by the calling one's own country, or a region that one particularly likes, "God's country"). With regard to the purported agreeable, temperate climate that prevails in the vicinity of the 45th parallel of north latitude, Cotton Mather, after having explained that it is due to "the posture of the earth's axis being inclined as it is and not perpendicular to the plane going through the center of the sun," noted that Mr. Ray had observed that "God has better chosen for us, than we could have done for ourselves."[96] "Us" must mean the English. In the *Magnalia*, Cotton Mather explained that "some things done since [Columbus' discovery] by Almighty God for the English in these regions [New England] have exceeded all that has been hitherto done for any other nation: if the new world were not first found out by the English; yet in those regards that are of the greatest, it seems to be found out more for them than any other."[97] Professors L. B. Wright and Perry Miller have collected other examples of seventeenth- and eighteenth-century claims that in America, Bermuda, and elsewhere God was on the side of the British in their empire-building and colonizational enterprises.[98] Professor Michael Kraus has pointed out that Professor John Winthrop of Harvard, who took a rationalistic view of earthquakes, had ridiculed (in a letter to Ezra Stiles, 1756) one reverend gentleman for being "pleased lately to observe to his audience that not one Protestant place had suffered the late European [Lisbon] earthquake."[99]

The belief that God was disposed to show special favors to the English in their early colonial ventures gave place in this country toward the close of the eighteenth century to similar interpretations by Americans of the Deity's attitude toward *them* with respect to the settlement of the North American continent and exploitation of its natural resources. Inspired, perhaps, by Bishop Berkeley's famous lines, Nathaniel Ames wrote in his *Astronomical Diary* for 1758:

> The Curious have observ'd, that the Progress of Humane Literature (like the Sun) is from the East to the West; thus it has travelled thro' Asia and Europe, and now is arrived at the Eastern Shore of *America*. As the Cœlestial Light of the Gospel was directed here by the Finger of GOD, it will doubtless, finally drive the long! long! Night of the Heathenish Darkness from *America*:—so Arts and Sciences will change the Face of Nature in their Tour from Hence over the Appalachian Mountains to the Western Ocean; and as they march thro' the vast Desert, the Residence of wild Beasts will be broken up, and their obscene Howl cease for ever. Huge Mountains of Iron Ore are already discovered; and vast Stores are reserved for future Generations: this Metal more useful than Gold and Silver, will employ millions of Hands, not only to form the martial Sword, and peaceful Share, alternately; but an infinity of Utensils improved in the Exercise of Art, and Handicraft amongst Men. Nature thro' all her Works has stamp'd Authority on this Law, namely, "That all fit Matter shall be improved to its best Purposes."[100]

Many variations on this theme were heard after 1776, giving expression to what Professor Miller called the "official faith" of the United States,[101] a faith that formed part of a larger orthodoxy of which Carl Ritter (1779–1859) and his Swiss-American disciple Arnold Guyot (1807–1884) were exponents. Here the essential dogmas were that God looks with partiality upon the white race, and in particular upon the Protestants, in their overseas enterprises of evangelism, colonization, settlement, and exploitation of natural resources. Ritter's philosophy of geographical history, as read by Americans during the mid-nineteenth century in Gage's translation of Ritter's lectures and as conveyed through Guyot's *Earth and Man* (1849), is drenched with such geotheological "humidity"[102] (see above, p. 253).

In an extraordinary little book, *The World a Workshop: or, the Physical Relationship of Man to Earth* (1855), the inventive Scottish-American, Thomas Ewbank, set forth a different facet of this "orthodoxy" by expressing convictions which any conservatively minded Christian theologian of the period would no doubt have held to be impious, had he read them. Ewbank, who visualized God as a super-factory-manager, espoused "characteristically American" doctrines of the sacrosanctity of mechanical ability and of the inexhaustibility of natural resources.

> This mundane habitation was designed and literally fitted up for the cultivation and application of chemical and mechanical science.—In no character does the Creator so prominently, constantly, and universally appear as in that of THE MECHANICIAN . . . —A first element of progress for all time, it is preposterous to suppose that the supplies of coal can ever be exhausted or even become scarce. The idea is almost blasphemous. It is a reflection on the Proprietor of the World.[103]

This is not how eighteenth-century Puritans thought of God, nor how present-day geologists and conservationists think of coal.

We have seen that Jared Eliot was piously inspired by contemplation of the continuous terrestrial circulation of iron and water. Ewbank was similarly inspired:

> God has made coal and ores for man to use — not to conceal them uselessly . . . [with increasing population] a deficiency [of coal and ores] would inevitably occur, were it not for the continual recomposition of decomposed bodies—a principle that prevents the establishments

from being closed for want of fresh stock to work up. There is no reason whatever to infer that the fertility of a surface layer can be indefinite in duration—that it can continue beyond certain periods to support vegetable or animal life. To recover or maintain its virtue, the materials must descend into the alembic again; for the heat of the central furnace is as necessary to prepare materials for man as to furnish the power that pushes them up to him.[104]

THE TELEOLOGY OF GEODIVERSITY AS SUCH

Up to this point in this paper the focus has been upon divinely ordained forces, processes, and events conceived as having produced (or as going to produce in the future) specific geographic effects. Let us now shift it to teleologic conceptions regarding the origins and nature of certain divinely-ordained arrangements of geographic actualities in space—in other words, concerning geodiversity as such.

Geodiversity, the essential subject of geographical knowledge and belief, is exhibited in two great overlapping classes of terrestrial conditions and processes: *nonhuman* and *human*. The former comprise the circumstances of man's physical and biological milieu (sometimes misleadingly called "the geographic environment") insofar as they have remained essentially unaltered and would appear to be unalterable by man. The human category comprises the bodies, minds, souls, institutions, and ideas of men. These two classes merge in an *intermediate* zone or class of phenomena due to the combined operation of human and nonhuman forces and consisting of mixtures or compounds of human with nonhuman elements. Nonhuman geodiversity is illustrated upon a map showing the distribution of volcanoes, human upon a map showing that of religions, and the intermediate type on a map showing the distribution of dams and reservoirs. After some brief remarks about the teleology of nonhuman I shall have somewhat more to say about that of human geodiversity. The intermediate kind will not be considered except incidentally. (The utility of distinguishing between these three kinds of geodiversity and the reality of the distinctions seem obvious enough. It is easy, however, to bog down in a semantic morass in trying to define their boundaries, as not a few geographers have discovered. Fortunately, such definition is not needed here.)

NONHUMAN GEODIVERSITY

Both the beginning and the continuing teleologic purposes of nonhuman geodiversity have been explained sometimes as punitive, sometimes as awardative, and sometimes as corrective. The smooth, egg-shaped, featureless earth of antediluvian times, as Burnet envisaged it, was characterized by such a high degree of geouniformity that geographers, had there been any in those times, would have been hard put to it to occupy themselves. The earth's postdiluvian pattern of nonhuman geographic actualities (mountains, oceans, plains), accordingly, was explained as a by-product or aftereffect of the punishment inflicted upon man by the Flood. Nonhuman geodiversity has also been interpreted as a continuing punitive affliction laid upon fallen man with a view to making it more difficult and disagreeable than it had been in antediluvian times for him to keep body and soul together—designed, in other words, to increase the punitive toilsomeness of his having to live by the sweat of his brow. This, however, was a somewhat out of the ordinary view of the nature of nonhuman geographic actuality, which more often was thought to be awardative or, at worst, corrective. Thus, Strachey and other English writers of the seventeenth century conceived of nonhuman geodiversity as a divinely ordained measure for providing different commodities in different parts of the world in the interests of commerce: "International trade was conceived in the bed of religion" (Perry Miller).[105]

Thomas Hutchinson wrote in his *History of Massachusetts* in the 1760's:

> The great creator of the universe . . . has so formed the earth that different parts of it, from the soil, climate, &c. are adapted to different produce, and he so orders and disposes the genius, temper, numbers, and other circumstances relative to the inhabitants as to render some employments peculiarly proper for one country, and others for another.

Hutchinson went on to illustrate this by saying that Virginians should not plant rice because South Carolina is "peculiarly designed for rice," and that New Englanders should keep to whaling, fishing, lumbering, and shipbuilding and not try to compete with the British in manufactures.[106] A century later we find Ritter explaining that each continent's shape is so planned and formed as to have its own special function in the progress of human culture, although the crowning thought of geographical science is of the unity and symmetry of the whole. "The nature of the parts is understood only from comprehension of the whole. That was the most just saying of Plato."[107] Guyot, following Ritter, believed that the differences in the "physical organization" of the northern continents were "intentional," and were prepared by God "to act different parts in the education of mankind."[108] Thomas Ewbank may well have read Guyot's *Earth and Man*, for he developed somewhat similar ideas not long afterward:

> It appears to have been a distinct trait in the Divine plan that the factory [that is, the earth] should not be equally and uniformly matured. The advantages of this were great: Man was sooner introduced than otherwise he could have been; while growing up on one part, another was preparing for him . . . Another advantage was that the occupants of one part were made mutually dependent on and beneficial to those of others. In a manufacturing point of view that plan commends itself as the most economical and productive in every conceivable respect.[109]

HUMAN GEODIVERSITY

There have also been acute theological differences of opinion regarding *human geodiversity*, notably as to how greatly men in different regions actually differ, why and how these differences came to be, and whether or not they are punitive, awardative, or corrective.[110]

American Indians and Negroes. Just as shells on the tops of mountains and other "marks of the Deluge" led to theological speculation about matters paleontological and geological, so the presence of Indians and Negroes in America led to similar speculation about matters of archeology, anthropology, and human history. Both kinds of speculation have had ethical implications. The Flood was a great moral lesson to mankind—though an ineffectual one, for despite the punitive drowning of most of the human race man seems to have remained as sinful as ever. But all this is rather vague, and the argument over whether the Flood or the initial Creation was responsible for mountains and other manifestations of nonhuman geodiversity raised geological and geographical rather than moral issues. But the clearly moral problems of how you treat your fellow man may depend substantially upon your ideas as to where his ancestors came from and why they came and what they were like—questions of a historico-geographical rather than of a more strictly theological nature and questions upon which the Bible has been believed to shed authoritative light.

We have seen that Cotton Mather and William Hubbard both mentioned *Mede's* theory that the Indians might eventually turn out to be the tribes of Gog and Magog of the medieval legends. In these legends the latter were frequently identified with the Lost Tribes of Israel, and the hypothesis that the Indians were the Lost Tribes or otherwise of Jewish origin appears often in Colonial American writings.[111] Hubbard, however, questioned this on the ground that the Indian languages disclose no trace of Hebrew influence: "no instance can bee given of any nation in the world that hath so fare degenerated from the purity of their originall tongue in 1500 or 2000 yeares, butt that there may be observed some rudiments of the ancient language, as may be seen in the Greeke and Latin tongues." Indians living within 200 miles of each other spoke utterly different languages.[112] William Penn, on the other hand, thought that the Indians might well be Jewish physically: they resembled people he had seen in the Jewish quarter of London and they had many of the same rites and ceremonies.[113] Hubbard likewise felt it most improbable that the Indians had come via the Straits of Magellan, because this would have entailed "a passage too near the frigid zone"; he was inclined to favor the view that they had come out of Tartary, "by the streights of Anian beyond California," on the ground that they "do in their manners more resemble the Salvage Tartar, than any other people whatsoever." At any rate, it must have been a long time after God first made man before his posterity found his way to America, for "the shortest cutt they can be supposed to take from Eden or Armenia, could not bee less than a journey of eight or ten thousand miles." Furthermore, it would not be easy to ascertain how they got here "unless the astrologers can find it in the starrs, or that it came from the motion [of] the celestial bodies, that lighted them hither; none of the inhabitants being knowne to have keept any annals or records of things done in fore past tymes."[114]

As the progress of exploration from Columbus' day onward through the sixteenth and seventeenth centuries gradually revealed the characteristics of the American Indian, there was a tremendous surge of interest in Europe in the problem of their origins. Many theories were developed and controversies raged.[115] Besides the tribes of Gog and Magog, the Lost Tribes, and the Tartars, Welshmen,[116] Scandinavians, Chinese, Moors, Carthaginians, Polynesians, Plato's Atlanteans, and no doubt others, were all given credit for having procreated the Red Man. Some held that all Indians were descended from a single racial stock of immigrants to America, others that the tribes in different parts of America came from different parts of the Old World, and echoes of the argument long reverberated. But, though opinions might differ as regards the more immediate places of origin, the overwhelmingly prevalent view was that the Americas had been peopled after the Flood by offspring of Noah migrating from the Old World to the New, and that, as Gookin put it, "they are Adam's posterity and therefore children of wrath."[117]

Hugh Jones in his book on Virginia expounded a characteristically geoteleological hypothesis of racial origins, though admitting that such speculation is "meer guesswork."[118] He thought it altogether possible that the Europeans and "western Asiaticks" stemmed from Japheth; that from Canaan, son of Ham, issued the Canaanites, from whom in turn might have sprung the Egyptians, Moors, Negroes, and other inhabitants of Africa; and that the American Indians might be the posterity of Shem. It is a little upsetting to find that Jones picked out the following verse for signal honor by putting it on his title page: "God shall enlarge Japheth, and he shall dwell in the tents of Shem; and Canaan shall be his servant." In the text this injunction of Noah is referred to as seeming "fulfilled in our possession of lands in the East and West-Indies, the tents of the sons of Shem, where Canaan or the Negroe is our servant and slave."[119] Thus, because Ham had seen the nakedness of Noah when the latter was drunk, and Noah for this reason, presumably, had cursed Ham's son, Canaan, an Anglican clergyman in the eighteenth century upheld the doctrines of colonialism and Negro slavery as foreordained by Noah, if not by God. Nor was Jones the last man of the cloth to employ similar arguments.

Gookin would hardly have taken pains to assert that the Indians were "Adam's posterity" had he not heard of theories to the contrary, such, for example as the "Pre-adamite" theory of Isaac de la Peyrère, published in 1655.[120] This was to the effect that the earth had been inhabited by human beings, some of whom had attained to a fairly high level of civilization before God had created Adam, and that the Flood of Noah, far from wiping out all mankind, had been merely local. Lewis Evans must have held similar views, in maintaining that "the brute Creation as well as the Human of America are originally of this continent,"[121] since he obviously did not think that Adam was an American Indian. He was, however, bold enough to suggest that the northern American Indians might "have given beginning to the Laplanders, Samoieds, Zemblans, and some other Tartar Nations, because the dark swarthy Complexion of these last mentioned Nations discover that they have not come from any beginning on the old Continent." Evans annotated this as follows: "This Hypothesis dont very well correspond with the Account Moses has given us of the creation [;] to some it may appear quite heretical."[122] It so appeared to the Father of American Geography, along with Evans's thoughts about the effects of the Flood in Pennsylvania and New Jersey (see above, pp. 262–263). Dr. Morse wrote:

> Those who call in question the authority of the sacred writings say, the Americans [that is, American Indians] are not descendants from Adam, that he was the father of the Asiatics only, and that God created other men to be the patriarchs of the Europeans, Africans and Americans. But this is one among the many weak hypotheses of unbelievers, and is wholly unsupported by history.[123]

Monogeny versus Polygeny. The particular "weak hypothesis" of Evans and others that offended Dr. Morse has been called the doctrine of human *polygeny*. This holds that different races of men were separately created in different regions and therefore constitute different species, as with the animals; whereas its antithesis, human *monogeny*, holds that all human beings are descended from a single pair (Adam and Eve, in the Judaeo-Christian tradition) and therefore constitute a single species of which the different races are merely varieties due chiefly to adaptations to the environment.[124]

As early as 1742, in answer to an inquiry addressed to him by Dr. Peter Collinson of London, Cadwallader Colden of New York wrote:

The Observation you made . . . that we have in America many different Species of Plants and Animals from those found in Europe or other parts of the World under the same Climate is certainly true & I think we may likewise add that we have different species of Men. This naturally leads to the Question . . . whether they be the effects of a different Creation. But, Dear Sir, I dare not pretend to give any answer in a matter so high and out of my reach. It is a subject fit to be treated only by first rate Philosophers and Divines.[125]

Others were not so reticent. According to Dr. Samuel Miller,

knowledge gained by modern voyagers and travellers of the manners, customs, and traditions of different countries, especially those of the Eastern Continent, has . . . furnished abundant and striking evidence in support of the Mosaic account of the common origin, the character, the dispersion, and the subsequent history of mankind.[126]

The "whole philosophy of man is confounded by the doctrine that divides man into different species radically different from one another." Were this the case, "the laws of morals would not apply universally to all men," argued Dr. Samuel Stanhope Smith at the close of the eighteenth century.[127] He rejected polygeny in no uncertain terms, both because it contradicts divine revelation and on the basis of factual evidence marshaled in support of his position. As stated in the "Advertisement" to his *Essay on the Causes of the Variety of Complexion and Figure in the Human Species,* he brought in "science to confirm the verity of the Mosaic history." Yet the verity of the Mosaic history would seem to have called for confirmation in the interests of morality rather more than of truth. Apparently, like many others before and later, Dr. Smith felt that, if a belief that something is true is thought to lead to immorality, such a belief should not be entertained. Dr. Smith was convinced that the earliest men must have been fairly civilized—intelligent, religious, law-abiding—and that savage life

seems to have arisen only from idle, or restless spirits, who shunning the fatigues of labor or spurning the restraints and subordinations of civil society, sought at once, liberty, and the pleasures of the chace in wild, uncultivated regions remote from their habitation.[128]

Savagery was equivalent to degeneracy and attributed in part to original sin and in part to the environment (many examples could, no doubt, readily be found of comparably unchristian beliefs concerning primitive people on the part of Christian missionaries[129]). The thought of "equivocal generation resulting from the united action of heat and moisture on the primitive mass of the world"[130] disgusted Dr. Smith, as the geologist J. D. Dana was disgusted a half century later by the "debasing association of Man with the Quadrumana."[131]

(The great botanist, Asa Gray, however, in lecturing before theological students at Yale in 1880 said: "We are sharers . . . with the higher brute animals in common instincts, and feelings and affections. It seems to me that there is a sort of meanness in the wish to ignore the tie. I fancy that human beings may be more humane when they realize that, as their dependents live a life in which man has a share, so they have rights which man is bound to respect."[132])

With respect to the white race, Guyot wrote that it is "the most pure, the most perfect type of humanity . . . Western Asia is not only the geographical center of the human race, but it is, moreover, the spiritual centre; it is the cradle of man's moral nature . . . If man came from the hands of the divine Author of his being, pure and noble, it was in those privileged countries where God placed his cradle, in the focus of spiritual light, that he had the best chance to keep himself such." The races of man show "the degeneracy of his type in proportion as he is removed from the place of his origin, and the focus of his religious traditions."[133] These are but scraps from the exposition of a philosophy of history upheld by Guyot but inspired by Ritter, in which a basically naïve geopiety shows through an elaborate façade of geotheology.

Ewbank, however, presented the opposite point of view. Convinced that "man made his appearance in different countries at different times"[134] and that the idea that "ancient and modern civilizations are debris of primeval science and refinement . . . is reversing matters," he proclaimed the view that human geodiversity was divinely planned for man's benefit with no less confidence than he did his similar views regarding nonhuman geodiversity:

As no factory involving a multiplicity of operations can be carried on without diversity in the workmen, so this diversity is a prominent feature of the plan on which this earthly workshop has been established. [The inequality from place to place in the mental capacity of men is a] beautiful provision, since, if the arts and sciences were equally cultivated, there would be no exchange of knowledge in them, and little to foster national intimacies and universal brotherhood . . . Universal progress can only arise from and be maintained by universal intercourse.[135]

Ewbank also felt that "the progress of science and art is furthered by the separation of men into national groups," for otherwise there would be "a universal heaviness, if not torpor."[136]

Here this essay, which began with the Psalms, comes abruptly to an end—with Thomas Ewbank! Yet the incongruity is not too great. We owe much to the Psalms for whatever poetic feeling we may have for the beauty and wonder of our terrestrial surroundings. If we owe less to Ewbank, whose book is rare and seldom read, his rather prosaic concept of God the Mechanician foreshadowed the even more prosaic one of God the Machine, the object of worship in many quarters today. On the other hand, Ewbank's more inspiring concept of universal brotherhood as stemming in part from human geodiversity also foreshadowed ideas of a similar sort that, despite the universal prevalence of hatred, are hopefully more widespread today than they were in the days of Cotton Mather and of Hugh Jones.

Epilogue

To every thing there is a season, and a time to every purpose under the heaven: A time to be born, and a time to die; a time to plant, and a time to pluck up that which is planted.[1]

THE PAPERS in this book are notes on scattered localities visited during more than a half-century of exploration in the Great South Seas of past human geographical awareness. While preparing them for publication or republication I have been tempted to linger nostalgically in these enchanted places, explore them more thoroughly, and branch out into wider unknowns. But there is a time to study and write, and also a time to send that which is written to the press, and unless a retired scholar sets his own deadlines and abides by them (more or less) he may find no time for anything. That is why this book, like most things human, is unfinished. In the different papers explicit hints are given now and then, with more implied, regarding routes that further exploration might follow. I want now to point to two possible main routes that have seemed particularly alluring and then to conclude with some remarks on the spirit in which voyaging along them (as also elsewhere) might be pursued.

Both routes lead in more or less the same direction. The first, in continuation of the last two papers, would be toward a better understanding of the relations in this country, past and present, between geography, mathematics, and religion—a matter of historical facts and interpretation. The second, suggested by a problem that has cropped up at many points in the writing of these papers, would be toward a better understanding of principles applicable to any and all relations between geography and other kinds of awareness—a matter of epistemological theory and technique. By "awareness" I have in mind mental consciousness of the actual or imagined existence of anything, whether the consciousness is naïve or sophisticated, unscientific or scientific, simple or complex, rational or irrational, emotional or unemotional, vaguely felt or clearly expressed. "Awareness," thus, is a blanket term embracing understanding, knowledge, wisdom, belief, religion, philosophy, geography, scholarship, magic, science, art, literature, et cetera (though obviously each of these may be conceived as including in varying degrees other components besides awareness alone; religion and geography, for example, could each be regarded as a body of awareness plus institutions for its acquisition, exposition, and dissemination).

THE TWO "HUMIDITIES" IN AMERICAN GEOGRAPHY

In the Introduction the religious "unmodernity" of Early American geography was contrasted with its mathematical "modernity." The former was likened to the lingering snows of a winter that has passed and the latter to fresh young growths, "harbingers of spring and of a summer that promises to be hot and thundery." On page 253, however, I observed that "the theological drought" of the period since the Civil War has "evaporated most of the patent (if not the latent) piety out of American geographical scholarship and education." This might seem inconsistent in a meteorological climate. In a metaphorical and intellectual one, mathematical humidity and theological aridity may increase simultaneously. Thus, while the American "air" has become desiccated of pietistic humidity since Puritan times it has also become ever more heavily charged with the vapors of mathematics and statistics, under whose fructifying influence the young growths have flourished and changed into mighty rain forests and tangled jungles.

Note, however, that it is the *patent* piety only that has so largely evaporated away. Patent piety is religious piety when openly expressed in speech or writing, and one certainly finds little evidence of it in American geographical publications and teachings of the last hundred years. Yet American geography is imbued with *latent* piety, consisting in part of religious piety *not* openly expressed in words but manifested in deeds and in part of what I venture to call *parapiety*. The latter could be defined as nonreligious or extrareligious piety lying outside the confines of religion. It is evidenced in the worship of nonreligious gods (for example, Science) and in the observance of codes of ethics, aesthetics, ritualism, and so forth, associated therewith.

(The delicate questions of where to draw the line between piety and parapiety and of which preceded which in the development of human awareness I leave to philosophers, sociologists, psychologists, anthropologists, archeologists, theologians, and experts in semantics. Yet it would seem undeniable that the parapiety now prevalent in American geography is largely a legacy of earlier American religious piety.)

Latent piety, both religious and parapious, has played a powerful role in the making of modern geography, a role to which historians of the subject have given but little heed. It has lent fervor to the geographer's faith in what he believes geography is or should or should not be. It has found expression in the credo of some geographers that there is no god but Science and Quantification is his prophet, and in the caution of others: "Take heed, and beware of the leaven of the Quantifiers and of the Scientifiers." It has accounted for much of the heat and thunder when geographical dogmas[2] and ethics have been discussed. If it has fostered bigotry, if not exactly fanaticism, on the part of some, so also has it fostered brotherhood and scholarly compassion on the part of others. How has it been related to the continued persistence, despite the theological drought, of cultural—even aesthetic—interests among a goodly number of geographers? Would the rain forests and jungles have crowded out all other vegetation in American geography had it not been for the latent piety that has infiltrated it from the circumambient intellectual atmosphere. These are interesting questions.

GEOGRAPHY'S FOREIGN RELATIONS

Both as a form of awareness and as an institution concerned with the development and dissemination of such awareness geography is inseparably related to many other forms of awareness and their associated institutions. This book deals in the main with such relations as they have existed in the past in specific respects. But, there are also attributes common to all such relations and analogies of varying degrees of strength and significance between the nature of geography and the natures of the other kinds of awareness with which it is related. Could these common attributes and analogies be studied more systematically than has been done hitherto? Would there be any point in such study? I should answer both questions: "Yes."

The kinds of awareness—or, to put it more definitely, the several disciplines such as those of history, science, theology, the law, the humanities, medicine—with which geography has been and is now related are extremely diverse in character, and there are countless different types of relation between them, from those of mere contact or juxtaposition to those of amalgamation and identification. Hence, it might be argued—especially by those who have an aversion to systems—that to try to systematize anything so heterogeneous and complex would be futile and that each particular relation or analogy should be treated separately and on its own merits. But even so, and while it would naturally be futile, misleading, and highly objectionable to carry any such system to extremes and to claim for it any unique and exclusive virtue, I am convinced that useful and interesting purposes could be served by attempts to examine geography's foreign relations in a considerably more systematic manner than they have been examined to date.

In order to do so tools are needed, and first and foremost a language in which the relations in question may be described and discussed concisely, intelligibly, and as unambiguously as possible. English as it exists is suited to this purpose *up to a certain point*, but beyond that point it yields diminishing returns, owing to a paucity of vocabulary and other semantic reasons—an obvious fact borne in on me over and over again by the necessity that has arisen while writing this book of discoursing about previously nameless concepts (such as *geopiety*) and about concepts that in normal English share their names ambiguously with others (as when the term *geography* is applied indiscriminately to a kind of awareness and to the terrestrial circumstances with which it is concerned). Hence, in order to carry on such discourse more or less comprehensibly, I have not hesitated to coin new terms where they have seemed needed; and in doing so my moral courage would have been reinforced by the following remark of George Perkins Marsh, had I known of it at the time: "The Adams of modern botany and zoology have been put to hard shifts in finding names for the multiplied organisms which the Creator brought before them, 'to see what they would call them:' and naturalists and philosophers have shown much moral courage in setting at naught the laws of philology in the coinage of uncouth words to express scientific ideas."[3] Since reading this I have felt better about "geopiety" and the other new terms that may have irritated you by breaking out in rashes, like the measles, in this book.

After coining new terms and specially defining old ones (as "geographical," "geographic") on an *ad hoc* basis over the years, I have become conscious of inconsistencies and ambiguities not only in the established terminology (for example, "religious geography," "American geography," "animal geographer," "woman geographer") but in my own coined terms (for instance, the *ge-* in "geosophy" refers to geographical knowledge and belief, in "geopiety" it refers to the earth). This has led to the development of a scheme of symbols and formulas for designating specific relations between geography and other kinds of awareness (and, indeed, between any two or more kinds of awareness), with an associated technique for cross-classifying the symbols and formulas to suggest new ones designating actual or possible concepts that might be overlooked otherwise.[4] The scheme thus has both a descriptive and a heuristic use. Simple in essence, it invites elaboration and refinement along intriguing lines. It has attracted me off and on for several years as a bone attracts a dog, and I have kept reverting to it and worrying it in canine fashion.

THE GEOGRAPHER AND THE PORPOISE

"Gee! that's interesting!" "It ain't necessarily so."

As slogans these are more meaningful than "Think!", and a geographer could put worse ones over his desk. The "Gee!" and the "ain't" might remind him that to be scholarly (even scientific) his language need not invariably be statistical, incomprehensible, dignified or stately. There is a light-hearted spirit of good humor and *joie de vivre* in both, and these have a place in times when there is abroad so much heavy-heartedness, bad humor, and distaste for life. "He had the glee of the porpoise then, pouring and leaping through strange seas,"[5] T. H. White wrote of "the Wart," or young King Arthur, when Merlin's magic had begun to open the boy's eyes to the wondrous world of science and lore. The Wart no doubt often responded with the Old English equivalent of "Gee! that's interesting!" and I am sure that Merlin did not reprove him for being so juvenile in his language. But Merlin must also have taught the Wart a healthy skepticism, for Arthur became quite a wise old man (in most respects). The magician developed in his pupil the counterpart of the delicate navigational sense that keeps the porpoise from running aground on reefs and rocks[6] and that prevents the enthusiastic but circumspect geographer from grounding himself on hazards to scholarly navigation. In the case of the geographer this, in the main, is the habit of confronting his own *idées fixes*, along with those that he borrows wittingly or unwittingly from others, with the tolerant, almost affectionate skepticism of Porgy's song: they ain't *necessarily* so, and the "necessarily" is the saving grace, precluding both the categorical negative fanaticism of total denial (" 'tain't so") and the categorical positive fanaticism of total affirmation (" 'tis so"). Children's arguments run along " 'tis-'tain't" lines. It needs maturity to insert "necessarily" after the "ain't."

Ruskin wrote *The Seven Lamps of Architecture*. "The Seven Banes of Geography"[7] surely must include erroneous or misleading *idées fixes* and boredom (you may name the other five for yourself). An *idée fixe* is an idea (whether true or false, helpful or misleading) that has "set" like concrete and is no longer amenable to change in the light of factual evidence or rational thought. *Idées fixes* may become clichés when shared by many persons.

The 1961 edition of the Merriam-Webster unabridged dictionary defines a cliché as "a trite or stereotyped phrase or expression; also the idea expressed by it . . . an overworked idea or its expression." Professor Whitaker has suggested that an interesting treatise could be written on the part that clichés, shibboleths, slogans, and other *idées fixes* have played in geography—for it is indeed a large part—and I hope that he will write it someday. Naturally, some clichés are bad, some good, and some harmless. Among the bad ones, I am inclined to class the cliché that all clichés are bad.[8] From a meteorological point of view, "it never rains but it pours" is untrue, but, since any fool knows that it often rains without pouring, the cliché is harmless and may even perform a good service by expressing concisely what we are often tempted to regard as the normal behavior of our misfortunes. Nevertheless, we should always be on our guard against bad clichés. Without attempting to classify them or assess their merits or demerits I shall list at random a few geographical clichés that I have encountered from time to time: one should not coin terms for use in one's own writings; one should never employ vague expressions like "meaningful," "significant," "interesting"; figures of speech are out of place in serious geographical works; geography is (is not) a science; it should (should not) be made more quantitative; it should (should not) be developed as a unified, autonomous discipline with a distinctive purpose differing from that of any other discipline; unless it is based upon fieldwork it is "geography with the *ge-* left out" (disastrous, since only "-ography" would remain); the nature of geography is independent of the nature of geographers; it is unseemly to mention "environmental determinism" in polite geographical circles. Of course, it ain't necessarily so that all these are always clichés. Some may be sincerely held convictions based on logical reasoning from well-established premises. It is for the reader to judge.

A writer for the *New York Times* recently quoted an English scholar "at the turn of the century" as having "declared with a yawn: 'Americans—they bore me as it is impossible to be bored in Europe.' "[9] Certain geographers have been known to be similarly affected by the works of certain other geographers, developing *idées fixes* that the latter are boresome. Boredom with incessant argument over "what is geography" has prompted the answers "geography is what I like" and "geography is what geographers do," and these harmless witticisms have become clichés in our discipline and profession, quoted when one wishes to express *ennui* at the mere thought of methodological discussion. "Geography is what I like," however, implies the less harmless "what I don't like is *not* geography" or, differently, "what bores me in geography is of no consequence." Porpoises do not look with lofty disdain upon one another.

Before we leave the disagreeable subject of clichés, one more point must be made. When a thought-cliché gains wide currency in any circle, geographical or otherwise, it tends to produce a reaction in the form of an anti-thought-cliché, and perhaps ultimately cliché and anticliché become reconciled according to Hegelian principles[10] in a syncliché (or else they mutually exterminate one another by boring all concerned). Among American geographers, for example, the normative cliché that geography should concern itself primarily with the question of environmental determinism produced the anticliché that it should do nothing of the kind, and these, in turn, engendered the syncliché that it might do so to some slight extent (and also the annihilating view that the whole matter might just as well be forgotten).

Interesting as are the porpoise's navigational abilities, his most appealing qualities are his exuberance, his friendliness, and his uninhibited leaping and pouring into strange seas; and, of course, the same is (or should be) true of geographers and historians of geography. If, as the years pass by, they cannot keep up quite such an energetic pouring and leaping as they used to, there is no rule of geographical or historical ethics that requires them to stay within well-charted waters or, as explained elsewhere in this book, to remain deaf to the sirens' song as they cruise along.[11] They could learn something both from Ulysses and from the porpoise.

Notes

Introduction

1. See J. K. Wright, "Some Boyhood Memories of William M. Davis." *Annals of the Association of American Geographers* 40 (1950), 179–180.

2. A. T. Wright, *Islandia* (1st ed. with introduction by Leonard Bacon; New York: Farrar and Rinehart, 1942; 2nd ed. with introduction by Sylvia Wright; New York: Rinehart, 1958). On the location of Islandia see the second edition, pp. 1017–1018. See also *An Introduction to Islandia, its History, Customs, Law, Language, and Geography* as prepared by Basil Davenport . . . (New York: Farrar and Rinehart, 1942).

3. Alexander von Humboldt, *Kosmos: Entwurf einer physischen Weltbeschreibung* (Stuttgart and Tübingen, 5 vols., 1845–1862); see also Chapter 9, n. 8.

4. O[scar] Peschel, *Geschichte der Erdkunde bis auf Alexander von Humboldt und Carl Ritter*, 2nd ed. by Sophus Ruge (Munich, 1877). Siegmund Günther, *Geschichte der Erdkunde* (Leipzig and Vienna: Deuticke, 1904).

5. See Chapter 1; also J. K. Wright, "The History of Geography: A Point of View," *Annals of the Association of American Geographers 15* (1925), 192–201.

6. J. K. Wright, *The Geographical Lore of The Time of the Crusades: A Study in the History of Medieval Science and Tradition in Western Europe* (New York: American Geographical Society, 1925; reprinted with Foreword by C. J. Glacken, New York: Dover, 1965), p. xix.

7. See also J. K. Wright and the late Elizabeth T. Platt, *Aids to Geographical Research* . . . (2nd ed.; New York: published by Columbia University Press for the American Geographical Society, 1947), pp. 4–9. It would be natural, though a mistake, to assume that Hume, Collingwood, and Thoreau inspired me to write about human nature in science and the place of the imagination in geography. I did not read Collingwood's book, which has sections pertinent to these subjects, until after the present

volume was nearly ready for the printer. See R. G. Collingwood, *The Idea of History* (New York: Oxford University Press, 1946; Galaxy paperback, 1956), pp. 82, 205–249; also, concerning Thoreau, below, Chapter 5, n. 6.

8. C. O. Paullin, *Atlas of the Historical Geography of the United States*, ed. by J. K. Wright (Carnegie Institution of Washington and American Geographical Society of New York, 1932). See also Chapter 11, n. 30.

9. *New England's Prospect: 1933*, by 27 authors, ed. by J. K. Wright (New York: American Geographical Society, 1933).

10. J. K. Wright, *Geography in the Making: The American Geographical Society, 1851–1951* (New York: American Geographical Society, 1952).

11. See J. K. Wright, "British Geographers and the American Geographical Society, 1851–1951," *Geographical Journal 118* pt. 2 (1952), 153–167; "AAG Programs and Program-Making, 1904–1954," *Professional Geographer 9* (1954), 6–11.

Chapter 1. A Plea for the History of Geography

1. Hugo Berger, *Geschichte der wissenschaftliche Erdkunde der Griechen* (2nd ed.; Leipzig: Viet, 1903), pp. 387–388, 443–453, 460, 504, 534–537, 576–577.

2. Marcel Dubois, *Examen de la Géographie de Strabon . . .* (Paris, 1891), pp. 169–180; A. Thalamas, *La Géographie d'Ératosthène* (Paris: Marcel Rivière, 1921), pp. 199–201.

3. The *Italia illustrata* of Flavio Biondo, written about 1451 and published in 1471, was perhaps the first study in classical geography of any importance (see J. C. Husslein, *Flavio Biondo als Geograph des Frühhumanismus* (Dissertation, Würzburg, 1901). On the broader aspects of this subject see Siegmund Günther, "Der Humanismus in seinem Einflusse auf die Entwicklung der Erdkunde," *Geographische Zeitschrift* 7 (1900), 65–89.

4. The translation of Ptolemy's *Geography* into Latin was begun by Emmanuel Chrysoloras and completed by Jacopus Anglicus at the beginning of the fifteenth century. This translation was the basis of subsequent Latin editions of the fifteenth and sixteenth centuries. See A. E. Nordenskiöld, *Facsimile Atlas to the Early History of Cartography*, trans. by J. A. Ekelöf and C. R. Markham from the Swedish original (Stockholm, 1889), pp. 9–10. Strabo's *Geography* was translated into Latin by Guarino of Verona and Gregory Tiferna and appeared in 1470 (see Dubois, *Examen*, pp. 7–8).

5. O[scar] Peschel, *Geschichte der Erdkunde bis auf Alexander von Humboldt und Carl Ritter* (2nd ed. by Sophus Ruge; Munich, 1877), pp. 654–655; Christian Sandler, *Die Reformation der Kartographie um 1700* (Munich and Berlin: Oldenbourg, 1905), p. 8.

6. See especially Armand Rainaud, *Le Continent austral: hypothèses et découvertes* (Paris, 1893).

7. See Max Böhme, *Die grossen Reisesammlungen des 16. Jahrhunderts und ihre Bedeutung* (Strassburg: Heitz, 1904). [See also G. R. Crone and R. A. Skelton, "English Collections of Voyages and Travels, 1625–1846," in *Richard Hakluyt and his Successors . . .*, ed. by Edward Lynam (London: Hakluyt Society, 1946), pp. 65–140. (1965)]

8. [For bibliographical guidance to works on these subjects published prior to about 1947 see J. K. Wright and the late Elizabeth T. Platt, *Aids to Geographical Research . . .* (2nd ed.; New York: published by Columbia University Press for the American Geographical Society, 1947), pp. 54, 104–111. For the subsequent period see *Research Catalogue of the American Geographical Society* (14 vols. and map supplement; Boston: G. K. Hall, 1962), II, 1596–1837, also the sections devoted to the history of geography in *Current Geographical Publications: Additions to the Research Catalogue of the American Geographical Society* and the bibliographies currently published in *Isis.* (1965)]

9. See note 5. [It should be noted, however, that Hettner regarded Peschel's history of geography as too brief for the earlier periods and as overemphasizing German work for the more modern; see Alfred Hettner, *Die Geographie: ihre Geschichte, ihr Wesen, und ihre Methoden* (Breslau: Hirt, 1927), p. 4. (1965)]

10. Siegmund Günther, *Geschichte der Erdkunde* (Leipzig and Vienna: Deuticke, 1904). [Despite Hettner's damnation of this book as "an indigestible compilation of not always reliable facts" (*Die Geographie*, p. 4), I have found its bibliographical data useful, especially for the period before 1800. (1965)]

11. Louis Vivien de St.-Martin, *Histoire de la géographie et des découvertes géographiques depuis les temps les plus reculés jusqu' à nos jours* (Paris, 1873; with atlas).

12. For the nineteenth century, see Siegmund Günther, *Entdeckungsgeschichte und Fortschritte der wissenschaftlichen Geographie im neunzehnten Jahrhundert* (Berlin, 1902). [For the period c.1850–c. 1960, see T. W. Freeman, *A Hundred Years of Geography* (Chicago: Aldine, 1962). (1965)]. In addition to the above-mentioned general survey, the volume of Karl Weule, *Geschichte der Erdkenntnis und der geographischen Forschung: zugleich Versuch einer Würdigung beider in ihrer Bedeutung für die Kulturentwicklung der Menschheit* (Berlin, 1904), deserves mention. Currents of geographical thought in the seventeenth, eighteenth, and early nineteenth centuries especially are discussed by Emil Wisotzki, *Zeitströmungen in der Geographie* (Leipzig, 1897). An important series of articles on the development of geographical studies in different countries will be found in the *Atti del X Congresso Internazionale di Geografia, Roma MCMXIII* (Rome, 1915). For Europe during the period immediately preceding 1922, see W. L. G. Joerg, "Recent Geographical Work in Europe," *Geographical Review 12* (1922), 431–484. (For the United States especially, see works cited below, Chapter 8, nn. 6, 11, and 13.)

13. [Notably, J. Scott Keltie and O. J. R. Howarth, *History of Geography* (London: Rationalist Press Association; New York: Putnam, 1913); R. E. Dickinson and O. J. R. Howarth, *The Making of Geography* (Oxford: Clarendon Press, 1933); G. R. Crone, *Background to Geography* (London: Museum Press, 1964). (1965)]

14. Emmanuel de Martonne, *Geography in France* (New York: American Geographical Society, 1924), p. 62.

15. W. H. Tillinghast, "The Geographical Knowledge of the Ancients Considered in Relation to the Discovery of America," in Justin Winsor, ed.,

Narrative and Critical History of America (Boston and New York, 1889), I, ch. 1.

16. G. E. Nunn, *The Geographical Conceptions of Columbus: a Consideration of Four Problems* (New York: American Geographical Society, 1924).

17. N. M. Crouse, *Contributions of the Canadian Jesuits to the Geographical Knowledge of New France, 1632–1675* (Ithaca, New York: dissertation, Cornell University, 1924).

18. Jean Brunhes, *La Géographie humaine* (3rd ed., 3 vols.; Paris: Alcan, 1925), II, 921. See also H. E. Barnes, ed., *The History and Prospects of the Social Sciences* (New York: Knopf, 1925), p. 100.

19. See above, p. 83, and Chapter 5, n. 3.

20. See W. Dröber, *Kartographie bei den Naturvölker* (dissertation, Erlangen, 1903); R. J. Flaherty, "The Belcher Islands of Hudson Bay: their Discovery and Exploration," *Geographical Review 5* (1918), 440–458 (on p. 440 there is a reproduction of a remarkably accurate map drawn from memory by an Eskimo). [See also works referred to in *Geographical Review 10* (1920), 414, *22* (1932), 491–492, and Leo Bagrow, *History of Cartography*, revised and enlarged by R. A. Skelton (Cambridge, Mass.: Harvard University Press, 1964), pp. 23–28; and Kevin Lynch, *The Image of the City* (Cambridge, Mass.: Massachusetts Institute of Technology, 1960; M.I.T. Press, paperback ed., 1964), pp. 133–134. (1965)]

21. See the delightful little book of essays by W. P. James, *The Lure of the Map* (London: Methuen, 1920).

22. D. G. Hogarth, *The Penetration of Arabia: A Record of the Development of Western Knowledge Concerning the Arabian Peninsula* (New York: Stokes, 1904), pp. 39–40 [this book, of which I bought a copy in September 1909, probably more than any other, sparked my interest in the history of exploration and in Arabia (1965)]. See also Thorkild Hansen, *Arabia Felix: The Danish Expedition of 1761–1767*, translated by James and Kathleen McFarlane (New York and Evanston: Harper and Row, 1964; first published in Danish, 1962).

23. See F. A. Golder, *Bering's Voyages: An Account of the Efforts of the Russians to Determine the Relation of Asia and America* (2 vols.; New York: American Geographical Society, 1922, 1925).

24. An interesting chapter on early scientific expeditions will be found in Peschel, *Geschichte der Erdkunde*, pp. 535–640.

25. [See Wright and Platt, *Aids*, p. 108; as regards travel in the United States and by Americans in other parts of the world, see *Literary History of the United States*, edited by R. E. Spiller, Willard Thorp, T. H. Johnson, and H. S. Canby (3 vols.; New York: Macmillan, 1948), vol. III, *Bibliography;* also *Bibliography, Supplement* (New York: Macmillan, 1959). (1965)]

26. [See the references in the preceding note; also many of the papers in the magazine *Landscape* (1951–) and R. I. Wolfe, "Perspective on Outdoor Recreation: A Bibliographical Survey," *Geographical Review 54* (1964), 203–238. (1965)]

27. See, especially, Brunhes, *Géographie humaine*, II, 831–858; H. E. Barnes, *The New History and the Social Studies* (New York: Century, 1925), pp. 53–75.

28. See Chapter 3.
29. See above, p. 134.
30. See J. K. Wright, "The History of Geography: A Point of View," *Annals of the Association of American Geographers 15* (1925), 192–201; also Chapter 2.
31. See J. K. Wright, *The Geographical Lore of the Time of the Crusades* . . . (New York: American Geographical Society, 1925; Dover, paperback, 1965), pp. 27–30, 59–60, 184–187; see also above, pp. 121, 251, and below, Chapter 2, n. 4.
32. Wright, *Geographical Lore*, pp. 213, 446.
33. See above, pp. 130–131.
34. See Nunn, *Geographical Conceptions*, pp. 31–53; also below p. 29.
35. See also above, p. 158, and Chapter 14.
36. John Ruskin, *Modern Painters* (5 vols.; London, 1843–1860), vol. IV, containing Part 5, "Mountain Beauty."
37. See Francis Gribble, *The Early Mountaineers* (London, 1899); W. W. Hyde, "The Development of the Appreciation of Mountain Scenery in Modern Times," *Geographical Review 3* (1917), 107–118; [and especially Marjorie Nicolson, *Mountain Gloom and Mountain Glory: The Development of the Aesthetics of the Infinite* (Ithaca, N.Y.: Cornell University Press, 1959, New York: Norton paperback, 1963). (1965)]
38. See Chapter 7.
39. H. R. Mill, *Guide to Geographical Books and Appliances* . . . (2nd ed.; London: Philip, 1910), pp. 58–63. [For references to more recent works of a similar nature, see *Geographical Review 14* (1924), 659–660; *28* (1938), 499–501; and Wright and Platt, *Aids*, pp. 62–63. (1965)]
40. [This phrase was suggested by Dr. Sarton's subtitle of *Isis*, namely, *International Review Devoted to the History of Science and Civilization*, in which this paper first appeared. Later those responsible for the publication of that admirable journal dropped the "civilization" out of the title, which seems a pity (see above, p. 53); see also A. C. Crombie, ed., *Scientific Change* (New York: Basic Books, 1963), pp. 772–773. (1965)]

Chapter 2. Where History and Geography Meet

1. No detailed bibliographical references were given in this paper as first published, and, since it is being reprinted for its methodological rather than its factual interest, no attempt has been made to complete it or bring it up to date in this respect. By consulting either the decennial indexes to the *Geographical Review* for the periods 1926–1935 and 1936–1945 or the catalogue of any large library the titles of most of the then-recent books by writers mentioned in the text may readily be found. For some general comments on the nature and history of exploration see my introduction to the Torchbook edition of Sir Percy Sykes, *A History of Exploration* (New York: Harper, 1961).
2. John Leighly, "Error in Geography," in Joseph Jastrow, ed., *The Story of Human Error* (New York: Appleton-Century, 1936), pp. 89–119:90.
3. See *Journal of the Washington Academy of Science*, Lewis and Clark Anniversary Number, *44* (11), 333–373.

4. For an interpretation of the history of geographical discovery, of medieval cartography, and of large aspects of the general history of geography—one which, to make an understatement, is out of the ordinary—see W. G. Niederland, "River Symbolism," *The Psychoanalytic Quarterly* *25* (1956), 469–504; *26* (1957), 50–75; "The Symbolic River—Sister Equation in Poetry and Folklore," *Journal of the Hillside Hospital 6* (1957), 91–99. This Freudian study in "sexual geography" shows by indirection how mistaken I was a few years ago in asserting that "the paucity of women . . . has almost completely deprived the history of exploration of sexual interest" (introduction to Torchbook edition of Sykes, *A History of Exploration*, p. xxi).

Chapter 3. Map Makers Are Human

1. See, however, Max Eckert, "On the Nature of Maps and Map Logic," *Bulletin of the American Geographical Society 40* (1908), 344–351:347 [and, much more recently, O. M. Miller and R. J. Voskuil, "Thematic-Map Generalization," *Geographical Review 54* (1964), 13–19. (1965)]

2. Hans Speier, "Magic Geography," *Social Research 8* (1941), 310–330: 310–311, 316.

3. Max Eckert, *Die Kartenwissenschaft: Forschungen und Grundlagen zu einer Kartographie als Wissenschaft* II (Berlin and Leipzig: Verlag Wissensch. Verleger, 1925), 145.

4. See R. S. Patton, "The Physiographic Interpretation of the Nautical Chart," *Geographical Review* 17 (1927), 115–127; on the "limitations to which all charts are subject" (p. 116).

5. [The ideas expressed in this and the two following sections of this paper and illustrated by Figs. 1 and 2 have been greatly developed and elaborated in recent years by American geographical cartographers; see, for example, G. F. Jenks, "Generalization in Statistical Mapping," *Annals of the Association of American Geographers 53* (1963), 15–26, and Miller and Voskuil, "Thematic Map-Generalization." (1965)]

6. R. G. Collingwood, *The Idea of History* (New York: Oxford University Press, 1946; Galaxy paperback, 1956), pp. 240–241. See also above pp. 45–46.

7. See Sten De Geer, "On the Definition, Method, and Classification of Geography," *Geografiska Annaler 5* (1923), 1–37:14–23, where this subject is analyzed from a somewhat different point of view; also J. K. Wright, "Problems in Population Mapping," in *Notes on Statistical Mapping, With Special Reference to the Mapping of Population Phenomena*, edited by J. K. Wright (mimeographed; American Geographical Society and Population Association of America, 1938), pp. 1–18; "A Proposed Atlas of Diseases, Appendix I—Cartographic Considerations," *Geographical Review 34* (1944), 649–652; "The Terminology of Certain Map Symbols," *ibid.*, 653–654; A. H. Robinson, *Elements of Cartography* (2nd ed.; New York: Wiley, 1960), pp. 172ff.

8. Term coined in 1938 (Wright, "Problems," p. 14).

9. Robinson, *Elements*, pp. 172ff.

10. See J. K. Wright, "A Method of Mapping Densities of Population, with Cape Cod as an Example," *Geographical Review 26* (1936), 103–110;

also in *Mélanges de géographie offerts . . . à M. Václav Švambera . . .* (Prague: privately published, 1936), pp. 143–151; R. S. Platt, editor, *Field Study in American Geography: the Development of Theory and Method Exemplified by Selections* (Chicago: University of Chicago, Department of Geography Research Paper No. 61, 1959), pp. 281–287.

11. Frank Debenham, *Map Making* (London and Glasgow: Blackie, 1936), p. 98.

12. A. C. Veatch and P. A. Smith, "Atlantic Submarine Valleys of the United States and the Congo Submarine Valley," *Geological Society of America Special Paper No.* 7 (1939), pp. 54–55.

Chapter 4. Human Nature in Science

1. [Commenting on a paper by T. S. Kuhn, "The Function of Dogma in Scientific Research," in A. C. Crombie, editor, *Scientific Change: Historical Studies in the Intellectual, Social and Technical Conditions for Scientific Discovery and Invention, from Antiquity to the Present—Symposium on the History of Science, University of Oxford 9–15 July, 1961* (New York: Basic Books, 1963), pp. 347–369, S. E. Tuolmin wrote: "And of course we know very well, both from Snow's novels and from our own experience that 'scientists are human too'—that they sometimes allow their personal emotions to distort their approach both to their colleagues and to their subject-matter" (*ibid.*, p. 382). This is one among several examples that could be given of the expression in *Scientific Change* of ideas not unlike certain previously or otherwise independently developed ideas expressed in the present volume. See also T. S. Kuhn, *The Structure of Scientific Revolutions* (Chicago: University of Chicago Press, 1962; Phoenix, paperback, 1964). (1965)]

2. [This theme is further developed in J. K. Wright and the late Elizabeth T. Platt, *Aids to Geographical Research . . .* (2nd ed.; New York: published for the American Geographical Society by the Columbia University Press, 1947), pp. 4–9. (1965)]

3. [W. M. Davis, "The Value of Outrageous Geological Hypotheses," *Science* 63 (1926), 463–468. (1965)].

4. [I further developed this idea in a paper, "Training for Research in Political Geography," *Annals of the Association of American Geographers* 34 (1944), 190–201. A periodical entitled *Conflict Resolution* was established in the United States in 1957. See also Baron de Montesquieu, *The Spirit of the Laws*, transl. by Thomas Nugent (New York: Hafner, 1949), bk. i, ch. 3, p. 5. (1965)]

Chapter 5. *Terrae Incognitae*

1. In a passage from his inaugural lecture delivered at the Sudan Cultural Centre, quoted by R. A. Hodgkin, *Sudan Geography* (Education Department of the Sudan Government, 1946), p. 147; see also *Geographical Review* 37 (1947):340.

2. See Derwent Whittlesey, "The Horizon of Geography," *Annals of the Association of American Geographers* 35 (1945), 1–38.

3. [This idea has been developed in a scholarly and thought-provoking paper by David Lowenthal, "Geography, Experience, and Imagination: Towards a Geographical Epistemology," *Annals of the Association of American Geographers 51* (1961), 241–260. (1965)]

4. [See J. K. Wright, "Communication" (letter to the editor), *Annals of the Association of American Geographers 39* (1949), 47. (1965)]

5. See R. H. Brown, *Mirror for Americans: Likeness of the Eastern Seaboard, 1810* (New York: American Geographical Society, 1943). pp. xix–xxxii.

6. [It would be natural enough, though erroneous, to assume that this was inspired by reading the "Conclusion" of Thoreau's *Walden*, where the exploration of *terrae incognitae*, real and metaphorical, is the theme of a marvelous passage that I read for the first time on December 8, 1964. Perhaps bushwacking explorations in the then miniature *terra incognita* of the northern Mahoosuc Range (Oxford County, Maine) during the years 1910–1914 were its principal inspiration (see J. K. Wright, "The Northern Mahoosucs, 1910–1911: Terra Incognita," *Appalachia* [periodical of the Appalachian Mountain Club] *35*, December 1965; also above, p. 148, Fig. 6, and p. 295, n. 7.) (1965)]

Chapter 6. The Open Polar Sea

1. No references are given below to books that can be readily identified under their authors' names in bibliographies or library catalogues. The voluminous scrapbooks kept by Henry Grinnell (1799–1874), now in the library of the American Geographical Society, New York City, contain many clippings concerning the Arctic from American and British newspapers and magazines of the period 1850–1885, together with a list of those bearing on the Open Polar Sea compiled through the courtesy of Mr. Andrew Taylor of Ottawa, Canada.

2. Vilhjalmur Stefansson, *Ultima Thule: Further Mysteries of the Arctic* (New York: Macmillan, 1940), pp. 234–265; J. K. Wright, *The Geographical Lore of the Time of the Crusades . . .* (New York: American Geographical Society, 1925), pp. 18, 156–157, 165, 179.

3. Stefansson, *Ultima Thule*, pp. 109–222.

4. *Ibid.*, pp. 158–159, 266–267.

5. See Cesare de Lollis, "Scritti di Cristoforo Colombo . . . ," *Raccolta di documenti e studi pubblicati dalla R. Commissione Colombiana pel quarto centenario dalla scoperta dell'America*, pt. I, vol. II (Rome, 1894), pp. LVI, 73.

6. Richard Hakluyt, *Divers Voyages Touching the Discovery of America and the Islands Adjacent*, Collected and Published . . . in the Year 1582, edited by J. W. Jones, *Hakulyt Society* [*Publs.*], ser. 1, vol. VII (London, 1850), pp. 27–54. See also F. T. McCann, *English Discovery of America to 1585* (New York: King's Crown Press, 1952), p. 60; and E. G. R. Taylor, *Tudor Geography, 1485–1583* (London: Methuen, 1930), pp. 10–11, 45–51, 57–58, 92.

7. Although the theory of an Open Polar Sea, as such, was developed at this period chiefly as an outgrowth of the search for either a Northeast Passage or a due-north passage across the Pole, the quest for a Northwest

Passage stimulated much lively speculation in England concerning the geography and climatology of the Arctic. See Taylor, *Tudor Geography*, pp. 34–35, 38–39, 79–82; and Vilhjalmur Stefansson's edition ("from the original 1578 text of George Best") of *The Three Voyages of Martin Frobisher* (2 vols.; London: Argonaut Press, 1938), I, lxxxiii–lxxxviii, xcix, 39–45, 135–145.

8. See G. M. Asher, ed., *Henry Hudson the Navigator*, in *Hakluyt Society* [*Publs.*], ser. 1, vol. 27 (London, 1860), pp. 246–250. On Plancius see "Petrus Plancius als geograf," in the introduction to Gerritt de Veer, *Reizen van Willem Barentz . . .*, ed. by S. P. L. 'Honoré Naber, Part II (The Hague, 1917), pp. III–XXXII (*Werken uitgegeven door de Linschoten Vereeniging*, No. 15); and J. Keuning, *Petrus Plancius: Theoloog en geograaf, 1552–1622* (Amsterdam: van Kampen, 1946).

9. Edward Heawood, *A History of Geographical Discovery in the Seventeenth and Eighteenth Centuries* (Cambridge, England: University Press, 1912), p. 401.

10. In this connection Dr. Stefansson kindly called my attention to the instructions issued to Captain David Buchan on the eve of his departure for his North Polar expedition of 1818. These implied that, in sailing from the Atlantic to Bering Strait, Buchan would encounter less ice in crossing the central Polar Basin than on entering or leaving it; see F. W. Beechey, *A Voyage of Discovery Towards the North Pole . . . under the Command of Captain David Buchan . . .* (London, 1843), pp. 7–10.

11. E. K. Kane, "Access to an Open Polar Sea along a North American Meridian," *Bulletin of the American Geographical and Statistical Society*, 1, no. 2 (1853): 85–102, p. 87.

12. Silas Bent, "Communication . . . upon Routes to be Pursued by Expeditions to the North Pole," *Journal of the American Geographical and Statistical Society* 2, pt. 2 (1870), 31–40; *Thermometric Gateways to the Pole*, address before the Saint Louis Historical Society, December 10, 1868 (Saint Louis, 1869); *Thermal Paths to the Pole*, address before the Saint Louis Mercantile Library Association, January 6, 1872 (Saint Louis, 1872).

13. T. B. Maury, "The Gateways to the Pole," *Putnam's Magazine* 4 (1869), 521–537; "The Dumb Guides to the Pole," *ibid.*, 727–740; "The New American Polar Expedition and its Hopes," *Atlantic Monthly* 26 (1870), 492–504.

14. C. P. Daly, "Annual Address: Review of the Events of the Year, and Recent Explorations and Theories for Reaching the North Pole," *Journal of the American Geographical and Statistical Society* 2, pt. 2 (1870), lxxxiii–cxxvi.

15. Fridtjof Nansen, "How Can the North Polar Region Be Crossed?" *Geographical Journal* 1 (1893), 1–32.

16. Taylor, *Tudor Geography*, p. 11.

17. Plancius' map of 1592 is reproduced in F. C. Wieder, editor, *Monumenta cartographica* (5 vols., The Hague: Nijhoff, 1925–1933), II, pl. 35; that of 1594 with Volume IV of "Itinerario . . . van Jan Huygen van Linschoten," *Werken . . . Linschoten-Vereeniging*, No. 43 (The Hague, 1939).

18. This appears on an inset on Mercator's great world map of 1569 (reproduced by the International Hydrographic Bureau, Monaco, 1931,

with text, 1932). See also A. E. Nordenskiöld, *Fascimile Atlas to the Early History of Cartography* . . . (Stockholm, 1889), p. 95; Taylor, *Tudor Geography*, p. 131; and below, p. 338.

19. John Barrow, *A Chronological History of Voyages into the Arctic Regions* (London, 1818), p. 376.

20. J. N. L. Baker, *A History of Geographical Discovery and Exploration* (London: Harrap, 1931), p. 456.

21. Kane, "Access," pp. 97, 99.

22. Joseph Moxon, *A Brief Discourse of a Passage by the North-Pole to Japan, China, &c.* (London, 1674), pp. 1–2. The American Geographical Society possesses a copy of this work, referred to as "a now very scarce tract" by Barrington, *The Possibility of Approaching the Pole Asserted* (new ed.; London, 1818), p. 33, note.

23. C. R. Markham, *The Threshold of the Unknown Region* (4th ed.; London, 1876 [1st ed., 1873]), p. 55.

24. Augustus Petermann, "On the Proposed Expedition to the North Pole," *Proceedings of the Royal Geographical Society 9* (1864–1865), 90–99: 92.

25. Hayes believed that his farthest north was about a degree north of the latitude of Cape Constitution; see, however, N. M. Crouse, *The Search for the North Pole* (New York: Richard R. Smith, 1947), pp. 76–78.

26. R. V. Hamilton, "On Open Water in the Polar Basin," *Proceedings of the Royal Geographical Society 13* (1868–1869), 234–243: 242.

27. See also Siegmund Günther, *Geschichte der Erdkunde* (Leipzig and Vienna: Deuticke, 1904), p. 152, n. 6.

28. See Augustus Petermann, *The Atlas of Physical Geography: . . . With Descriptive Letter-press* . . . by . . . Thomas Milner (London, 1850), pl. 7.

29. *The Polar Exploring Expedition: A Special Meeting of the American Geographical & Statistical Society, Held March 22, 1860*, pamphlet on meeting held to discuss plans for Dr. I. I. Hayes's expedition, 1861–1862 (New York, 1860), pp. 7–8.

30. T. L. Kane, *Alaska and the Polar Regions: Lecture . . . before the American Geographical Society* . . . , May 7, 1868, pamphlet (New York, 1868), pp. 20–21.

31. Markham, *Threshold*, p. 60.

32. M. F. Maury, *The Physical Geography of the Sea* (new ed.; London, 1859 [1st ed., 1855]), pp. 159–160. See also the edition by John Leighly entitled *The Physical Geography of the Sea, and its Meteorology* (Cambridge, Massachusetts: Harvard University Press, 1963), p. 234.

33. The *Report of the Committee on Recent Discoveries and Publications on Sub-Oceanic Geography* of the American Geographical and Statistical Society, January 8, 1857, presents extracts from this letter (see pp. 2–5).

34. "A Plan of Search Proposed by Mr. Petermann," [Great Britain, Admiralty], *Arctic Expedition: Further Correspondence and Proceedings* . . . (London, 1852), pp. 142–147.

35. E. Weller, "August Petermann: Ein Beitrag zur Geschichte der geographischen Entdeckungen und der Kartographie im 19. Jahrhundert," *Quellen und Forschungen zur Erd- und Kulturkunde 4* (Leipzig, 1911), 147, n. 25.

36. *Ibid.*, pp. 135–139.

37. Petermann's conceptions of the distribution of land and water in the Arctic are shown on the map accompanying the work cited in note 34 above; in *Petermanns Mitteilungen* (1865), pl. 5; in *Petermanns Mitteilungen Ergänzungsheft No. 26* (1869), pl. 1; and on the frontispiece to S. R. Van Campen, *The Dutch in the Arctic Seas* (2 vols.; London, 1876). The first two of these items also show ocean currents. De Long planned to follow the coast line of Petermann's hypothetical land mass.

38. Weller, *Petermann*, p. 134.

39. E. V. Blake, editor, *Arctic Experiences: Containing Capt. George E. Tyson's Wonderful Drift on the Ice-Floe, A History of the Polaris Expedition . . .* (New York, 1874), p. 388.

40. See Markham, *Threshold*, pp. 84, 126.

41. T. C. Chamberlin, "The Method of Multiple Working Hypotheses," *Journal of Geology 5* (1897), 837–848; reprinted in 39 (1931), 155–165. See also T. S. Kuhn, *The Structure of Scientific Revolutions* (Chicago: University of Chicago Press, 1962; Phoenix, paperback, 1964), p. 144.

For a recent study, which came to my attention after this book was in press, see E. L. Towle, "The Myth of the Open North Polar Sea," *Actes du Dixième Congrès International d'Histoire des Sciences*, vol. 2 (Paris: Hermann, 1964), pp. 1037–1041. Professor Towle informs me that he is "completing a study of the peculiar revival of the open polar sea hypothesis in the United States in the 1840's and 1850's." He is particularly interested in the applications of Kuhn's ideas to the history of geographical thought.

42. W. M. Davis, *The Coral Reef Problem* (New York: American Geographical Society, 1928).

43. See the stimulating and methodologically sophisticated discussion of this principle in Kuhn's *Structure of Scientific Revolutions*, pp. 143–158.

44. W. M. Davis, "The Value of Outrageous Geological Hypotheses," *Science 63* (1926), 463–468.

Chapter 7. From *Kubla Khan* to Florida

1. Insofar as it deals with Dr. Francis Harper's identifications of localities mentioned in the 1791 edition of Bartram's *Travels*, the present version of this paper is not "documented" in detail, as was the first version (1956). This is because, for all of the localities in question, page references both to the text of the *Travels* and to Dr. Harper's commentary thereon, where the localities are mentioned, described, and identified, may now easily be found by consulting the index to *The Travels of William Bartram: Naturalist's Edition*, edited with commentary and an annotated index by Francis Harper (New Haven: Yale University Press, 1958).

2. J. L. Lowes, *The Road to Xanadu: A Study in the Ways of the Imagination* (Boston and New York: Houghton Mifflin, 1927), pp. 367–370. Although he "came upon all the evidence independently," Lowes points out (p. 587) that he was not the first to note the links between *Kubla Khan* and these places as mentioned in Bartram's *Travels*. E. H. Coleridge had mentioned the "Manate" Spring and the Alligator Hole in 1906 and Georg Barsch in 1909 had referred to these and also to "der bezaubernde und erstaunliche Krystalquell."

3. William Bartram, *Travels Through North and South Carolina, Georgia, East and West Florida* . . . (Philadelphia, 1791; many subsequent editions). See also note 1, above.

4. Bartram, *Travels*, p. 165.

5. William Bartram, *Travels in Georgia and Florida, 1773–74: A Report to Dr. John Fothergill*, annotated by Francis Harper, *American Philosophical Society, Transactions* (n.s.) 33, pt. 2 (Philadelphia, 1943), 123–242. Cited below as Bartram, *Report*.

6. *Ibid.*, p. 123.

7. See above, note 2.

8. Bartram, *Report*, pp. 161–162 (MS., vol. 2, p. 32).

9. In the Naturalist's edition (see above, note 1), however, brief mention is made of some of these links.

10. Notably, Lowes, *The Road to Xanadu;* Ernest Earnest, *John and William Bartram* (Philadelphia: University of Pennsylvania Press, 1940); N. B. Fagin, *William Bartram, Interpreter of the American Landscape* (Baltimore: The Johns Hopkins University Press, 1933).

11. Owing to a typographical error, the essential words "but hitherto no one seems to have thought of crediting Salt Springs Run" were omitted in the first version of this paper, *American Quarterly 8* (1956), 78. See also above, p. 20.

12. Bartram, *Report*, p. 149 (MS., vol. 1, p. 80).

13. Lowes, *Road to Xanadu*, p. 587, n. 21; p. 588, n. 28.

14. *Ibid.*, pp. 8, 47.

15. *Ibid.*, p. 368.

16. *Ibid.*, 365–368.

17. Earnest, *John and William Bartram*, p. 125. I find no mention in Lowes (*Road to Xanadu*) of this passage in Bartram. It is possible that the image in *Kubla Khan* may have been the result of a blending of the two widely separated passages in the *Travels* (1791) relating to the Isle of Palms (pp. 156–157) and to the "spacious high forests and flowery lawns" (p. 364).

18. Lowes, *Road to Xanadu*, pp. 369, 370, 387.

19. Bartram, *Report*, p. 161 (MS., vol. 2, p. 29).

Chapter 8. What's "American" about American Geography?

1. This paper was first written in 1956 for the Columbia University Seminar in American Civilization which, at the suggestion of Dr. John A. Kouwenhoven, Professor of English at Barnard College, was considering from divers angles the question "What's 'American' about America?" See J. A. Kouwenhoven, *The Beer Can by the Highway: Essays on What's "American" about America* (Garden City, N.Y.: Doubleday, 1961), pp. 37–73.

2. For an annotated bibliography comprising "the principal writings [in English] of social scientists and historians on culture and personality, national character and American character since 1940," see Michael McGiffert, editor, *The Character of Americans: A Book of Readings* (Homewood, Illinois: Dorsey Press, 1964), pp. 361–367.

3. "So quintessentially American a performance as 'Porgy and Bess.'" C. L. Sulzberger, *New York Times*, December 28, 1955.

4. See R. B. Perry, *Characteristically American* (New York: Knopf, 1949), p. 36.

5. See John Gillin, "National and Regional Cultural Values in the United States" (*Social Forces*, 14 [Dec., 1955]: 107–113, as reprinted in McGiffert, ed., *Character*, pp. 217–223, especially pp. 221–223, where six cultural regions are differentiated: Northeast, Southeast, Middle States, Southwest, Northwest, and Far West.

6. *American Geography: Inventory and Prospect*, P. E. James and C. F. Jones, editors; J. K. Wright, consulting editor; published for the Association of American Geographers by Syracuse University Press (1954).

7. See Merle Curti, *The Growth of American Thought* (2nd ed.; New York: Harper, 1951), pp. 149–152; R. H. Brown, *Mirror for Americans: Likeness of the Eastern Seaboard, 1810* (New York: American Geographical Society, 1943), pp. xiii–xxii.

8. The first edition, which appeared in one volume under the title *The American Geography* (Elizabethtown, New Jersey, 1789) and approximately four-fifths of which were devoted to the United States, was followed by six official American editions entitled *The American Universal Geography* (Boston, 1793, 1796, 1802, 1805, 1812, 1819), each in two volumes. The work was also published in London, Edinburgh, and Dublin, and in French and Dutch translations; see R. H. Brown, "The American Geographies of Jedidiah Morse," *Annals of the Association of American Geographers 31* (1941), 145–217: 214–217.

9. *Ibid.* See also R. H. Brown, "A Letter to the Reverend Jedidiah Morse . . . ," *Annals of the Association of American Geographers 41* (1951), 188–198.

10. On Morse's British background see J. K. Wright, "Some British 'Grandfathers' of American Geography," *Geographical Essays in Memory of Alan G. Ogilvie*, edited by R[onald] Miller and J. W. Watson (London: Nelson, 1959), pp. 144–165.

11. See J. K. Wright, *Geography in the Making: The American Geographical Society, 1851–1951* (New York: American Geographical Society, 1952), pp. 123–127.

12. See H. E. Gregory, "A Century of Geology: Steps of Progress in the Interpretation of Land Forms," *American Journal of Science* (4th ser.), *46* (1918), 104–132.

13. W. M. Davis, "The Progress of Geography in the United States," *Annals of the Association of American Geographers 14* (1924), 159–215: 170 quoting J. P. Lesley, *A Manual of Coal and its Topography* (Philadelphia, 1856), p. 124.

14. Davis, "Progress," pp. 182–183. Davis used "simplified spelling."

15. Henri Baulig, "William Morris Davis: Master of Method," *Annals of the Association of American Geographers*, 40(1950):188–195, p. 188.

16. Davis, "Progress," p. 173.

17. Charles Poore, "Books of the Times," *New York Times*, December 15, 1955, quoting from *Places: A Volume of Travel in Space and Time . . .* , edited by Geoffrey Grigson and C. H. Gibbs-Smith (New York: Hawthorn, 1935).

18. See F. V. Hayden, *Annual Report of the U.S. Geological and Geographical Survey of the Territories . . . 1873* (Washington, 1874), p. 7 and plates facing pp. 136, 192; W. M. Davis, "Biographical Memoir of Grove

Karl Gilbert, 1843–1918," *Memoirs of the National Academy of Sciences 21* (1926), 5th Memoir, pp. 73, 76, 77. For the broader background see Wallace Stegner, *Beyond the Hundredth Meridian: John Wesley Powell and the Second Opening of the West* (Boston: Houghton Mifflin, 1953; Sentry paperback, 1962), pp. 146–191; see also R. J. Chorley, A. J. Dunn, and R. P. Beckinsale, *History of the Study of Landforms, or the Development of Geomorphology* (to be in 3 vols.; London: Methuen; New York: Wiley), I (1964), 467–662.

19. Isaiah Bowman, *Geography in Relation to the Social Sciences* (New York: Scribner, 1934), pp. 220–222.

20. Personal letter to J. K. Wright, July 29, 1949; see also G. M. Wrigley, "Isaiah Bowman," *Geographical Review 41* (1951), 7–65: 30.

21. *Ibid.*, pp. 13–17, 30–35; Wright, *Geography in the Making*, pp. 257–260.

22. Isaiah Bowman, *The Pioneer Fringe* (New York: American Geographical Society, 1931).

23. See *American Geography: Inventory and Prospect*, pp. 125–141.

24. *Ibid.*, p. 17.

25. See Kouwenhoven, *The Beer Can by the Highway*, pp. 72–73.

26. Wrigley, "Bowman," pp. 21–25.

27. Isaiah Bowman, *The New World: Problems in Political Geography* (Yonkers-on-Hudson and Chicago: World Book Co., 1921; supplements, 1923, 1924; 4th ed., 1928; Chinese trans., Shanghai, 1927, 2nd ed., 1928).

28. See above, note 6.

29. See Perry, *Characteristically American*, pp. 9, 13, 28–29, 56.

30. Ironical (lest you misunderstand, as others have, dear literal-minded reader).

31. *American Geography: Inventory and Prospect*, pp. 246–247.

32. *Ibid.*, 145.

33. *Ibid.*, 215.

34. "Through American geography runs a reasonably adequate emphasis on natural factors in man's relation to the earth; but relatively slight attention has been given to cultural factors" (J. R. Whitaker in *American Geography*, p. 238).

35. Based mainly on my own observation and experience, this nowise remarkable generalization is supported by the replies (donated to the archives of the American Geographical Society) to a questionnaire sent in 1965 to some ten geographers in different parts of the United States. "America is notable for the number of her scientists of the second rank, and for her development of useful applications, rather than for individual genius and epoch-making originality" (Perry, *Characteristically American*, p. 24).

36. E. G. R. Taylor, "Whither Geography? A Review of Some Recent Geographical Texts," *Geographical Review 27* (1937), 129–135: 129.

37. Michael McGiffert, editor, *The Character of Americans: a Book of Readings* (Homewood, Illinois: The Dorsey Press, 1964), p. x.

Chapter 9. The Heights of Mountains: "An Historical Notice"

1. Alexander von Humboldt, *Aspects of Nature in Different Lands and Different Climates . . .*, trans. Mrs. Sabine (Philadelphia, 1849), p. 89.

2. See *Guiness Book of Superlatives* (London: Superlatives Inc., c. 1956); *Guiness Book of Records* (New York: Sterling, 1956; Bantam paperback, 1963).

3. Translation in Helmut De Terra, *Humboldt: The Life and Times of Alexander von Humboldt, 1769–1859* (New York: Knopf, 1955), p. 128.

4. Edward Whymper, *Travels Among the Great Andes of the Equator* (New York, 1892), pp. 76–77.

5. A. H. Bent, "Early American Mountaineers," *Appalachia 13* (1913), 45–67: 46.

6. S. I. Bailey, "Peruvian Meteorology, 1888–1890," *Annals of the Astronomical Observatory of Harvard College 39* (1899), pt. 1, pp. 10–11.

7. *Bulletin of the American Geographical Society 39* (1907), 108; *42* (1910), 55.

8. See, for example, Humboldt, *Aspects of Nature*, pp. 51–64, 73–94, 217–223, 247–252; also Alexander von Humboldt, *Cosmos: essai d'une description physique du monde*, trans. H. Faye (3 vols.; Paris, 1848–1851), I, 9–10, 435–442; *Cosmos: A Sketch of the Physical Description of the Universe*, trans. E. C. Otte (5 vols.; New York, 1851–1860), I, 28–33. See also above, Introduction, n. 3.

9. Florian Cajori, "History of Determinations of the Heights of Mountains," *Isis 12*, no. 39 (1929), 482–515: 496.

10. O[scar] Peschel, *Geschichte der Erdkunde bis auf Alexander von Humboldt und Carl Ritter* (2nd ed. by Sophus Ruge; Munich, 1877).

11. Knut Lundmark, "Kunskapen om bergens höjder . . ." (with English summary, "Knowledge of Heights of Mountains"), *Svensk Geografisk Årsbok: The Swedish Geographical Yearbook 18* (1942), 127–156.

12. Cajori, "History," pp. 484–485.

13. Peschel, *Geschichte der Erdkunde*, p. 687, n. 2.

14. See W. E. K. Middleton, *The History of the Barometer* (Baltimore: Johns Hopkins Press, 1964).

15. Peschel, *Geschichte*, pp. 688–697; Cajori, "History," pp. 499–512. On Caswell (1656–1712), who also measured the height of Cader Idris, see E. W. Gilbert, *Geography as a Humane Study: An Inaugural Lecture Delivered Before the University of Oxford on 12 November 1954* (Oxford: Clarendon Press, 1955), p. 5.

16. Samuel Williams, *The Natural and Civil History of Vermont* (Walpole, N.H., 1794), p. 21; 2nd ed., 2 vols. (Burlington, Vermont, 1809), I, 27–29.

17. Peschel, *Geschichte*, p. 702, n. 1. Peschel gives Humboldt's total as 122, but his figures add to only 62.

18. J. S. T. Gehler, *Physikalisches Wörterbuch* (Leipzig, 1829), 5, 339–397 (reference from Cajori, "History," p. 504).

19. See R. Benini, "Origine, sito, forma, e dimensioni del Monte del Purgatorio e dell'Inferno dantesco," *Rendiconti della Reale Accademia dei Lincei, Classe di scienze morali, storiche e filologiche* (ser. 5) *25* (Rome, 1917), 1015–1129.

20. J. A. Lawson, *Wanderings in New Guinea* (London, 1875). Neither the *Journal of the Royal Geographical Society* nor *Petermanns Mitteilungen* deigned to notice this book.

21. This and the material in the preceding paragraph come from Cajori, "History," pp. 482–493.

22. Aristotle, *Meteorologica*, I, 13, trans. by E. W. Webster (Oxford: Clarendon Press, 1923), 350a.

23. Scipio Claramontius, *De altitudine Caucasi* (Paris, 1649; Bologna, 1653); I have not seen this work; it is cited in J. B. Riccioli, *Geographiae et hydrographiae reformatae, libri duodecim* (Venice, 1672), p. 202 (the first edition of Riccioli was published in Bologna, 1661).

24. Riccioli's *Geographiae* contains a wealth of historical and bibliographical data. On pp. 210–211 two tables are presented, one showing what different authorities had believed to be the world's highest mountain and the other various estimates by different authorities with regard to the altitudes of some two dozen mountains.

25. *Ibid.*, pp. 203, 211.

26. *Ibid.*, p. 210. For other examples of excessive estimates based on at least a semblance of mathematical reasoning, see Robert Hues, *Tractatus de globis*, edited by C. R. Markham (London, 1889), pp. 8–14; also Cajori, "History," pp. 493–494. Cajori here quotes Hues to the effect that Cardan (Geronimo Cardano, 1501–1576) and other mathematicians estimated that mountains rise to an altitude of 288 miles; but, read in its context, Hues's statement would appear to have reference to the maximum height to which vapor rises.

27. Riccioli, *Geographiae*, p. 211. See also Wilhelm Capelle, "Berges- und Wolkenhöhen bei griechischen Physikern," Στοιχεῖα: *Studien zur Geschichte des antiken Weltbildes und der griechischen Wissenschaft herausgegeben von Franz Boll* (Leipzig and Berlin, 1916), V, 13.

28. Peschel, *Geschichte der Erdkunde*, pp. 426, 697–699.

29. José de Acosta, *Historia natural y moral de las Indias* (Seville, 1590), III, 9; (Madrid, 1894), I, 208); quotation from Edward Grimston's trans. (London, 1604), edited by C. R. Markham (2 vols.; London: Hakluyt Society, 1880), I, 131.

30. Peschel, *Geschichte der Erdkunde*, p. 545.

31. *Asiatick Researches* (publication of the Asiatic Society of Bengal) *12* (1816), 266.

32. Maurice Herzog, *Annapurna: Conquest of the First 8000-metre Peak [26,493 feet]*, trans. from the French by Nea Morin and Janet Adam Smith (London: Jonathan Cape, 1952). There is no mention in this volume that Dhaulagiri was once believed to be the world's highest mountain.

33. *Proceedings of the Royal Geographical Society 1* (1856–1857), 345–347. Peschel, *Geschichte der Erdkunde*, p. 699. *Briefwechsel Alexander von Humboldt's mit Heinrich Berghaus . . . 1825–1858* (2nd ed.; Jena, 1869), III, 109. Alexander von Humboldt, *Aspects of Nature* (Mrs. Sabine's trans.), p. 88.

34. *Proceedings of the Royal Geographical Society 1* (1856–1857), 350.

35. See B. L. Gulatee, "Mount Everest: Its Name and Height," *Himalayan Journal 17* (1952), 131–142.

36. B. L. Gulatee, "The Height of Mount Everest . . . ," *Survey of India, Technical Paper No. 8* (1954), pp. i–ii. See also *Geographical Journal 122* (1956), 141–142.

37. See *Geographische Zeitschrift 36* (1930), 300; *Life* (October 3, 1949), 92, 104.

38. Humboldt, *Aspects of Nature*, pp. 63–64, 217–219.

39. *The Military Engineer 48* (1956), 384.

40. J. D. Whitney, "Which is the Highest Mountain in the United States, and Which in North America," *Proceedings of the California Academy of Natural Sciences 2* (1858–1862), 219–224. See also E. T. Brewster, *Life and Letters of Josiah Dwight Whitney* (Boston: Houghton Mifflin, 1909), in which there is much of interest concerning early mountain measurements and estimates in the western United States.

41. Jedidiah Morse, *The American Universal Geography* . . . (3rd ed.; Boston, 1796), I, 401.

42. Jeremy Belknap, *The History of New Hampshire* . . . (2nd ed.; Boston, 1813), III, 28–29.

43. Williams, *Vermont* (1794 ed.), pp. 22–24; (1809 ed.), I, 27–28.

44. John Drayton, *A View of South Carolina as Respects her Natural and Civil Concerns* (Charleston, 1802), p. 12.

45. Thomas Jefferson, *Notes on the State of Virginia* (Paris, 1785; many later eds.), query iv (edited by William Peden; Chapel Hill, N.C.: University of North Carolina Press, 1955, p. 20).

46. *Ibid.*, Peden ed., p. 262.

47. Peschel, *Geschichte der Erdkunde*, pp. 687–688.

48. "Note on Diversities in Published Estimates of the Height of Mt. Washington, New Hampshire," *American Journal of Science* (2nd ser.) *44* (1867), 377–379. This "Note" may well have been contributed by D. C. Gilman (see above, pp. 173, 178).

49. See Frederick Tuckerman, "Early Visits to the White Mountains," *Appalachia 15* (1921), 111–127; L. S. Mayo, "The White Mountains in Three Centuries," in *Three Essays by Lawrence Shaw Mayo: An Appreciation* (Boston: privately printed, 1948), pp. 23–60.

50. Robert Rogers, *A Concise Account of North America* . . . (London, 1765), p. 47.

51. Belknap, *New Hampshire*, III, 37–39. See also Belknap, "A Description of the White Mountains," *Transactions of the American Philosophical Society 2* (1786), 42–49; *Journal of a Tour of the White Mountains in July 1784* . . . (Boston: Massachusetts Historical Society, 1876), pamphlet.

52. Williams, *Vermont* (1794 ed.), pp. 23–24, n.; (1809 ed.), I, 28–29, n.

53. Morse, *American Universal Geography* (6th ed., 1812), I, 231; (7th ed., 1819), p. 257.

54. Timothy Dwight, *Travels in New England and New York* (New York, 1802), II, 163; (New Haven, 1821), p. 164; (London, 1823), p. 150. Such generous estimates prompted Robert Frost to comment wryly on how an early map had disillusioned him by showing the White Mountains as twice their actual height. Robert Frost, "New Hampshire," in *Complete Poems* (New York: Holt, 1949), pp. 199–212:209.

55. Jacob Bigelow, "Some Account of the White Mountains of New Hampshire," *New England Journal of Medicine and Surgery 5* (1816), 321–358.

56. Arnold Guyot, "On the Appalachian Mountain System," *American Journal of Science* (2nd ser.) *31* (1861), 157–187. *Report of the Superintendent of the United States Coast Survey*, 1853, p. 29; 1854, p. 40. C. H. Hitchcock, *The Geology of New Hampshire* (3 vols.; Concord, 1874–1878), I, 64–69.

57. Morse, *American Universal Geography* (3rd ed., 1796), I, 383.
58. *Massachusetts Historical Society, Collections* (2nd ser.) *8* (1826), 112–116.
59. T. G. Bradford and S. G. Goodrich, *A Universal Illustrated Atlas* (Boston, 1842), p. 35.
60. Moses Greenleaf, *A Survey of the State of Maine* (Portland, 1829), p. 47.
61. W. C. Redfield, "Some Account of Two Visits to the Mountains in Essex County, New York, in the Years 1836 and 1837 . . . ," *American Journal of Science* (1st ser.) *33* (1838), 301–323:322–323. Before 1837 the Catskills were regarded as the highest mountains in New York State (*ibid.*, p. 314).
62. *State of New York No. 200. In Assembly, February 20, 1838. Communication from the Governor Relative to the Geological Survey of the State*, pp. 185–250 ("Report of E. Emmons, Geologist of the 2nd Geological District of the State of New York"); "Visit to the Mountains of Essex" (pp. 240–250).
63. E[lisha] Mitchell, "Notice of the Height of Mountains in North Carolina," *American Journal of Science* (1st ser.) *35* (1839), 377–380.
64. Guyot, "Appalachian Mountain System," pp. 159–161.
65. See D. C. Gilman, "The Last Ten Years of Geographical Work in this Country," *Journal of the American Geographical Society 3* (1873), 113–133: 122–123; W. H. Brewer, "Explorations in the Rocky Mountains and the High Peaks of Colorado," *ibid.*, 193–215: 196–204.
66. Mount Hood "was formerly called the 'monarch of American mountains'—16,000, 17,000, and 18,000 feet high, of which the latter number was by far the most popular belief; and, indeed, one called it 21,000 feet high." Brewer, "Explorations," p. 197.
67. *Ibid.*, p. 203. See also Brewster, *Josiah Dwight Whitney*, pp. 268–269.
68. Gilman, "Last Ten Years," p. 132.

Chapter 10. On Medievalism and Watersheds in the History of American Geography

1. H. S. Commager, *The American Mind: An Interpretation of American Thought and Character Since the 1880's* (New Haven: Yale University Press, 1950; Yale Paperbound, 1959), p. 41.
2. See J. K. Wright, *The Geographical Lore of the Time of the Crusades . . .* (New York: American Geographical Society, 1925; Dover reprint, New York, 1965).
3. S. E. Morison, *Harvard College in the Seventeenth Century* (2 vols.; Cambridge, Mass.: Harvard University Press, 1936); *The Puritan Pronaos: Studies in Intellectual Life in New England in the Seventeenth Century* (New York: New York University Press, 1936). Perry Miller, *The New England Mind:* [I] *The Seventeenth Century* (New York: Macmillan, 1939; Cambridge, Mass.: Harvard University Press, 1954); [II] *From Colony to Province* (Cambridge, Mass.: Harvard University Press, 1953); paperback ed., 2 vols. (Boston: Beacon Press, 1961); *Errand into the Wilderness* (Cambridge, Mass.: Harvard University Press, 1956). Stow Persons, *American Minds: A History of Ideas* (New York: Holt, 1958).

4. Lewis Evans, *Geographical . . . Essays: The First, containing an Analysis of a General Map of the Middle British Colonies in America . . .* (Philadelphia, 1755), facsimile reproduction of text and map in L. H. Gipson, *Lewis Evans* (Philadelphia: Historical Society of Pennsylvania, 1939), pp. 140–176, map VI (following p. 219).

5. Jedidiah Morse, *The American Universal Geography . . .* (3rd ed.; Boston, 1796), I, 383. Agamenticus is the only mountain in Maine mentioned in the first edition, *The American Geography* (Elizabethtown, N.J., 1789), p. 196.

6. Morse, *The American Universal Geography* (1796), I, 794. William Guthrie, *A New System of Modern Geography . . .* (1st American ed., Philadelphia, 1794–1795), II, 629–630.

7. E. D. Fite and Archibald Freeman, *A Book of Old Maps Delineating American History . . . to the Close of the Revolutionary War* (Cambridge, Mass.: Harvard University Press, 1926), pp. 117–119. Bernard De Voto, *The Course of Empire* (Boston: Houghton Mifflin, 1952), pp. 55–59. William Bradford, *Of Plymouth Plantation, 1620–1647*, edited by S. E. Morison (New York: Knopf, 1952), p. 79, n. 7.

8. See Alan Moorehead, *The White Nile* (New York: Harper, 1960, 1961); Armand Rainaud, *Le continent austral: hypothèses et découvertes* (Paris, 1893); and, on the Open Polar Sea, Chapter 6.

9. C. R. Beazley, *The Dawn of Modern Geography* (3 vols.; London: Frowde, 1897, 1901; Oxford, Clarendon Press, 1906).

10. See J. K. Wright, "Some British 'Grandfathers' of American Geography," *Geographical Essays in Memory of Alan G. Ogilvie*, edited by R[onald] Miller and J. W. Watson (London: Nelson, 1959), pp. 144–165: 153–154.

11. Morse, *The American Universal Geography* (Boston, 1793), II, 18–19 (this passage in Morse is almost word for word as in Guthrie, *Modern Geography* [London, 1771 ff.], p. 74; see also 1st American ed., I, 88). See Herman Melville, *Moby Dick* (New York: Random House, 1940), chap. 59, p. 403.

12. Thomas Jefferson, *Notes on the State of Virginia* (Paris, 1785, and many later eds.), query vi (ed. by William Peden [Chapel Hill, N.C.: University of North Carolina Press, 1955], pp. 43–44).

13. Morse, *American Universal Geography* (1793), II, 396.

14. John Drayton, *A View of South Carolina as Respects her Natural and Civil Concerns* (Charleston, 1802), pp. 57, 59.

15. Jeremy Belknap, *The History of New Hampshire . . . Containing also a Geographical Description of the State . . .* (2nd ed.; Boston, 1813), III, 29. Morse, *American Universal Geography* (Boston, 1796), II, 363. C. H. Hitchcock, *The Geology of New Hampshire* (3 vols.; Concord, N.H., 1874–1878), I, 178.

16. Loyal Durand, Jr., "'Mountain Moonshining' in East Tennessee," *Geographical Review 46* (1956), 168–181. Wilbur Zelinsky, "The New England Connecting Barn," *ibid. 48* (1958), 540–553. Fred Kniffen, "The American Covered Bridge, *ibid. 41* (1951), 114–123. Eric Sloane, "The First Covered Bridge in America," *ibid. 49* (1959), 315–321. Eugene Mather and J. F. Hart, "The Geography of Manure," *Land Economics 32* (1956), 25–38. C. F. Bennett, Jr., "Cultural Animal Geography: An Inviting Field of Research," *The Professional Geographer 12:* 5 (1960), 12–14.

17. A. J. Toynbee, *East to West* (New York: Oxford University Press, 1958), as quoted in *Geographical Journal 124* (1958), 571.

18. George Sarton, *The Study of the History of Science* (Cambridge, Mass.: Harvard University Press, 1932), p. 5, as quoted in *Isis 48* (1957), 305. T. S. Kuhn discusses the question "Why is progress a perquisite reserved almost exclusively for the activities we call science?" (*The Structure of Scientific Revolutions* [Chicago: University of Chicago Press, 1962; Phoenix paperback, 1964], pp. 159–172).

19. Morse, *American Universal Geography* (1793), I, preface. See also above, p. 128.

20. J. P. Thompson, "The Value of Geography to the Scholar, the Merchant, and the Philanthropist," *Journal of the American Geographical and Statistical Society 1* (1859), 98–107: 98; see also J. K. Wright, *Geography in the Making: The American Geographical Society, 1851–1951* (New York: American Geographical Society, 1952), pp. 44–45.

21. William Libbey, "The Life and Scientific Work of Arnold Guyot," *Journal of the American Geographical Society 20* (1888), 194–221: 213–214.

22. C. R. Dryer, "A Century of Geographic Education in the United States," *Annals of the Association of American Geographers 14* (1924), 117–149: 124.

23. I. C. Russell, "Report of a Conference on Geography," *Journal of the American Geographical Society 27* (1895), 30–41: 40 (Russell's report was a discussion of *Report of the Committee on Secondary School Studies* appointed at the meeting of the National Educational Association, July 7, 1892 [Washington: U.S. Bureau of Education, 1893], which was republished with the title *Report of the Committee of Ten . . .* [New York: American Book Co., 1894]).

24. *Journal of the American Geographical Society 31* (1899), 382.

25. W. S. Tower, "Scientific Geography: The Relation of its Content to its Subdivisions," *Bulletin of the American Geographical Society 42* (1910), 801–825: 825.

26. J. N. L. Baker, "Geography and its History," *Advancement of Science 46* (1955), 188–198: 189; also in *The History of Geography: Papers by J. N. L. Baker* (Oxford: Blackwell; New York: Barnes and Noble, 1963), pp. 84–104: 86–87. The arrogance in question is sometimes accompanied by execrable manners. It has been reported to me by a reliable witness that at a dinner party recently a geographer in his early thirties told two distinguished geographers now in their late sixties that the older generation of geographers had contributed *nothing* upon which the younger generation could build.

27. See above, p. 193; also David Riesman, "The Found Generation," as reprinted from *Abundance for What?* by David Riesman (Garden City, New York: Doubleday, 1956) in Michael McGiffert, ed., *The Character of Americans: A Book of Readings* (Homewood, Illinois: The Dorsey Press, 1964), pp. 265–276.

28. Kuhn, *Structure of Scientific Revolutions*, p. 137. For some striking examples of the "law of the disparagement of the past," see Giorgio di Santillana, "On Forgotten Sources in the History of Science," in A. C. Crombie, ed., *Scientific Change . . .* (New York: Basic Books, 1963), pp. 813–828, and comments by Peter Laslett (*ibid.*, pp. 861–862).

29. Richard Hartshorne, "The Nature of Geography: A Critical Survey

of Current Thought in the Light of the Past," *Annals of the Association of American Geographers 29* (1939), 171–658 (also paginated [i–vi], [1–482]): 198 [22], "second printing" (photolith, 1946), with additional data.

30. Richard Hartshorne, *Perspective on the Nature of Geography* (Chicago: published for the Association of American Geographers by Rand McNally, 1959).

31. *American Geography: Inventory and Prospect*, ed. P. E. James and C. F. Jones; consulting editor, J. K. Wright (published for the Association of American Geographers by the Syracuse University Press, 1954), p. vi (the quoted sentences are paraphrased from a letter from J. K. Wright to J. R. Whitaker, July 11, 1953).

32. Estimated from inspection of a graph showing the years of birth for 1273 members of the Association of American Geographers, compiled by Herman R. Friis in 1956 and reproduced in the Association's *Handbook-Directory* (Washington, 1956), p. 37.

33. A. M. Wright, *The American Short Story in the Twenties* (Chicago: University of Chicago Press, 1961).

34. See above, p. 193; below, Chapter 12, n. 11.

35. See S. R. Packard, "A Medievalist Looks at the Renaissance," reprinted from *Smith College Studies in History 44* (1964), 3–12.

36. See J. K. Wright, *Geography in the Making*, pp. 172–175; *American Geography: Inventory and Prospect*, chapter on "The Regional Concept and the Regional Method" (pp. 19–68), for which the original draft was prepared by the late Derwent Whittlesey; also, for a divergent opinion, G. H. T. Kimble, "The Inadequacy of the Regional Concept," *London Essays in Geography . . .* , ed. L. D. Stamp and S. W. Wooldridge (Cambridge, Mass.: Harvard University Press), pp. 151–174.

Chapter 11. Daniel Coit Gilman: Geographer and Historian

1. See, for example, S. E. Morison, *Harvard College in the Seventeenth Century*, in *Tercentennial History of Harvard College and University, 1636–1936* (2 parts; Cambridge, Mass.: Harvard University Press, 1936), pp. 214–251; Theodore Hornberger, *Scientific Thought in the American Colleges, 1638–1800* (Austin, Tex.: University of Texas Press, 1945), pp. 37–40, 54, 74–75; William Warntz, *Geography Now and Then: Some Notes on the History of Academic Geography in the United States* (New York: American Geographical Society, 1964). The present paper is based almost entirely on publications readily available in large libraries. Quotations and statements for which no references are given below are derived from the following books: Abraham Flexner, *Daniel Coit Gilman: Creator of the American Type of University* (New York: Harcourt Brace, 1910); Fabian Franklin, *The Life of Daniel Coit Gilman* (New York: Dodd, Mead, 1910); D. C. Gilman, *University Problems in the United States* (New York: Century, 1898), *The Launching of a University and other Papers: a Sheaf of Reminiscences* (New York: Dodd Mead, 1906); "Daniel Coit Gilman: First President of the Johns Hopkins University, 1876–1901," *The Johns Hopkins University Circular* (n. s., 1908), no. 10 (whole no. 211).

2. Warntz, *Geography Now and Then*, pp. 3–4, in his interesting study of early American academic geography, maintains that this assertion is unjustified.

3. R. H. Brown, "A Plea for Geography, 1813 Style," *Annals of the Association of American Geographers 41* (1951), 233–236: 233 (this paper, one of "Four Papers by the Late Ralph H. Brown," *ibid.*, pp. 187–236, is a reprint, with a brief explanatory note, of an essay on "Geography" written by Sparks when an undergraduate at Harvard.

4. Franklin, *Gilman*, p. 339.

5. See Gladys M. Wrigley, "Isaiah Bowman," *Geographical Review 41* 1951, 7–65; J. C. French, *A History of the University Founded by Johns Hopkins* (Baltimore: Johns Hopkins Press, 1946), pp. 456–462.

6. Personal letter from Dr. Fulmer Mood, University of California, October 29, 1952.

7. The first address, read January 31, 1871, covered the years 1861–1870 (*Journal of the American Geographical Society 3* [1872], 111–133); the second, read January 30, 1872, covered 1871 (*ibid. 4* [1873], 119–144).

8. Council Minutes, American Geographical and Statistical Society (MS. in the Society's library), vol. 4, p. 73.

9. J. D. Whitney, *Geographical and Geological Surveys* (Cambridge, Mass., 1875), p. 53 (from *North American Review*, July and October, 1875).

10. See J. K. Wright, *Geography in the Making: The American Geographical Society, 1851–1951* (New York: American Geographical Society, 1952), p. 296.

11. Fulmer Mood, "The Rise of Official Statistical Cartography in Austria, Prussia, and the United States, 1855–1872," *Agricultural History 20* (1946), 209–225: 221–222.

12. D. C. Gilman, *On the Structure of the Earth, with Some Reference to Human History: A Synopsis of Twelve Geographical Lectures Delivered Before the Senior and Junior Classes in the College of New Jersey, Feb. 1871* (Princeton, 1871).

13. G. P. Marsh, *Man and Nature, or Physical Geography as Modified by Human Action* (New York, 1864; ed. by David Lowenthal, Cambridge, Mass.: Harvard University Press, 1965); *The Earth as Modified by Human Action: A New Edition of Man and Nature* (New York, 1874). Quotation from David Lowenthal, *George Perkins Marsh: Versatile Vermonter* (New York: Columbia University Press, 1958), p. 246.

14. D. C. Gilman, "The Last Ten Years of Geographical Work in this Country," *Journal of the American Geographical Society 3* (1872), 116. Wright, *Geography in the Making*, p. 87.

15. J. D. Whitney, *The United States: Facts and Figures Illustrating the Physical Geography of the Country, and its Material Resources* (Boston, 1889). N. S. Shaler, ed., *The United States of America* (2 vols., New York, 1894). W. L. G. Joerg, "The Geography of North America: A History of its Regional Exposition," *Geographical Review 26* (1936), 640–663: 647, described Whitney's book as "in some respects the first adequate geography of the country by an American in the period under discussion [1877–1934]."

16. On this subject as it was discussed in the United States during the last half of the nineteenth century see H. B. Adams, "The Study of History in American Colleges and Universities," *U.S. Bureau of Education Circular of Information* (1887), no. 2; "The Study of History in Schools, Being the Report . . . by the Committee of Seven," *Annual Report of the American Historical Association for the Year 1898* (Washington, 1899), pp. 427–564;

G. Stanley Hall, *Educational Problems* (2 vols.; New York: Appleton, 1911), II, chap. 16, "Pedagogy of History," and chap. 21, "School Geography" (pungent discussion and useful bibliographical data); J. F. Jameson, *An Historian's World: Selections from the Correspondence of John Franklin Jameson*, ed. Elizabeth Donnan and L. F. Stock (*Memoirs of the American Philosophical Society*, vol. 42 [Philadelphia, 1956]); A. D. White, *Autobiography of Andrew Dickson White* (2 vols.; New York: Macmillan, 1905; reprint, Century, 1922).

17. White, *Autobiography*, I, 257–259; Adams, "The Study of History," pp. 134–136.

18. D. C. Gilman, *The Life of James Dwight Dana* (New York and London, 1899), p. 187. See also above, p. 284.

19. See P. R. Fossum, "The Anglo-Venezuelan Boundary Controversy," *Hispanic-American Historical Review 8* (1928), 299–329; D. R. Dewey, "National Problems, 1885–1897," *The American Nation: A History* (28 vols.; New York: Harper, 1904–1918), XXIV (1907), pp. 305–313.

20. Michael Kraus, *A History of American History* (New York: Farrar & Rinehart, 1937), pp. 16–20.

21. Wright, *Geography in the Making*, pp. 199–200.

22. White, *Autobiography*, II, 120–122.

23. Justin Winsor, ed., *Narrative and Critical History of America* (8 vols.; Boston and New York, 1884–1889).

24. See Jameson, *An Historian's World*, pp. 83–84.

25. United States Commision on Boundary between Venezuela and British Guiana (Commission to Investigate and Report upon the True Division Line between Venezuela and British Guiana), *Report and Accompanying Papers* . . . (4 vols.; Washington, 1897), III, *Geographical*, pp. 4–6. Vol IV, "Maps of the Orinoco-Essequibo Region," contains 61 facsimile reproductions of critical maps and 15 maps especially compiled for the purpose and reproduced on a uniform base, lithographed by A. Hoen and Company of Baltimore. The maps show boundaries proposed or claimed, drainage basins, geology, and facts relating to European occupation.

26. White, *Autobiography*, II, 122.

27. Franklin, *Gilman*, p. 326.

28. French, *History of the University Founded by Johns Hopkins*, pp. 228–231.

29. Adams, "The Study of History," pp. 191–193.

30. C. O. Paullin, *Atlas of the Historical Geography of the United States* (published jointly by the Carnegie Institution of Washington and the American Geographical Society of New York, 1932). See also J. K. Wright, "J. Franklin Jameson and the 'Atlas of the Historical Geography of the United States,'" in *J. Franklin Jameson: A Tribute*, ed. Ruth Anna Fisher and W. L. Fox (Washington: Catholic University of America Press, 1965), pp. 66–79. The statement in the present paper as first published in the *Geographical Review 51* (1961), 398, n. 37, concerning Jameson's part in the publication of George Sarton's *Introduction to the History of Science* (3 parts in 5 vols.; 1927, 1937) would seem to be incorrect.

31. S. E. Morison, ed., *The Development of Harvard University Since the Inauguration of President Eliot, 1869–1929* (The Tercentennial History of Harvard College and University, 1636–1936; Cambridge, Mass.: Harvard University Press, 1930), p. 170.

32. J. F. Jameson in the Introduction to American Historical Association, Committee on Planning and Research, *Historical Scholarship in America: Needs and Opportunities* (New York: Long and Smith, 1932), p. 5.

33. On these developments between about 1875 and about 1925, and especially on the "impact of German scholarship" and criticism that it invoked, see Merle Curti, *The Growth of American Thought* (New York: Harper, 1943), pp. 581–593. On how the seminar method originated in Germany in the early eighteenth century and its subsequent development there, see J. H. Wright [my father], "An Address on the Place of Original Research in College Education" (extract from *Transactions of the National Educational Association, 1882*, Boston, 1883), pp. 6–13.

34. Curti, *Growth*, p. 586; see also French, *History*, pp. 85–87.

35. See G. Donald Hudson, "Professional Training of the Membership of the Association of American Geographers," *Annals of the Association of American Geographers 41* (1951), 97–115: 103–109; C. F. Jones, "Status and Trends of Geography in the United States, 1952–1957", ed. Ruth E. Baugh, *Professional Geographer 11*, no. 1, pt. 2 (1959), 40.

Chapter 12. Miss Semple's "Influences of Geographic Environment"

1. George Tatham, "Environmentalism and Possibilism," *Geography in the Twentieth Century*, ed. Griffith Taylor (New York: Philosophical Library; London: Methuen, 1951), pp. 128–162: 147.

2. See, for example, E. S. Bates, *Biography of the Bible* (New York: Simon & Schuster, 1937).

3. C. van Paassen, *The Classical Tradition of Geography* (Groningen: Wolters, 1957), warns against interpreting "passages in Homer and Herodotus as though they had been written by Friedrich Ratzel or Ellen Churchill Semple," to construe his thought as in my review of his book, *Geographical Review 49* (1959), 300.

4. See, for example, C. J. Glacken, "Changing Ideas of the Habitable World," *Man's Role in Changing the Face of the Earth*, ed. W. L. Thomas, Jr. (Chicago: University of Chicago Press, 1956), pp. 70–92; Richard Hartshorne, "The Nature of Geography, A Critical Survey of Current Thought in the Light of the Past," *Annals of the Association of American Geographers 29* (1939), 171–658 (also paginated i–vi, 1–482), pp. 296–302, and *Perspective on the Nature of Geography* (Chicago: published for the Association of American Geographers by Rand McNally, 1959), pp. 55–63.

5. E. C. Semple, *American History and its Geographic Conditions* (Boston and New York: Houghton Mifflin, 1903; revised in collaboration with C. F. Jones, Boston: Houghton Mifflin, 1933); *The Geography of the Mediterranean Region: Its Relation to Ancient History* (New York: Holt, 1931).

6. For references to, and discussions of, early American (c. 1750–1810) speculations on geographic environmentalism (especially climatic), see R. H. Brown, *Mirror for Americans: Likeness of the Eastern Seaboard, 1810* (New York: American Geographical Society, 1943), pp. 13–24, 266–267; Merle Curti, *The Growth of American Thought* (2nd ed., New York: Harper, 1951), pp. 122–123, 165–169; and, especially, D. J. Boorstin, *The Lost World of Thomas Jefferson* (New York: Holt, 1948; Boston: Beacon

paperback, 1960); E. T. Martin, *Thomas Jefferson, Scientist* (New York: Schuman, 1952), and three papers by Gilbert Chinard: (1) "The American Philosophical Society and the World of Science (1768–1800)," *Proceedings of the American Philosophical Society 87* (1943–1944), 1–11:9; (2) "The American Philosophical Society and the Early History of Forestry in America," *ibid. 89* (1945), 444–488: 451–470; and (3) "Eighteenth Century Theories on America as a Human Habitat," *ibid. 91* (1947), 27–57: 39–54.

7. G. G. Chisholm, "Miss Semple on the Influences of Geographical Environment," *Geographical Journal 39* (1912), 31–37: 31.

8. C. O. Sauer, "Foreword to Historical Geography," *Annals of the Association of American Geographers 31* (1941), 1–24: 5, also in C. O. Sauer, *Land and Life: A Selection from the Writings of Carl Ortwin Sauer*, ed. John Leighly (Berkeley and Los Angeles: University of California Press, 1963), pp. 351–379: 356.

9. Harriet Wanklyn, *Friedrich Ratzel: A Biographical Memoir and Bibliography* (Cambridge [England]: University Press, 1961), pp. 32–33.

10. See Lucien Febvre, *La Terre et l'évolution humaine: Introduction géographique à l'histoire* (Paris: La Renaissance du Livre, 1922), pp. 334–338.

11. A. C. Crombie, ed., *Scientific Change . . .* (New York: Basic Books, 1963), p. 861. This is related to what the Germans call the *Generationenproblem;* see Crane Brinton, *Isis 50* (1959), 181. See also above, pp. 161, 314, n. 27.

12. Reviews in seven American periodicals, all laudatory, are quoted in the *Book Review Digest* 7 (1911), 422; the first quotation is from an unsigned review in the *Journal of Geography 10* (1911), 33; the second from a review by R. H. Whitbeck, *Bulletin of the American Geographical Society 43* (1911), 937–939.

13. G. G. Chisholm, "Miss Semple." In "Some Recent Contributions to Geography," *Scottish Geographical Magazine 27* (1911), 561–573; on pp. 569–573 Chisholm discusses *Influences* as disproving an assertion made by Colonel (later Sir) Charles F. Close that geographers cannot hope to do original work; see also *Bulletin of the American Geographical Society 44* (1912), 27–39.

14. Chisholm, "Miss Semple," pp. 31–32.

15. *Journal of Geography 10* (1911), 33.

16. Harold and Margaret Sprout, *Man-Milieu Relationship Hypotheses in the Context of International Politics* (Princeton, N.J.: Center of International Studies, 1956), p. 34.

17. See above, Chapter 8, n. 30.

18. *American Historical Review 17* (1911–1912), 355–357.

19. Febvre, *La Terre*, p. 206, as translated by Mark Jefferson, *Geographical Review 13* (1923), 147.

20. *Geographical Review 13* (1923), 144–146: 146.

21. See, for example, "Theory and Practice in Historical Study," *Social Science Research Council Bulletin 54* (New York, 1946); "The Social Sciences and Historical Study," *ibid. 64* (1954); K. E. Bock, *The Acceptance of Histories: Toward a Perspective in Historical Study*, constituting *University of California Publications in Sociology and Social*

Institutions, 3, no 1. (1956); *Generalization in the Writing of History*: A Report of the Committee on Historical Analysis of the Social Science Research Council, ed. by Louis Gottschalk (Chicago: University of Chicago Press, 1963); and Page Smith, *The Historian and History* (New York: Knopf, 1964).

22. A. C. Crombie, ed., *Scientific Change*, pp. 604–605.

23. Tatham, "Environmentalism," p. 148.

24. Emrys Jones, "Cause and Effect in Human Geography," *Annals of the Association of American Geographers 46* (1956), 369–377 ("Sheer numbers of examples in works such as Miss Semple's are impressive," p. 370); Hartshorne, "Nature," p. 300.

25. Whereas most American geographers would seem to have become bored with the problem of environmental determinism and inclined to join Hartshorne in wishing "a plague o' both your houses" (*Perspective*, p. 57), some of the British have taken up the cudgels with renewed verve since World War II; see O. H. K. Spate, "The End of an Old Song? The Determinism-Possibilism Problem," *Geographical Review 48* (1958), 280–282, for discussion and references.

26. See the articles by William Warntz, "Progress in Economic Geography," *New Viewpoints in Geography*, ed. P. E. James (*29th Yearbook of the National Council for the Social Studies*, Washington, 1959), pp. 54–75, and "Geography at Mid-Twentieth Century," *World Politics 11* (1958–1959), 442–454. See also below, p. 338.

27. O. H. K. Spate, "Quantity and Quality in Geography," *Annals of the Association of American Geographers 50* (1960), 377–394: 380.

28. Sprout, *Man-Milieu Hypotheses;* R. S. Platt, *Field Study in American Geography: The Development of Theory and Method Exemplified by Selections*, constituting *University of Chicago, Department of Geography, Research Paper No. 61* (Chicago, 1959), pp. 61–62; also G. R. Lewthwaite, "The Nature of Environmentalism," *Proceedings of the Second New Zealand Geography Conference*, Christchurch, 1958 (Christchurch: New Zealand Geographical Society, 1958), pp. 5–12; also Hartshorne, *Perspective*, pp. 55–60.

29. Hartshorne, *Perspective*, p. 149. The distinction between the terms "nomothetic" and "idiographic" as applying to natural science and history, respectively, was made by the German historian of philosophy, Wilhelm Windelband, in 1894. See R. G. Collingwood, *The Idea of History* (New York: Oxford University Press, 1946; Galaxy paperback, 1956), p. 166.

30. See David Lowenthal, *George Perkins Marsh: Versatile Vermonter* (New York: Columbia University Press, 1958), pp. 246–276, and "George Perkins Marsh on the Nature and Purpose of Geography," *Geographical Journal 126* (1960), 413–417. See also the symposium, *Man's Role in Changing the Face of the Earth* cited in note 4, above.

31. See Tatham, "Environmentalism," pp. 151–159; O. H. K. Spate, "How Determined is Possibilism?" *Geographical Studies 4* (1957), 3–12; Sprout, *Man-Milieu Hypotheses*, pp. 39–49.

32. Sprout, *Man-Milieu Hypotheses*, pp. 24, 32–34.

33. Harold and Margaret Sprout, "Environmental Factors in the Study of International Politics," *Journal of Conflict Resolution 1* (1957), 309–328: 312.

34. Platt, *Field Study*, pp. 61–62; see also pp. 200–202.

Chapter 13. Notes on Measuring and Counting in Early American Geography

1. *The Columbia-Lippincott Gazetteer of the World* (New York: Columbia University Press, 1952), p. 924.

2. J. K. Wright, "'Crossbreeding' Geographical Quantities," *Geographical Review 45* (1955), 52–65. See also Maurice Daumas, "Precision of Measurement and Physical and Chemical Research in the Eighteenth Century," in A. C. Crombie, ed., *Scientific Change . . .* (New York: Basic Books, 1963), pp. 418–430: 418–426.

3. G. P. Marsh, *Man and Nature; or, Physical Geography as Modified by Human Action,* ed. David Lowenthal (Cambridge, Mass.: Harvard University Press, 1965), p. 225 (1st ed., New York, 1864).

4. S. E. Morison, *Harvard College in the Seventeenth Century* (Cambridge, Mass.: Harvard University Press, 1936), p. 157.

5. J. H. Alstedius, *Scientiarum omnium encyclopaediae* (4 vols.; Lyons, 1649 [1st ed., Herborn, 1630]), II, 379.

6. Cotton Mather, *The Christian Philosopher: A Collection of the Best Discoveries in Nature with Religious Improvements* (London, 1721), p. 283.

7. *Economic Geography 38* (1962), 38–55; *Annals of the Association of American Geographers 53* (1963), 505–515.

8. See, for example, G. V. T. Matthews, *Bird Navigation* (Cambridge, England: University Press, 1955); J. P. Scott, *Animal Behavior* (Chicago: University of Chicago Press, 1958; Doubleday-Anchor paperback, 1963), ch. 10, "Behavior and the Environment." See also below, Epilogue, n. 6.

9. Jedidiah Morse, *The American Geography* (Elizabethtown, N.J., 1789), p. 11.

10. Morse, *The American Universal Geography* (3rd ed., 2 vols.; Boston, 1796), I, 17. See, regarding the medieval Aristotelian background for this concept, J. K. Wright, *The Geographical Lore of the Time of the Crusades . . .* (New York: American Geographical Society, 1925; Dover reprint, 1965), p. 129.

11. Bernardus Varenius, *Geographia generalis* (Amsterdam, 1664), p. 1.

12. See J. N. L. Baker, "The Geography of Bernhard Varenius," *Transactions and Papers, Institute of British Geographers 21* (1955), 51–60, reprinted in J. N. L. Baker, *The History of Geography* (Oxford: Blackwell; New York: Barnes and Noble, 1963), pp. 105–118; also J. K. Wright, "Some British 'Grandfathers' of American Geography," *Geographical Essays in Memory of Alan G. Ogilvie* (London: Nelson, 1959), pp. 144–165: 155–156.

13. Morison, *Harvard College,* p. 157.

14. Bartolomaeus Keckermann, *Systema compendiosum totius mathematices, hoc est geometriae, astronomiae, geographiae . . .* (Hannover, 1612, 1617, 1621, and later eds.), p. 52 (1621 ed.). Medieval minds, both during the Middle Ages and subsequently, have tended to seek for logical and symmetrical subdivisions of the sum of all knowledge, a matter that has practical applications today in the problem of subject classifications for libraries of universal scope. Until the nineteenth century, geography was frequently subordinated to mathematics in such classifications; J. K. Wright, *Geographical Lore,* pp. 127–129; Alstedius, *Encyclopaedia,* I, un-

numbered pages at beginning of the volume; L. W. McKeehan, *Yale Science: The First Hundred Years, 1701–1801* (New York: Schuman, 1947), pp. 7–9 (on the American Dr. Samuel Johnson's scheme); *Catalogue of the Library of Thomas Jefferson,* compiled with annotations by E. Millicent Sowerby (Washington: Library of Congress), I (1952): plates following p. xv (show Jefferson's scheme, in which geography appears as a branch of the physiocomathematical branch of moral philosophy).

15. See Wright, "'Crossbreeding'", p. 56.

16. Harry Woolf, in *Isis 52*, pt. 2 (1961), 133.

17. See, for example, Herodotus, *History*, II, 109.

18. Leonard Whibley, ed., *A Companion to Greek Studies* (Cambridge, England: University Press, 1905), p. 202.

19. See E. G. R. Taylor, *The Mathematical Practitioners of Tudor and Stuart England* (Cambridge University Press, 1954).

20. See Wright, "Crossbreeding," Table II, pp. 62–63.

21. On the height of the atmosphere see Charles Morton, *Compendium physicae,* with a biographical memoir by S. E. Morison and an introduction by Theodore Hornberger, *Publications of the Colonial Society of Massachusetts, 33* (1940: *Collections*), p. 48; Benjamin Martin, Φιλότεχνος: *The Philosophical Grammar* . . . (7th ed.; London, 1769), p. 183.

22. My distinction between *bathymetry* on the one hand and *hypsometry* and *altimetry* on the other accords with established usage. For the distinction between hypsometry and altimetry, as well as for the other terms pertaining to geomensuration, I alone am responsible. The measurement of the heights of vegetational zones on mountains would be classed as hypsometric, that of zones of temperature, pressure, and so forth in the atmosphere as altimetric. See Carl Troll, "Die dreidimensionale Landschaftsgliederung der Erde," in *Herman Von Wissman-Zeitschrift,* ed. Adolf Laidlmair (Tübingen: Geographisches Institut der Universität Tübingen, 1962), pp. 54–80, "a historical and critical discussion of the vertical dimension, elevation, in geographical literature from [and including] Humboldt to the present" (John Leighly, *Geographical Review 54* (1964), 142). In his comprehensive *Climatology of the United States* (Philadelphia, 1857), Lorin Blodget devoted considerable attention to the compilation and analysis of data concerning altitudes as bearing upon climatic differences. Among the early geographical undertakings of the Smithsonian Institution was the systematic compilation of data concerning altitudes in the United States. This resulted in the publication of a small hypsometric map of the country (by C. A. Schott) in the Census Bureau's *Statistical Atlas of the United States* of 1874 (G. D. Hubbard, "Geography," in *The Smithsonian Institution, 1846–1896: The History of its First Half Century,* ed. G. B. Goode [Washington, 1897], p. 784).

23. Acts, xxvii, 28.

24. What may have been the earliest contour maps of a land area in the United States accompanied a report on engineering and geology submitted to the governor of Maryland in 1836; *Military Engineer 33* (1941), 69. On the European background, concerning the development of contour mapping see Max Eckert, *Die Kartenwissenschaft: Forschungen und Grundlagen zu einer Kartographie als Wissenschaft* (2 vols.; Berlin and Leipzig: Verlag Wissensch. Verlager, 1921, 1925), I, 438; and especially

François de Dainville, "De la profondeur à l'altitude," *International Yearbook of Cartography 2* (1962), 151–161 (on the origins in hydrographic maps of the concept of contour mapping as applied to land surfaces). A monograph or even a popular book could well be written around the central theme of the development of the representation of topographical and submarine relief on American maps from the seventeenth century onward. Many studies in the history of cartography are disappointing because they deal separately, one by one, with individual map makers and their works. This scatters in a multitude of places the treatment of topics that could be discussed interestingly as units.

25. See A. P. Middleton, *Tobacco Coast: A Maritime History of Chesapeake Bay in the Colonial Era* (Newport News: Mariners' Museum, 1953), pp. 27, 72–76.

26. Marsh, *Man and Nature*, p. 51.

27. Lewis Evans, ["Analysis"] *Geographical, Historical, Political, Philosophical and Mechanical Essays: the First, containing an Analysis of a General Map of the Middle British Colonies in America . . .* (Philadelphia: printed by B. Franklin and D. Hall, 1755); facsimile reproduction in L. H. Gipson, *Lewis Evans* (Philadelphia: Historical Society of Pennsylvania, 1939), pp. 141–176.

28. *Ibid.*, pp. 1, 10 (Gipson, pp. 145, 154); see also comment on this passage in G. W. White, "Early American Geology," *Scientific Monthly 76* (1953), 134–141: 139.

29. John Smith, *A Map of Virginia, with a Description of its Commodities, Government, and Religion* (Oxford, 1612), p. 10 (citation from E. D. Fite and Archibald Freeman, *A Book of Old Maps Delineating American History* . . . [Cambridge, Mass.: Harvard University Press, 1926], p. 118). See also William Strachey, *The Historie of Travell into Virginia Britania (1612)*, ed. by L. B. Wright and Virginia Freund (London: Hakluyt Society, 1953), p. 50, and W. C. Ford, "Captain John Smith's Map of Virginia, 1612," *Geographical Review 14* (1924): 433–443.

30. That a bibliography of the Mason and Dixon line could list more than 2000 manuscripts and other documents suggests the superabundance of source material concerning early American boundaries and the related surveys; E. M. Douglas, "Boundaries, Areas, Geographic Centers, and Altitudes of the United States and the Several States," *U. S. Geological Survey, Bulletin 817* (2nd ed., 1939), p. 122. Many secondary studies have also been published, though most of them as contributions to political or engineering history rather than to the history of geography as such. Douglas, "Boundaries," gives useful references; see also Brooke Hindle, *The Pursuit of Science in Revolutionary America, 1735–1789* (Chapel Hill, N.C.: The University of North Carolina Press, 1956), pp. 32–33, 174–180, 242–243, 337–338.

31. See especially W. D. Pattison, *Beginnings of the American Rectangular Land Survey System, 1784–1800* (constituting *University of Chicago, Department of Geography, Research Paper No. 50*, 1957).

32. See J. K. Wright, *Early Topographical Maps: Their Geographical and Historical Value as Illustrated by the Maps of the Harrison Collection of the American Geographical Society* (New York: American Geographical Society, 1924). See also below, p. 338.

33. See, however, Christopher Colles, *A Survey of the Roads of the United States of America, 1789*, ed. W. W. Ristow (Cambridge, Mass.: Harvard University Press, 1961), pt. I, by W. W. Ristow.

34. See G. S. Dunbar, "Thomas Jefferson, Geographer," *Special Libraries Association, Geography and Map Division, Bulletin No. 40* (1960), pp. 11–16.

35. Evans, "Analysis," p. 5 (Gipson, p. 149).

36. Jeremy Belknap, *The History of New-Hampshire . . . Containing also a Geographical Description of the State* (3 vols.; Philadelphia, 1784–1792; 2nd ed., Dover, N. H., 1812; Boston, 1813), III (1813), 57–58.

37. Samuel Williams, *The Natural and Civil History of Vermont* (Walpole, N. H., 1794; 2nd ed., 2 vols., Burlington, Vt., 1809), I (1809), 22; Belknap, *New-Hampshire*, III (1813), 10.

38. Williams, *Vermont*, I (1809), 473–474.

39. See George Graham, "On the Variation of the Horizontal Needle," *Philosophical Transactions 33 (No. 383;* London, 1724), 96 (reference from O[scar] Peschel, *Geschichte der Erdkunde . . .* [2nd ed., edited by Sophus Ruge; Munich, 1877], p. 730).

40. Peschel, pp. 730–731.

41. Williams, *Vermont*, I (1809), p. 474.

42. Evans, "Analysis," p. iv (Gipson, p. 144). Further references regarding American interest in terrestrial magnetism in the eighteenth century will be found in Hindle, *Pursuit*, pp. 180–181, 349–353. For the broader background see Heinz Balmer, *Beiträge zur Geschichte der Erkenntnis des Erdmagnetismus* (Veröffentlichungen der Gesellschaft für Geschichte der Medizin und der Naturwissenschaften, 20; Aarau: Sauerländer, 1956), 892 pp.; G. Hellmann, *Die ältesten Karten der Isogonen, Isoklinen, Isodynamen* (Berlin, 1895).

43. Williams, *Vermont*, I (1809), 23.

44. See L. S. Mayo, "The Forty-Fifth Parallel: A Detail of the Unguarded Boundary," *Geographical Review 13* (1923), 255–265; 263.

45. *William Byrd's Histories of the Dividing Line Betwixt Virginia and North Carolina*, ed. W. K. Boyd (Raleigh, N. C.: North Carolina Historical Commission, 1929), p. xxi.

46. Mayo, "Forty-Fifth Parallel," p. 260.

47. See Hindle, *Pursuit*, pp. 175–176; T. D. Cope, "Charles Mason and Jeremiah Dixon," *Scientific Monthly 62* (1946), 541–544; also see note 30, above.

48. Douglas, "Boundaries," p. 121.

49. As located by the United States Geological Survey (*ibid.*, p. 122) the line at long. 77° 29′05.6 W is about 7″ south of its starting point at the northeast corner of Maryland.

50. See article on Charles Mason (1730–1787), *Dictionary of National Biography*, XII, 1302.

51. Cope, "Mason and Dixon," pp. 547–549.

52. T. S. Eliot, "Morning at the Window," in *Complete Poems* (New York: Harcourt Brace, 1952), p. 16 (see also p. 156).

53. Williams, *Vermont*, I (1809), 24, 128.

54. The history of the measurement of geographical areas does not appear to have been treated comprehensively until the early years of the

twentieth century, when Walther Schmiedeberg, "Zur Geschichte der geographischen Flächenmessungen bis zur Erfindung des Planimeters," *Zeitschrift der Gesellschaft für Erdkunde zu Berlin* (1906), 152–176, 233–256, discussed developments down to about 1850, and Th. Willers, "Zur Geschichte der geographischen Flächenmessung," *Petermanns Mitteliungen Ergänzungsheft 170* (1911), dealt with subsequent developments in Europe. M. J. Proudfoot summarized these papers and contributed more besides, especially with regard to progress in the United States, in *Measurement of Geographic Area*, U. S. Department of Commerce, Seventeenth Census of the United States, 1940; Washington, n.d. [c 1945].

55. Mather, *Christian Philosopher*, p. 75.

56. Morse, *American Universal Geography*, I (1793), 61–62; see also I (1796), 61; I (1812), 85–86; I (1819), 67.

57. *Géographie d'Aboulféda, traduite de l'arabe en français*, by J. T. Reinaud, II, pt. i (Paris, 1848), pp. 17–19. Schmiedeberg, "Flächenmessungen," p. 158.

58. Carl Schoy, "Erdmessung bei den Arabern," *Zeitschrift der Gesellschaft für Erdkunde zu Berlin* (1917), 431–445:445.

59. Schmiedeberg, "Flächenmessungen," p. 161, n. 1.

60. Peschel, *Geschichte der Erdkunde*, p. 396.

61. See Proudfoot, *Measurement of Area*, pp. 10–14, for reproduction of Malynes' title page and the entire chapter in question.

62. See the discussion of this subject, *ibid.*, pp. 16–20.

63. Evans, "Analysis," pp. 31–32 (Gipson, pp. 175–176).

64. Thomas Jefferson, *Notes on the State of Virginia* (Paris, 1784, and many later eds.), queries i, viii, xiii (ed. by William Peden [Chapel Hill, N.C.: University of North Carolina Press], pp. 4, 83, 119, 283).

65. *Ibid.*, query i. (Peden's ed., p. 3).

66. Byrd, *Histories* (see above, n. 45), Boyd's introduction, pp. xvii–xviii.

67. See above, n. 56.

68. Notably those of Malynes (see above, p. 218, and n. 61).

69. Proudfoot, *Measurement of Area*, p. 16 (see also p. 248, note a).

70. William Guthrie, *A New System of Modern Geography* . . . (London, 1770; at least 25 later eds. until 1843; 1st American ed., Philadelphia, 2 vols., 1794, 1795).

71. *Ibid.* E. A. W. Zimmermann, *A Political Survey of the Present State of Europe* . . . (London, 1787). John Pinkerton, *Modern Geography* (1st ed., 2 vols., London, 1802; new ed., 3 vols., London, 1807; abridged ed., 1 vol., London, 1806). On Morse's use of these and other European sources, see J. K. Wright, " 'Grandfathers,' " (cited in note 12 above).

72. J. F. [James Freeman], *Remarks on the American Universal Geography* (Boston, 1793), p. 7. C. D. Ebeling, *Erdbeschreibung und Geschichte von Amerika: Die Vereinten Staaten von Nordamerika* (7 vols.; Hamburg, 1793–1816), V, 206, criticizes Morse for estimating the area of Maryland by multiplying the length by the breadth.

73. Morse, *American Universal Geography*, I (1793), 392–393.

74. Proudfoot, *Measurement of Area*, p. 7.

75. Morse, *American Geography* (1789), pp. 35–36.

76. *Ibid.*, p. 33.

77. Schmiedeberg, "Flächenmessungen," pp. 233–234; Proudfoot, *Measurement of Area*, p. 16. Büsching explains that, inspired by Templeman's example, he employed a friend, J. F. Hansen, to calculate the areas of the European states and regions (*Staaten und Länder*) in square German miles. The data are presented in his text and also in a table, *Neue Erdbeschreibung* . . . , I (Hamburg, 1758), 144; French trans., *Géographie universelle*, I (Strasbourg, 1785), 121–122. See also A. F. Büsching, *Vorbereitung für grundlichen und nützlichen Kenntnis der geographischen Beschaffenheit und Staatsverfassung der europäischen Reiche und Republiken* . . . (5th ed.; Hamburg, 1776), pp. 14–15.

78. A. F. Büsching, *A New System of Geography* . . . (6 vols.; London, 1762), I, 58; here the states are listed in order of size "in square geographical miles," but with no figures.

79. This information is presented in a little thumbnail description of the geographic features of his part of Virginia which Jefferson sent to Jean Baptiste Say, March 2, 1812, to be found in A. A. Lipscomb and A. L. Bergh, eds., *The Writings of Thomas Jefferson* (Washington, 1903), XIV, 260–263; also in *Thomas Jefferson's Garden Book, 1766–1824*, ed. E. M. Betts, *Memoirs, American Philosophical Society, 22* (Philadelphia, 1944), pp. 542–543.

80. Jedidiah Morse, *The American Gazetteer . . . on The American Continent* (2nd ed.; "printed in Boston, New England"; London, 1798), p. 163.

81. Morse, *American Geography* (1789), pp. 172–173, 284–285.

82. Proudfoot, *Measurement of Area*, pp. 27–29.

83. *Ibid.*, pp. 31–51.

84. See Wright, "'Crossbreeding,'" p. 57 (here the term "substantive" rather than "topical" is used).

85. *Ibid.*, p. 55.

86. Peschel, *Geschichte der Erdkunde*, pp. 747, 749.

87. Siegmund Günther, *Geschichte der Erdkunde* (Leipzig, 1904), pp. 153–155, 219–223.

88. On the development of climatology in the United States during the nineteenth and twentieth centuries, see John Leighly, "Climatology," in *American Geography: Inventory and Prospect*, ed. P. E. James and C. F. Jones (published by Syracuse University Press for the Association of American Geographers, 1954), pp. 335–361 (many bibliographical references). The "Preliminary Chapter" of Lorin Blodget, *Climatology of the United States* . . . (Philadelphia, 1857), pp. 17–29, is "a historical sketch of climatology from Theophrastus and Aristotle," and many more items of historical interest will be found here and there through the volume.

89. A. J. Henry, "Early Individual Observers in the United States," *Weather Bureau Bulletin 11* (1893), 291–302.

90. See R. H. Brown, "The First Century of Meteorological Data in America," *Monthly Weather Review 68* (1940), 130–133, which deals with the period 1738–1838; also Hindle, *Pursuit*, pp. 182–184, 347–349, and index *sub* "Meteorology."

91. See Henry, "Early Individual Observers," pp. 295–296.

92. R. H. Brown, *Mirror for Americans: Likeness of the Eastern Seaboard, 1810* (New York: American Geographical Society, 1943), pp. 13, 265.

93. Brown, "The First Century," p. 131; Leighly, "Climatology," pp. 335–336.

94. In Betts' edition of *Jefferson's Garden Book, 1766–1824*, many quotations will be found from the *Garden Book*, from Jefferson's correspondence, and from other sources that shed light on his lifelong interest in climate and weather (see especially pp. 124–129, 157, 255, 462, 515, 542, 567, 578, 579, 622–628). See also F. J. Randolph and F. L. Francis, "Thomas Jefferson as Meteorologist," *Monthly Weather Review 23* (1895), 456–458; E. T. Martin, *Thomas Jefferson: Scientist* (New York: Schuman, 1952), pp. 131–147; Dunbar, *Jefferson* (see note 34, above), pp. 12–13 (Professor Dunbar's paper put me on the track of the *Garden Book*).

95. Jefferson, *Garden Book* (1944 ed.), p. 69.

96. Jefferson, *Notes on Virginia*, query vi (Peden's ed., 1955, pp. 73–81). The late R. H. Brown deemed the essay on climate to be perhaps geographically the most influential part of the *Notes;* "Jefferson's 'Notes on Virginia,'" *Geographical Review 33* (1943), 467–473:471.

97. Alexander McAdie, "Simultaneous Meteorological Observations in the United States During the Eighteenth Century," *Weather Bureau Bulletin 11* (1893), 303–304.

98. *Jefferson's Garden Book* (1944 ed.), p. 579; also in Lipscomb and Bergh, *Writings of Jefferson*, XIX, 259–261.

99. See Peschel, *Geschichte der Erdkunde*, p. 752; Hindle, *Pursuit*, pp. 332, 348.

100. Williams, *Vermont*, I (1809), 58, 61, 63, 69.

101. John Drayton, *A View of South-Carolina as Respects her Natural and Civil Concerns* (Charleston, 1802), p. 23.

102. See Arnold Court, "Temperature Extremes in the United States," *Geographical Review 43* (1953), 39–49:39.

103. Peschel, *Geschichte der Erdkunde*, pp. 750–751.

104. Bibhutibhusan Datta, "On the Origin and Development of the Idea of 'Per Cent,'" *American Mathematical Monthly 34* (1927), 530–531.

105. D. E. Smith, *History of Mathematics* . . . (2 vols.; Boston, New York: Ginn, 1923, 1925; Dover reprint, 1958), II, 245–248.

106. George Sarton, "The First Explanation of Decimal Fractions and Measures (1585) together with a History of the Decimal Idea and a Facsimile of Stevin's Disme," *Isis 23* (1935), 153–244:174–176, 184.

107. Isaac Greenwood, *Arithmetick Vulgar and Decimal* (Boston, 1729).

108. L. C. Karpinski, *The History of Arithmetic* (Chicago, New York: Rand McNally, 1925), p. 135.

109. William Barton, "Observations on the Possibilities of the Duration of Human Life, and the Progress of Population, in the United States of America," *Transactions, American Philosophical Society 3* (1793), 25–62, table on p. 60.

110. According to information furnished by courtesy of Professor A. D. Bradley, Hunter College, New York City.

111. Zimmermann, *Political Survey*, Table [chap.] i.

112. Williams, *Vermont*, I (1809), 53–55, 73–74, 89–97. G. P. Marsh, many-sided American scholar of the mid-nineteenth century and pioneer advocate of conservation, quoted Williams in this connection in *Man and Nature*, ed. David Lowenthal, pp. 141–142, 151. See also David Lowenthal,

George Perkins Marsh: Versatile Vermonter (New York: Columbia University Press, 1958), pp. 259–260.

113. Lewis Evans, *A Brief Account of Pennsylvania*, 1753, in Gipson, *Lewis Evans*, pp. 87–137:109 (see above, n. 27).

114. *Jefferson's Garden Book* (Betts' ed., 1944), p. 628; see also p. 579.

115. Williams, *Vermont*, I (1809), 96–97.

116. See Brown, *Mirror for Americans*, pp. 19–21.

117. Williams, *Vermont*, I (1809), pp. 72–75.

118. *Ibid.*, pp. 475–478.

119. J. D. Whitney, *The United States: Facts and Figures Illustrating the Physical Geography of the Country and its Material Resources—Supplement I* (Boston, 1894), p. 294.

120. Martin, *Thomas Jefferson*, pp. 148–159. The "miserable America" doctrine was to reappear, though in a somewhat divergent form, in the lectures of Carl Ritter and Arnold Guyot; see Ritter, *Comparative Geography*, translated . . . by W. L. Gage (Philadelphia: 1865; Cincinnati and New York, 1881), pp. 56–57; Guyot, *The Earth and Man* (Boston, 1850), pp. 180–184, 215–218. See also above, p. 284.

121. Jefferson, *Virginia*, query vi (Peden, pp. 47–65).

122. *Ibid.*, pp. 50–52.

123. Concerning the sources from which Jefferson obtained many of these data, see Marie Kimball, *Jefferson: War and Peace 1776 to 1784* (New York: Coward-McCann, 1947), pp. 270–273.

124. Jefferson, *Virginia*, query vi (Peden, p. 49).

125. Williams, *Vermont*, I (1809), 130–131.

126. Jefferson, *Virginia*, query vi (Peden, p. 58).

127. Williams, *Vermont*, I (1809), 129.

128. See "Historical Review of Census Development," *Census of the Commonwealth of Australia*, I (1911), 2–35.

129. See W. S. Rossiter, *A Century of Population Growth from the First Census of the United States to the Twelfth, 1790–1900* (Washington: Bureau of the Census, 1909), pp. 3–15, 149–185; also E. B. Greene and V. D. Harrington, *American Population Before the Federal Census of 1790* (New York: Columbia University Press, 1932).

130. For example, C. O. Paullin, *Atlas of the Historical Geography of the United States* (Carnegie Institution of Washington and American Geographical Society, 1932), pls. 60–61; H. R. Friis, "A Series of Population Maps of the Colonies and the United States, 1625–1790," *Geographical Review 30* (1940), 463–470.

131. Jefferson, *Virginia*, query viii (Peden, 1955, pp. 82–87).

132. Rossiter, *Century of Population Growth*, p. 9.

133. Jefferson, *Virginia*, query ix (Peden, p. 89).

134. Rossiter, *Century*, pp. 42–50.

135. See below, n. 158.

136. Morse, *American Universal Geography*, I (1796), 229.

137. Among others besides Jefferson, the following based estimates of the increase in population, past or future, on assumptions with regard to the period of doubling: Sir William Petty, *Political Arithmetick* (London, 1690), see Lewis Bonar, *Malthus and His Work* (London: Allen and Unwin, 2nd ed., 1924), p. 68; J. P. Süssmilch, *Die göttliche Ordnung in denen Veränderungen des menschlichen Geschlechtes* . . . (1st ed.; Berlin, 1741);

Benjamin Franklin, *Observations Concerning the Increase of Mankind* (Philadelphia, 1751); T. R. Malthus, *An Essay on the Principle of Population* (London, 1798); and Samuel Williams, *Vermont*, II (1809), 416–425. On Petty as a geographer see Y. M. Goblet, *La Transformation de la géographie politique de l'Irlande au XVIII*e *siècle dans les cartes et essais anthropogéographiques de Sir William Petty* (Nancy: Berger-Levrault, 1930); "Un précurseur anglais de la géographie humaine au XVIIe siècle: Sir William Petty," *Mélanges de géographie offerts . . . à M. Václav Švambera* (Prague; privately published, 1936), pp. 60–71.

138. For other early predictions concerning the population of the United States, see Brown, *Mirror for Americans*, pp. 30–31.

139. Jefferson, *Virginia*, query viii (Peden, p. 84).

140. According to A. M. Carr-Saunders, *The Population Problem . . .* (Oxford: Clarendon Press, 1922), pp. 26–27, Sir Matthew Hale in *The Primitive Origination of Mankind* (London, 1677) had "calculated that the numbers of mankind must increase in a geometrical ratio unless hindered by checks," and "Sir William Petty deals with the geometrical ratio at some length."

141. Belknap, *New-Hampshire*, III (1813), 176–178.

142. *Ibid.*, pp. 344–352.

143. *Ibid.*, pp. 178–179; Williams, *Vermont*, II (1809), 419–420.

144. Williams, *Vermont*, I (1809), 237–241.

145. Jefferson, *Virginia*, query xi (Peden, p. 93).

146. Süssmilch, *Göttliche Ordnung* (1742), ch. iii. According to an eighteenth-century commentator on Süssmilch's work (Christian Johann Baumann), this was the first attempt ever made to estimate the potential population of the earth; *ibid.*, III (5th ed.; Berlin, 1776), 325.

147. On Malthus' debt to Süssmilch, see Bonar, *Malthus*, pp. 29, 124–126, 369, 414; see Süssmilch, *Göttliche Ordnung*, II (5th ed.; Berlin, 1775), 171–177, 233–234.

148. Süssmilch, *Göttliche Ordnung*, II (1775), 173; in the first edition, however, though acknowledging his debt to the earlier English pioneers in demography, John Graunt and Sir William Petty, Süssmilch made no mention of Templeman.

149. Süssmilch, *Göttliche Ordnung* (1742), pp. 75, 98.

150. Morse, *American Universal Geography*, I (1793), 353.

151. *Ibid.*, I (1796), 433.

152. *Ibid.*, I (1819), 209–210.

153. Hermann Wagner, *Lehrbuch der Geographie*, I, pt. 3 (Hannover, 1923), pp. 875–876.

154. A. F. W. Crome, *Ueber die Grösse und Bevölkerung der sämtlichen europäischen Staaten . . .* (Leipzig, 1785), p. 40.

155. Morse, *American Geography* (1789), p. 491. See above, n. 72.

156. Zimmermann, *Political Survey* (see above, n. 71), pp. 6–7.

157. Wagner, *Lehrbuch*, p. 875.

158. On early efforts to map densities of population (period 1830–1860) see Wagner, *Lehrbuch*, p. 876; Max Eckert, *Die Kartenwissenschaft* (see above, note 24), II, 160–161; and especially A. H. Robinson, "The 1837 Maps of Henry Harness," *Geographical Journal 121* (1955), 440–450.

159. Ebling, *Erdbeschreibung* (see above, n. 72), IV (1797), 197.

160. *Ibid.*, II (1794), 218; IV (1797), 197.

Chapter 14. Notes on Early American Geopiety

1. Psalms, cxiv, 4; cxlviii, 8–11, 13; civ, 18, 32, 35.

2. I am grateful to my brother-in-law, the Rev. Dr. A. C. McGiffert, Jr., for suggesting that I look both to the Psalms and to Jonathan Edwards in this connection.

3. Jonathan Edwards, *Images, or 'Shadows of Divine Things,'* ed. Perry Miller (New Haven: Yale University Press, 1948), pp. 68, 91. On medieval antecedents of these two rather inconsistent attitudes toward high places, see J. K. Wright, *The Geographical Lore of the Time of the Crusades . . .* (New York: American Geographical Society, 1952; Dover reprint, 1965), pp. 216–217 and index sub "Mountains"; also Jacob Burckhardt, *The Civilisation of the Renaissance in Italy*, trans. S. G. C. Middlemore (London: Swan Sonnenschein; New York; Macmillan, 1909), pp. 301–302 (Petrarch's ascent of Mont Ventoux, 1336).

4. Jared Eliot, *Essays upon Field Husbandry in New England and other Papers, 1748–1762*, ed. H. J. Carman and R. G. Tugwell (New York: Columbia University Press, 1934), pp. 170–171.

5. This book goes to press at a time when considerable interest is being developed in the United States concerning divers aspects of the relation between geography and religion, but more particularly concerning *theogeography*. Reed F. Stewart (Executive Council, Episcopal Church, 815 Second Avenue, New York, N.Y.) in a circular letter dated April 29, 1965, lists some 30 persons "known or reported to have written on the general area or . . . currently doing work in it." E. S. Gaustad, in the Preface to his *Historical Atlas of Religion in America* (New York: Harper and Row, 1962), p. x, uses the term "geotheology" (tentatively, as explained in a personal letter) with reference to the "anthropogeography of religion." According to my (equally tentative) terminology, this would be a branch of theogeography.

6. See Erich Isaac, "The Act and the Covenant: The Impact of Religion on the Landscape," *Landscape 11*, 2 (1961–62), 12–17; "God's Acre: Property in Land, a Sacred Origin?", *Landscape 14*, 2 (1964–65), 28–32; Lynn White, Jr., "What Accelerated Technological Progress in the Western Middle Ages?" in A. C. Crombie, ed., *Scientific Change . . .* (New York: Basic Books, 1963), pp. 272–291: 282–283, with commentary on pp. 328, 332.

7. See A. N. Whitehead, *Science and the Modern World* (New York: Macmillan, 1925 and later), pp. 269–273, (Mentor paperback, 1960), pp. 168–169.

8. On the similar "detheologicization" that has been going on in American historiography, see R. H. McNeal, "History vs. Theology," *Columbia University Forum 6* (1963), 45–48. On the larger subject of the impact of theology upon the writing of history see Page Smith, *The Historian and History* (New York: Knopf, 1964), especially chs. 3, 7, and 15.

9. See A. C. Crombie, ed., *Scientific Change*, p. 204.

10. Charles Morton, *Compendium physicae*, with a biographical memoir by S. E. Morison and an introduction by Theodore Hornberger, *Publications of the Colonial Society of Massachusetts 33* (1940: *Collections*), pp. 207–208.

11. *Ibid.*, p. 210.

12. From an undated letter from John Bartram to Jared Eliot, in Eliot, *Essays*, pp. 201–206: 205.

13. Edwards, *Images*, p. 44.

14. Adam Seybert, "Experiments and Observations on the Atmosphere of Marshes," *Transactions of the American Philosophical Society* (1st ser.) *4* (1799), 429, as quoted by D. J. Boorstin, *The Lost World of Thomas Jefferson* (New York: Holt, 1948; Beacon paperback, Boston, 1960), pp. 45–47, 260.

15. Cotton Mather, *Winter Meditations: Directions How to employ the Leisure of the WINTER for the Glory of God . . .*, (Boston, 1693), sect. 4, p. 40, as quoted by Josephine K. Piercy, *Studies in Literary Types in Seventeenth Century America (1607–1710)* (New Haven: Yale University Press; London: Oxford University Press, 1939), pp. 190–191.

16. Cotton Mather, *Diary*, in *Massachusetts Historical Society Collections* (7th ser.) 7, *8* (Boston, 1911, 1912), *8* (1912), 131, 152–153.

17. See, however, C. C. Gillispie, *Genesis and Geology: A Study in the Relation of Scientific Thought to Natural Theology and Social Opinion in Great Britain, 1790–1850* (Cambridge, Mass.: Harvard University Press, 1951); and R. J. Chorley, A. J. Dunn, and R. P. Beckinsale, *The History of the Study of Landforms, or The Development of Geomorphology*, vol. I, *Geomorphology Before Davis* (London, Methuen; New York: Wiley, 1964), index, *sub* "Theology."

18. See D. C. Allen, *The Legend of Noah: Renaissance Rationalism in Art, Science, and Letters* (University of Illinois Studies in Language and Literature, 33, Urbana, Ill., 1949).

19. Marjorie Hope Nicolson, *Mountain Gloom and Mountain Glory: The Development of the Aesthetics of the Infinite* (Ithaca, N.Y.: Cornell University Press, 1959; Norton Library paperback, 1963), p. 187.

20. Anne Bradstreet, *The Works of Anne Bradstreet in Prose and Verse*, J. H. Ellis, ed. (Charlestown, Massachusetts, 1867), p. 118.

21. Nicolson, *Mountain Gloom*, pp. 83–84.

22. Thomas Burnet, *Telluris theoria sacra . . .* (2 vols., London, 1681); *The Sacred Theory of the Earth . . .* (London, 1684 and later; last, 1826). John Woodward, *Essays Towards a Natural History of the Earth . . .* (London, 1695 and later). William Whiston, *A New Theory of the Earth* (London, 1696). For discussion of the theories of these men from the point of view of a historian of geography, see E. G. R. Taylor, "The English Worldmakers of the Seventeenth Century and their Influence upon the Earth Sciences," *Geographical Review 38* (1948), 104–112; "The Origin of Continents and Oceans: a Seventeenth Century Controversy," *Geographical Journal 116* (1950), 193–198; from that of a historian of geology, Ruth Moore, *The Earth We Live On: The Story of Geological Discovery* (New York: Knopf, 1956), chap. 2; and from that of literary historians, Allen, *Legend of Noah*, chap. 5; Nicolson, *Mountain Gloom*, chaps. 5, 6; and Perry Miller, *The New England Mind:* [II] *From Colony to Province* (Cambridge, Mass.: Harvard University Press, 1953; Beacon paperback, Boston, 1961), pp. 186–190; and *Errand into the Wilderness* (Cambridge, Mass.: Harvard University Press, 1956), chap. x. I am grateful to Dr. Jean Gottmann for first calling to my attention the geographical interest of *Errand*.

23. Nicolson, *Mountain Gloom*, pp. 74–95, 176–177.

24. Taylor, "English Worldmakers," p. 107.

25. Samuel Miller, *A Brief Retrospect of the Eighteenth Century: Part First, in Two Volumes, Containing a Sketch of the Revolutions and Improvements in Science, Arts, and Literature During that Period* (New York, 1803), I, 156–189. (Part Second, which was to have covered theology, morals, and politics, was never published.) Chap. 5 (I, 326–357) is on geography and constitutes one of the earliest American studies in the history of geography. The chapter on the United States (II, 330–410) also contains much of interest to the historian of geography and was reprinted with an Introduction by L. H. Butterfield, *William and Mary Quarterly* (3rd ser.) *10* (1953), 579–627.

26. Benjamin Martin, *Φιλοτέχνος: The Philosophical Grammar* . . . (London, 1735 and later; seventh ed., 1769), pp. 16–18 (1769).

27. See, however, on American recognition of Burnet and Whiston in the eighteenth century the references to Perry Miller in note 22, above.

28. Samuel Miller, *Brief Retrospect*, II, 339.

29. Cotton Mather, *The Christian Philosopher: A Collection of the Best Discoveries in Nature with Religious Improvements* (London, 1721), p. 97.

30. T. G. Wright, *Literary Culture in Early New England* (New Haven: Yale University Press, 1920), p. 253.

31. F. G. Kilgour, "The First Century of Scientific Books in the Harvard College Library," *Harvard Library Notes*, No. 29 (March 1939), 217–225: 221–223.

32. Perry Miller, *Errand*, p. 223.

33. John Lawson, *The History of North Carolina: Containing the Exact Description and Natural History of that Country* . . . (London, 1714), ed. F. L. Harriss (Richmond, Va.: Garrett and Massie, 1937), p. 179. There was also an edition entitled *A New Voyage to Carolina* (London, 1709).

34. Mark Catesby, *The Natural History of Carolina, Florida, and the Bahama Islands* . . . (2 vols.; London: I, 1731; II, 1743; later eds., 1754, 1771), I (1754), vii.

35. Hugh Jones, *The Present State of Virginia, From Whence is Inferred a Short View of Maryland and North Carolina* . . . (London, 1724), ed. R. L. Morton (Chapel Hill, N.C.: University of North Carolina Press, 1956), p. 50.

36. Eliot, *Field Husbandry* (1934 ed.), pp. 43–46.

37. *Ibid.*, p. 221.

38. "Seratas" in L. H. Gipson, *Lewis Evans* (Philadelphia: Historical Society of Pennsylvania, 1939), p. 12.

39. Transcribed from map as reproduced in Gipson, *Evans*, map II, following p. 219 (see also pp. 11–12). On the geological observations and ideas of Evans and other British and early American writers on the eastern part of North America from 1588 to the latter half of the eighteenth century, see G. W. White, "Early American Geology," *Scientific Monthly 76* (1953), 134–141, and Chorley, Dunn, and Beckinsale, *Landforms*, I, 236–279.

40. Jedidiah Morse, *The American Geography* (Elizabethtown, N.J., 1789), p. 51.

41. Marie Kimball, *Jefferson: War and Peace, 1776 to 1784* (New York: Coward-McCann, 1947), p. 300.

42. Thomas Jefferson, *Notes on the State of Virginia* (Paris, 1785, and many later eds.), query vi (ed. by William Peden [Chapel Hill, N.C.; 1955], p. 33).

43. *Ibid.*, query vi (p. 31).

44. On the diversity of ways in which the Bible has been "literally" interpreted, see Perry Miller's Introduction to Jonathan Edwards, *Images* (see above, note 3).

45. Samuel Miller, *Brief Retrospect*, I, 501; see also p. 186.

46. See H. W. Ahlmann, *Glacier Variations and Climatic Fluctuations* (New York: American Geographical Society, 1953), pp. 37–41.

47. Professor C. J. Glacken has kindly informed me that the expression *officina gentium* occurs in the sixth-century work of Jordanes, *De origine actibusque Getarum*, IV, 25.

48. Hugh Williamson, *Observations on the Climate in Different Parts of America Compared with the Climate in Corresponding Parts of the Other Continent, to which are added Remarks on the Different Complexions of the Human Race* . . . (New York, 1811), pp. 107–112.

49. Ellsworth Huntington, *The Pulse of Asia: A Journey in Central Asia Illustrating the Geographic Basis of History* (Boston and New York: Houghton Mifflin, 1907).

50. See Merle Curti, *The Growth of American Thought* (2nd ed.; New York: Harper, 1951), pp. 320–321; W. M. and M. S. C. Smallwood, *Natural History and the American Mind* (New York: Columbia University Press, 1941), p. 245; and Chorley, Dunn, and Beckinsale, *Landforms*, I, 241–247.

51. J. P. Lesley, *Manual of Coal and its Topography* (Philadelphia, 1856), pp. 167, 175.

52. See the discussion of this subject in Allen, *Legend of Noah*, chap. 6.

53. José de Acosta, *Historia natural y moral de las Indias*, (Saville, 1590). Translation by Edward Grimston, *The Natural and Moral History of the East and West Indies* (London, 1604); ed. C. R. Markham (2 vols., London: Hakluyt Society, 1880), I, 278.

54. *Ibid.*, I, 30; Smallwood, *Natural History*, p. 13.

55. Athanasius Kircher, *De arca Noë* (Amsterdam, 1675).

56. Allen, *Legend of Noah*, pp. 184–185.

57. Lawson, *History of North Carolina* (1937 ed.), pp. xii, 130.

58. Lewis Evans, "A Brief Account of Pennsylvania, 1753" (as collated from two MSS.), in Gipson, *Evans*, pp. 87–137: 117.

59. Abraham Milius, *De origine animalium et migratione populorum* (Geneva, 1667), as cited by Allen, *Legend of Noah*, p. 131.

60. Perry Miller, *Errand*, p. 224.

61. Edwards, *Images*, p. 83.

62. Cotton Mather, *Magnalia Christi Americana, or The Ecclesiatical History of New-England from its Plantation to the Year of our Lord, 1620* (London, 1702; Hartford, Conn., 2 vols., 1820), I(1820), 44.

63. Perry Miller, *Errand*, p. 235.

64. Morton, *Compendium physicae*, p. 74.

65. Perry Miller, *New England Mind*, II, 438–439.

66. Noah Webster, *A Brief History of Epidemic and Pestilential Diseases, with the Principal Phenomena of the Physical World which Precede and Accompany Them* (2 vols.; Hartford, Conn., 1799), II, 281.

67. C.-E. A. Winslow, "The Epidemiology of Noah Webster," *Transactions, Connecticut Academy of Arts and Sciences 32* (1934), 21–109: 64.

68. See Perry Miller, *New England Mind*, II, 345–366; R. H. Shryock, *The Development of Modern Medicine: An Interpretation of the Social and Economic Factors Involved* (Philadelphia: University of Pennsylvania Press; London: Oxford University Press, 1936), p. 134; for references, see W. J. Bell, Jr., *Early American Science: Needs and Opportunities for Study* (Williamsburg, Va.: Institute of Early American History and Culture, 1955), pp. 64–65.

69. Cotton Mather, *Christian Philosopher*, p. 102.

70. Perry Miller, *New England Mind*, II, 445, attributed 20 or more sermons to the earthquake of 1727 and L. C. Wroth, *An American Bookshelf, 1755* (Philadelphia: University of Pennsylvania Press, 1934), p. 73, attributed 13 sermons and one poem to that of 1755. See also below, p. 338.

71. See notably T. D. Kendrick, *The Lisbon Earthquake* (London: Methuen, 1956). For this reference I am indebted to Mr. R. A. Skelton of the British Museum.

72. Increase Mather, *An Essay for the Recording of Illustrious Providences* (Boston, 1684; as reprinted under the title *Remarkable Providences* . . . , with an introductory paper by George Offer, London, 1856), pp. 233–234.

73. Cotton Mather, *Christian Philosopher*, p. 52.

74. G. L. Kittredge, *Witchcraft in Old and New England* (Cambridge, Mass.: Harvard University Press, 1929), pp. 161–162. Daniel Defoe wrote a work on this catastrophe (*The Storm* . . . , 1704) while serving a term of imprisonment; see J. N. L. Baker, *The History of Geography* (Oxford: Blackwell; New York: Barnes and Noble, 1963), pp. 159–161.

75. T. H. White, *The Bestiary: A Book of Beasts: Being a Translation from a Latin Bestiary of the Twelfth Century* (New York: Putnam, 1954; Capricorn Books, 1960), p. 167.

76. Increase Mather, *Illustrious Providences*, p. 119.

77. *Ibid.*, p. 88.

78. Cotton Mather, *Magnalia*, I, 312–313.

79. Cotton Mather, *Wonders of the Invisible World* (1693; London, 1862), pp. 9ff, as quoted by Piercy, *Studies* (see note 15, above), p. 21.

80. Cotton Mather, *Magnalia*, I, 321.

81. George Alsop, *A Character of the Province of Mary-Land, Wherein is . . . Also A Small Treatise on the Wilde and Naked Indians (or Susquehanokes) of Mary-Land, their Customs, Manners, Absurdities, & Religion* . . . (London, 1666; reprinted, New York, 1869; reissued by the Maryland Historical Society, Baltimore, 1880), pp. 485–486 (1869 ed.). See also R. M. Dorson, *American Folklore* (Chicago: University of Chicago Press, 1959), pp. 16–20, 129–130; there is much "parageography" of interest in this volume; see above, p. 288, and below, p. 338.

82. Cotton Mather, *Magnalia*, I, 61; see also S. E. Morison, *The Maritime History of Massachusetts, 1783–1860* (Boston: Houghton Mifflin, 1921, and later; Sentry paperback, 1961), p. 13 (1961 ed.).

83. Jeremy Belknap, *Journal of a Tour to the White Mountains in July 1784*, printed from the Original Manuscript, with a Prefatory Note by the Editor, C. D. [Charles Deane?] (Boston: Massachusetts Historical Society, 1876), p. 15.

84. Cotton Mather, *Magnalia*, I, 44.

85. William Hubbard, *A General History of New England from the Discovery to MDCLXXX*, published for the first time from MS., in *Collections, Massachusetts Historical Society* (2nd ser.), V, VI (Cambridge, Mass., 1815; 2nd ed., Boston, 1848), V, p. 26 (1815 ed.).

86. See J. K. Wright, *The Geographical Lore of the Time of the Crusades . . .* (New York: American Geographical Society, 1925; Dover reprint, 1965), pp. 287–288.

87. F. D. Pastorius, *Umständige geographische Beschreibung der zu allerlezt erfundenen Provinz Pennsylvaniae* (Frankfurt and Leipzig, 1700), as translated under the title *Circumstantial Geographical Description of Pennsylvania*, in A. C. Myers, ed., *Narratives of Early Pennsylvania . . .*, 1630–1707 (Original Narratives of Early American History; New York: Scribner, 1912), p. 419.

88. Cotton Mather, *Christian Philosopher*, p. 92.

89. Morton, *Compendium physicae*, p. 62.

90. Cotton Mather, *Christian Philosopher*, pp. 93, 76, 77, 84, 99.

91. Jones, *Virginia*, p. 58.

92. Cotton Mather, *Christian Philosopher*, p. 136.

93. Eliot, *Field Husbandry*, p. 46.

94. J. K. Wright, *Geography in the Making: The American Geographical Society, 1851–1951* (New York: American Geographical Society, 1952), p. 40.

95. T. S. King, *The White Hills: their Legends, Landscape, and Poetry* (Boston, 1859), p. 293.

96. Cotton Mather, *Christian Philosopher*, p. 77.

97. Cotton Mather, *Magnalia*, I, 42.

98. L. B. Wright, *The Cultural Life of the American Colonies, 1607–1763* (New York: Harper, 1957), pp. 72–74; Perry Miller, *Errand*, pp. 115–121, esp. p. 115.

99. Michael Kraus, *Intercolonial Aspects of American Culture on the Eve of the Revolution, with Special Reference to Northern Towns* (New York: Columbia University Press, 1928), p. 188. See above, n. 71.

100. Quoted from Max Savelle, *Seeds of Liberty: The Genesis of the American Mind* (New York: Knopf, 1948), p. 575.

101. Perry Miller, *Errand*, p. 207.

102. Carl Ritter, *Allgemeine Erdkunde*, ed. H. A. Daniel (Berlin, 1862); trans. under the title *Comparative Geography* by W. L. Gage (Philadelphia, 1865; Cincinnati and New York, 1881). See also Richard Hartshorne, *The Nature of Geography . . .*, p. [63] (see above, Chapter 10, n. 29); Arnold Guyot, *The Earth and Man* (Boston, 1850; New York, 1890).

103. Thomas Ewbank, *The World a Workshop: or, the Physical Relation of Man to the Earth* (New York, 1855). On Ewbank see J. A. Kouwenhoven, *The Beer Can by the Highway: Essays on What's "American" about America* (Garden City, N.Y.: Doubleday, 1961), pp. 171–172.

104. Ewbank, *The World a Workshop*, pp. 38–39. Professor C. J. Glacken has called my attention to the following in this connection: F. D. Adams, *The Birth and the Development of the Geological Sciences* (Baltimore: Williams & Wilkins, 1938; New York: Dover, 1954), ch. ix and p. 440.

105. Perry Miller, *Errand*, p. 117.

106. Thomas Hutchinson, *The History of the Colony & Province of Massachusetts-Bay* . . . (Boston, 1764–1828; London, 1765–1828); ed. L. S. Mayo (3 vols.; Cambridge, Mass.: Harvard University Press, 1936), II (1936), 343. Quoted in Savelle, *Seeds of Liberty*, p. 210.

107. Ritter, *Comparative Geography*, p. 184.

108. Guyot, *Earth and Man*, pp. 299–300.

109. Ewbank, *The World a Workshop*, p. 47.

110. See especially Allen, *Legend of Noah*, pp. 113–129. "Herder (in his *Ideen zur Philosophie der Menschengeschichte*, 4 vols., 1784–1791), so far as I know, was the first thinker to recognize in a systematic way that there are differences between different kinds of men, and that human nature is not unified but diversified"; R. G. Collingwood, *The Idea of History* (New York: Oxford University Press Galaxy paperback, 1956), pp. 90–91.

111. Allen, *Legend of Noah*, pp. 125–128; also James Adair, *The History of the American Indians* (London, 1775), which presents arguments in favor of the Jewish origin of the Indians. Clark Wissler, "The American Indian and the American Philosophical Society," *Proceedings, American Philosophical Society 86*, 1 (1942), 190–204: 196, calls Adair's book "factually a classic."

112. Hubbard, *History of New England*, p. 28.

113. William Penn, *Letter to Committee of the Free Society of Traders, 1683*, in A. C. Myers, ed., *Narratives of Early Pennsylvania*, pp. 217–242: 236.

114. Hubbard, *History of New England*, pp. 26–27. Hubbard may well have been influenced by Acosta, who discussed the origin of the Indians in his *Historia*, bk. 1, chs. 15–25 (Grimston's trans., 1880 ed., I, pp. 42–72). See Saul Jarcho, "Origin of the American Indian as Suggested by Fray Joseph de Acosta (1589)," *Isis 50* (1959), 430–438.

115. See Allen, *Legend of Noah*, pp. 119–129.

116. See Bernard De Voto, *The Course of Empire* (Boston: Houghton Mifflin, 1952), pp. 68–73, 568–570.

117. Daniel Gookin, *Historical Collections of the Indians in New England* . . . first printed from the original MS, *Collections, Massachusetts Historical Society*, 1 (1792), 144–226: 145–147.

118. Jones, *Virginia*, p. 49 (see above, note 35).

119. *Ibid.*, p. 52.

120. Allen, *Legend of Noah*, p. 134.

121. Evans, *A Brief Account of Pennsylvania*, 1753, in Gipson, *Lewis Evans*, p. 117, (see above, note 58).

122. *Ibid.*, p. 91.

123. Jedidiah Morse, *The American Universal Geography* (3rd. ed., 2 vols.; Boston, 1796), I, 78.

124. See G. W. Stocking, Jr., "French Anthropology in 1800," *Isis 55* (1964), 134–150: 148–149.

125. Cadwallader Colden, *The Letters and Papers of Cadwallader Colden, 1711–1775*, in *New York Historical Society Collections* (7 vols.; New York, 1917–1923), II, 278.

126. Samuel Miller, *Brief Retrospect*, I, 357.

127. S. S. Smith, *An Essay on the Causes of the Variety of Complexion and Figure in the Human Species* (Philadelphia, 1787; Edinburgh and London, 1788; 2nd American ed., New Brunswick, N. J., 1810), p. 11 (1810 ed.).

128. *Ibid.*, p. 17.

129. See Bronislaw Malinowski, *Argonauts of the Western Pacific . . .* (New York: Dutton; London: Routledge & Kegan Paul, 1922), p. 10, note.

130. Smith, *Essay*, p. 30.

131. D. C. Gilman, *The Life of James Dwight Dana* (New York, 1899), p. 328.

132. Asa Gray, *Natural Science and Religion: Two Lectures Delivered at the Theological School of Yale University* (New York, 1880), p. 54.

133. Guyot, *Earth and Man*, pp. 228, 266–267.

134. Ewbank, *The World a Workshop*, pp. 129, 134.

135. *Ibid.*, 124, 128.

136. *Ibid.*, 165–166.

Epilogue

1. Ecclesiastes, iii, 1–2.

2. See T. S. Kuhn, "The Function of Dogma in Scientific Research," in A. C. Crombie, ed., *Scientific Change . . .* (New York: Basic Books, 1963), pp. 347–369.

3. G. P. Marsh, *Man and Nature; or, Physical Geography as Modified by Human Action*, ed. David Lowenthal (Cambridge, Mass.: Harvard University Press, 1965), pp. 68–69, n. 36.

4. See J. K. Wright, "'Crossbreeding' Geographical Quantities," *Geographical Review 45* (1955), 52–65, and, especially, "Geography and History Cross-Classified," *The Professional Geographer 12*, no. 5 (1960), 1–3. In the latter the studies of geography and of history are designated by G and H and the circumstances with which those studies have to do by g and h, respectively. These letter symbols are cross-classified to yield divers combinations suggestive of actual and conceptual relations. By using other letters—such as R for religion, GL for geology, X for any unspecified study—to designate related elements and a diversity of signs to specify different kinds of relation, the scope and flexibility of the scheme may be indefinitely increased. For example, whereas H(G) signifies history of geography (according to the notation suggested in the paper cited), X(G) would signify geosophy, G(R) theogeography, R(e) geopiety—religious belief (R), in or concerning ( ), the earth (e).

5. T. H. White, *The Once and Future King* (New York: Putnam, 1939; Dell, paperback, 1960), p. 44.

6. See Loren Eiseley, "The Long Loneliness: Man and the Porpoise: Two Solitary Destinies," *The American Scholar 30* (1960–1961), 57–64: 63. J. C. Lilly, *Man and Dolphin* (Garden City, N.Y.: Doubleday, 1961), pp. 220–224.

7. On "Seven Lamps of Geography" (according to the British geographer, E. W. Gilbert) see *Geography 36* (1951), 21–43; reviewed, *Geographical Review 43* (1953), 130–131.

8. See J. D. Adams, *The Magic and the Mystery of Words* (New York: Holt, 1963), p. 33.

9. L. H. Hannon, "American Culture: Qu'est-ce que c'est?" *New York Times Magazine* (February 14, 1964), p. 10.

10. "The Hegelian dialectical progression becomes enlightening if it is not treated metaphysically but is translated into terms of the psychology of fashion or the homely concept of the swing of the pendulum, a cardinal principle of the British political philosophy but equally relevant to the history of science" (C. A. Mace, in Crombie, ed., *Scientific Change*, p. 605).

11. See above, p. 76.

ADDENDA

To p. 304, n. 18: See also R. E. Skelton, T. E. Marston, and G. D. Painter, *The Vinland Map and the Tartar Relation* (New Haven and London: Yale University Press, 1965), pp. 179–182 and Pl. XV.

To p. 320, n. 26: See also Peter Gould, "Joshua's Trumpet: the Crumbling Walls of the Social and Behavioral Sciences," *Geographical Review 58* (1965), pp. 599–602.

To p. 232, n. 32: See also W. W. Ristow, "Historical Cartography in the United States, 1959–1963," *Imago Mundi: A Review of Early Cartography 17* (1963), pp. 106–114 (includes a helpful bibliography of selected references).

To p. 334, n. 70: See also C. E. Clark, "Science, Reason, and an Angry God: The Literature of an Earthquake," *New England Quarterly 38* (1965), pp. 340–362.

To p. 334, n. 81: By "parageography" I mean that which resembles but is *not* geography. The distinction between geography and parageography depends upon how one cares to define the former.

Index

Matter in the Notes (pp. 295–338) that can be found readily by means of the reference numbers in the text is not indexed (for example: under "Davis, W. M." no reference is given in the Index to p. 295). Other matter in the Notes is indexed rather fully. Since, however, most of the notes mean next to nothing when not considered in conjunction with the passages in the text to which they belong, a key is furnished below (p. 361) to assist the reader in locating the latter passages. Look up the references given in the Index under "Bible" and you will see the need for this key.

For many of the persons mentioned in the text (and also for some of the geographers mentioned in the Notes only) dates are given.

An asterisk (*) after a term appearing as a heading means that I either coined the term or have used it in a specially defined sense (see p. 290).

KEY TO REFERENCE NUMBERS

Reference numbers to the notes are listed here, together with the numbers (in parentheses) of the text pages on which they appear.

